T5-CVT-172

LAND TREATMENT OF HAZARDOUS WASTES

LAND TREATMENT OF HAZARDOUS WASTES

Edited by

JAMES F. PARR
PAUL B. MARSH
JOANNE M. KLA

Agricultural Environmental Quality Institute
Agricultural Research Service
U.S. Department of Agriculture
Beltsville, Maryland

NOYES DATA CORPORATION

Park Ridge, New Jersey, U.S.A.

1983

Library of Congress Catalog Card Number: 82-14402
ISBN: 0-8155-0926-X
Printed in the United States

Published in the United States of America by
Noyes Data Corporation
Mill Road, Park Ridge, New Jersey 07656

10 9 8 7 6 5 4 3 2 1

Library of Congress Cataloging in Publication Data
Main entry under title:

Land treatment of hazardous wastes.

Bibliography: p.
Includes index.
1. Hazardous wastes. 2. Waste disposal in the
ground. I. Parr, J. F. (James Floyd), 1929-
II. Marsh, Paul Bruce, 1914- III. Kla, Joanne M.
TD811.5.L365 1983 628.4'456 82-14402
ISBN 0-8155-0926-X

Foreword

Every segment of our society has benefited from the products of modern industrial chemistry. But production of a useful chemical usually creates a residue or by-product that must be disposed of. Often such disposal has been done on land by way of chemical dumps and landfills. In recent years, chemical dumpsites have become the subject of increased public concern, and in some instances, they have caused apprehension with respect to possible related health risks. Though few of these risks have been scientifically validated with indisputable evidence, the allegations of unacceptable risk have nevertheless remained. Although toxicological and epidemiological evidence is often inadequate in such situations, the public has called for remedial action and, consequently, the gathering of scientific evidence to assess the risks.

Better ways must be found to dispose of and detoxify industrial wastes in an environmentally safe and acceptable manner. One possibility currently under consideration is to detoxify these wastes by land treatment/landfarming, wherein the toxic constituents are subjected to inactivation, adsorption, and chemical or biological degradation. Such methods are already being practiced to some degree for petroleum and food processing wastes, and for municipal sewage sludge.

The authors examine the possibilities for improving the efficacy of land treatment and for its extension to other industrial chemical disposal situations. In the utilization of land as a treatment system for industrial wastes, the primary objective is to use the chemical and microbiological properties of a soil to enhance the detoxification, degradation, and inactivation of waste constituents, while preventing or minimizing the contamination of surface water, groundwater and the food-chain with these chemicals.

This book presents a critical review and assessment of current knowledge and management practices for ultimate disposal of a number of hazardous waste chemicals. It provides useful information and strategies for the design, development, and effective management of land treatment systems for hazardous wastes, and identifies areas of research that are needed to maximize the potential value of land treatment systems and to minimize associated risks.

Acknowledgments

Part of the material used in this book resulted from an Interagency Agreement (No. Ad-12-F-0-055-0) between the U.S. Department of Agriculture and the U.S. Environmental Protection Agency.

We gratefully acknowledge our Technical Information Specialists, Ms. M.M. Leech and Ms. N.K. Enkiri, for conducting literature searches, obtaining documents, and for their patience in the editing of our drafts; Mr. C. Bebee of the National Agricultural Library for instructing our staff and conducting literature searches and retrievals; and members of our laboratory staff, Dr. S.B. Sterrett, Mr. D. Hussong, Mr. M.A. Ramirez, Ms. T.I. Munitz, and Mr. C.F. Smith for their professional and technical assistance to the authors.

We are especially grateful to our typist Ms. M.C. Rolf who despite severe time constraints worked diligently through several drafts to obtain this final text.

A draft of this text has been reviewed by the U.S. Environmental Protection Agency and the U.S. Department of Agriculture, and approved for publication. Approval does not signify that the contents necessarily reflect the views and policies of these agencies. Mention of trade names or commercial products is made only for clearer communication, and does not constitute endorsement or recommendation for use over equivalent products.

Acknowledgments

Contributors*

W.D. Burge	Microbiologist, Biological Waste Management and Organic Resources Laboratory, Agricultural Environmental Quality Institute
R.L. Chaney	Research Agronomist, Biological Waste Management and Organic Resources Laboratory, Agricultural Environmental Quality Institute
R.H. Fisher	Microbiologist, Biological Waste Management and Organic Resources Laboratory, Agricultural Environmental Quality Institute
S.B. Hornick	Soil Scientist, Biological Waste Management and Organic Resources Laboratory, Agricultural Environmental Quality Institute
D.D. Kaufman	Research Soil Microbiologist, Pesticide Degradation Laboratory, Agricultural Environmental Quality Institute
J.M. Kla	Technical Information Specialist, Maryland Environmental Service, Biological Waste Management and Organic Resources Laboratory, Agricultural Environmental Quality Institute
P.B. Marsh	Microbiologist, Biological Waste Management and Organic Resources Laboratory, Agricultural Environmental Quality Institute
P.A. Paolini	Soil Scientist, Biological Waste Management and Organic Resources Laboratory, Agricultural Environmental Quality Institute

J.F. Parr — Chief, Biological Waste Management and Organic Resources Laboratory, Agricultural Environmental Quality Institute

L.J. Sikora — Microbiologist, Biological Waste Management and Organic Resources Laboratory, Agricultural Environmental Quality Institute

J.M. Taylor — Plant Physiologist, Biological Waste Management and Organic Resources Laboratory, Agricultural Environmental Quality Institute

G.B. Willson — Agricultural Engineer, Biological Waste Management and Organic Resources Laboratory, Agricultural Environmental Quality Institute

---

*Contributors are all located at the U.S. Department of Agriculture, Agricultural Research Center, Beltsville, Maryland.

Contents

Introduction

Present-day Americans enjoy many comforts and conveniences unknown a century ago. Our homes are more uniformly heated in winter and cooled in summer; our foods are available in greater variety in the supermarket; our accessibility to national and international news is broader and more prompt; our medical care is vastly improved, and our life expectancy has increased. Much of this progress has come, either directly or indirectly, from basic advances in science and from the genius of American industry in implementing these advances for improving the lot of mankind.

Many industries have been involved in this progress, but none more significantly than the chemical industry. Intertwined as it is with all other phases of U. S. industry, the chemical industry exerts a profound and pervasive influence on all of American life. The basic chemical-industrial goal of "a better life through chemistry" is backed by a long list of validated accomplishments.

But time and progress itself have brought with them new challenges. We have become increasingly aware, especially in the past two decades, of potential environmental hazards and health risks from undue exposure to certain chemicals. Much recent information suggests that cancer may be largely of environmental origin. Animal data which indicate that some chemicals may cause birth defects are readily available. Not surprisingly, then, the public has an increased sense of apprehension about chemicals.

In part, environmental pollution with chemicals is a direct result of progress in the sense that a large and ever growing list of new industrial chemicals is produced each year and with it the possibility of new byproducts and chemical wastes. But even more relevant is the fact that our concept of what constitutes a hazardous pollutant has undergone considerable change during the past two decades. Chemicals once thought to be relatively innocuous must now be regarded as potentially hazardous and toxic, and as possible causes of cancer, birth defects, and/or other health problems. Clearly, better methods are needed for the ultimate disposal of these wastes than have traditionally been used.

Two methods that appear to control the risk from disposal of hazardous wastes are 1) disposal in well designed hazardous waste landfills, and 2) high-temperature incineration. Though these methods can protect the environment, disposal of many wastes in this fashion is very expensive. Less expensive methods for disposing of these wastes and at the same time protecting the environment are badly needed.

Land treatment is a less expensive alternative method for ultimate disposal of many industrial wastes. In land treatment, soil is used to hold toxic chemicals while microbes degrade the compounds. For bulk organic sludges with high water content, land treatment may be the disposal method of choice; whereas for small volumes of still-bottoms containing very high concentrations of toxic organic chemicals, another method may be best.

The present book is a critical review and evaluation of available information relevant to land treatment of hazardous wastes. This information comes from basic research on organic and inorganic chemicals in the environment, from research and the practice of land application of sewage sludge, crop residues, food processing wastes, etc., and from the limited research conducted on hazardous wastes. The authors note processes by which land treatment degrades wastes, and by which environmental impacts could result from land treatment. Ways for using this information in designing and managing a land treatment site are also presented.

An important goal in writing of this book was to identify research needs for land treatment of hazardous wastes. The authors feel that if the suggested research is conducted, many industries and local, state, and federal agencies will have a better scientific basis on which to design and permit land treatment systems.

PART I

PROCESSES THAT INFLUENCE THE FATE AND EFFECTS OF WASTE

The Interaction of Soils with Waste Constituents

S.B. Hornick

The amount of soil contained in the upper six inches of a "plow layer" is estimated to be 2×10^6 lb/acre or 2.2×10^6 kg/ha. The ability of this large soil mass to absorb nutrients and hold sufficient water to sustain plant life is dependent on certain physical and chemical properties. Some of these are: texture, infiltration and permeability, water holding capacity, bulk density, organic matter content, cation exchange capacity, macronutrient content, salinity, and micronutrient content.

Mineral soils are composed of approximately 45% mineral material (varying proportions of sand, silt, and clay), 25% air, 25% water, and 5% organic matter. Any significant change in the balance of these components could affect the aforementioned physical and chemical properties. This, in turn, may alter the soil's ability to (a) support the necessary chemical and biological reactions to degrade, detoxify, and inactivate toxic waste constituents and (b) function as a natural filter in the adsorption and retention of waste constituents, thereby preventing or minimizing transport, leaching, and contamination of surface and groundwaters.

Effect of Waste Additions on the Physical Properties of Soil

Soil Texture--

An important aspect of site selection for waste application is the soil type or textural composition and parent material to which the waste or wastewater will be applied. Soil composition influences infiltration rate and permeability, water holding capacity, and adsorption capacity for various waste components.

Due to the very small pore size resulting from the predominance of clay and silt particles in finer textured soils, the water infiltration rate tends to be slow. If the clay portion is composed of minerals like montmorillonite that have high shrink-swell tendencies, added moisture or water will cause swelling which will result in the blockage of any further water movement in the profile. If the amount of irrigation or wastewater applied exceeds the soil infiltration rate, runoff or possibly flooding will occur depending on slope. In these instances, anaerobic conditions will be induced with the possibility of odors occurring. Slow or very slow permeability rates are less than 0.5 cm/hr (SCS USDA, 1973).

Coarse soils, which are mostly composed of sand and gravel particles, allow rapid water movement due to their large interconnecting pores. However, these soils also have severe limitations, in that, if a site is excessively drained, nutrients in wastes and wastewaters will move too rapidly to be sufficiently absorbed by plants or adsorbed on soils. Groundwater contamination can result if a more restrictive layer is not between the coarse layer and the water table.

Bulk Density--

The bulk density of a soil is a measure of soil weight per unit volume that determines pore space through which water can move. The frequent use of heavy machinery to either work the soil or apply wastes compacts the soil and, thus, increases the bulk density. The addition of organic wastes, such as animal manure and sewage sludge compost, decreases soil bulk density and increases infiltration and permeability since organic wastes tend to increase soil aggregation and porosity (Unger and Stewart, 1974; Hornick, Murray, and Chaney, 1979).

Water Holding Capacity--

The water holding capacity of a soil is directly related to its bulk density and soil texture. Soils having either very fine or very coarse textures or high bulk densities will not be able to maintain an adequate supply of water available for plant use. The water content determines available oxygen, redox potential, and microbial activity of a soil system. Although the addition of organic wastes or material will increase the water content of a soil, it may not necessarily increase the available water that a plant needs (Epstein, 1975). The effect of organic matter additions on water movement and water potential in a soil-plant system is discussed more fully in a later section.

Effect of Waste Additions on Chemical Status and Properties of Soils

Chemical Reactions in the Soil Profile--

Some factors which affect the chemical reactions that can occur in soils are the type and amount of inorganic and organic fractions which interact with the soil solution, the pH of a soil system, the relative concentrations of macro- and micronutrients, and the redox potential. These factors, coupled with others such as climate and microbial population, determine whether a soil is able to assimilate applications of industrial waste. Due to the exchange and adsorption processes in the soil, ions which might pollute the groundwater can be retained.

Soil exchange capacity--A net negative charge on soil colloid particles arises from (1) dissociation of OH^- radicals from the edge of the silica and alumina sheets comprising the silicate clay particles and from iron, aluminum, and manganese oxides; (2) the disassociation of phenolic and carboxylic groups on organic matter; and (3) isomorphous substitution,which is the substitution by one atom for another within the crystal lattice.

These charges are either permanent or pH dependent. Permanent charges are due to electrostatic forces resulting from isomorphous substitution within the clay structure. Cations attracted from the soil solution by these charges are subject to exchange or replacement at all pH levels.

Other charges such as the dissociation of H^+ from the edges of clays and humic material are directly affected by soil pH, with the CEC increasing as the pH increases. At low pH, the covalently bonded H is not dissociated, but at high pH the H can be replaced by cations which are themselves exchangeable with other cations. This replacement or interchange of a cation on the soil surface with one from the soil solution is termed cation exchange. The dominant exchangeable cations in a soil system are usually Ca^{+2}, Mg^{+2}, K^+, Na^+, Al^{+3} and H^+.

The term 'cation exchange capacity' or 'CEC' refers to the total amount of cations which are held exchangeably by a unit mass or weight of a soil (meq/100 g soil). The CEC of a soil is dependent on its clay and humus contents, i.e., kaolinites, 1-10 meq/100 g; illites, 10-40 meq/100 g; montmorillonites, 80-100 meq/100 g; vermiculites, 120-150 meq/100 g; and humus, 150-300 meq/100 g. But of course, soils are usually a mixture of several different clays and, except for peats, a small amount of organic or humic material ($< 5\%$). Although the organic matter usually comprises less than 5% of a soil, it usually accounts for 30-65% of the CEC in mineral soils and more than 50% in sandy and organic soils (Schnitzer and Khan, 1978).

Acid soils result from the domination of H^+, Al^{+3}, and Al hydroxypolymers on a soil's exchange sites, and the resultant contribution of H^+ ions to the soil solution either directly or indirectly through hydrolysis. In calcareous soils Ca >Mg >K = Na predominate and are called exchangeable bases. The proportion of the exchange sites or CEC which these bases occupy is called the percent base saturation. For example, when the percent base saturation is 50, half of the exchange capacity is satisfied by bases like Ca^{+2} and Mg^{+2}, and the other half by Al^{+3} and H^+.

Nonexchangeable cations which comprise a large percent of the soil solids do not exchange with cations from the soil solution. Instead, these cations are released slowly to the soil solution by physical, chemical, or biological actions such as weathering or decomposition.

The soil cation exchange capacity, therefore, allows the soil to act as a filter or buffer for many undesirable elements which can be added to it with organic and industrial wastes. The addition of organic wastes increases the CEC of a soil. Once additions are ceased, the CEC will slowly decrease as the organic matter decomposes.

Anion exchange--The anion and molecular retention capacity of soils is smaller than that for cations and is governed by more complex mechanisms than the simple electrostatic attraction responsible for most cation exchange reactions. Generally, silicate clay particles have a net negative charge which repels anions from a mineral's surface. However, soils contain a mixture of minerals, so that the positive charges arising from clay edges, amorphous material, and hydrous oxides, and the negatively charged surfaces affect the anions approaching the particle surface. Anions such as Cl^-, NO_3^-, and SO_4^{-2} are usually retained in soils by electrostatic forces from positively charged sites and are considered to be nonspecifically adsorbed. In soils with pH-dependent charge, lowering the pH decreases the net negative charge and, thus, decreases anion repulsion or increases anion adsorption.

Sorption and precipitation reactions--Hydrous oxides of Fe, Al and Mn comprise the largest non-silicate fraction of soils that are important in sorption reactions. Surfaces of these oxides contain ions which are not fully coordinated and, hence, the surfaces are electrically charged. In the soil solution, the resulting net charge that develops through amphoteric dissociation on the hydroxylated surfaces is pH dependent and can provide either anion or cation exchange properties. In addition to simple exchange reactions with the soil solution, hydrous oxides can specifically adsorb certain anions such as arsenate, molybdate and phosphate through ligand exchange (Hingston et al., 1967, 1968; Jenne, 1968). Ligand exchange is the strong adsorption capacity of an oxide for an anion over that predicted from electroneutrality.

Fixation of ions occurs in vermiculite soils and on weathered edges of mica soils. Ions such as K^+ and NH_4^+ have low energies of hydration and, thus, are easily dehydrated and retained or entrapped in the hexagonal holes of the vermiculite surfaces. Fixation increases as the pH of the soil increases.

Precipitation usually occurs when the solubility of a compound is exceeded and a sparingly soluble solid phase is formed. Usually both adsorption and precipitation reactions occur simultaneously in soils, with the predominant process being dependent on the concentration levels of the ions in both the solution and solid phase.

In addition to ion exchange, specific adsorption, and precipitation, the formation of organo-metal complexes through the organic matter chelation of metals is an important factor governing metal availability (Chaney and Hornick, 1978; Ellis and Knezek, 1972; Jenne, 1968; Schnitzer and Khan, 1978). In waste-amended soils, the addition of high amounts of organic matter insures a predominance of organic matter reactions. In turn, the mobility of heavy metals added by wastes is related to the organic matter content of soils, pH, hydrous oxide reactions, and the oxidation-reduction or redox potential of a soil. Specific metal reactions with organic matter will be discussed in a later section.

The presence of CO_2 in soil systems also affects the solubility of cations in the soil solution. The CO_2 reacts with many cations to form very slightly soluble carbonates. Calcium is one of the most abundant cations in the soil solution which forms carbonates and thus solid calcium carbonate is widely found in many soils.

Oxidation-reduction--Almost all chemical and biological soil reactions involve oxidation and reduction reactions which are the transfer of electrons from one ion or molecule to another. Oxidation is the loss or donation of electrons from other substances, while reduction is the gain or acceptance of an electron from other substances.

When oxygen accepts electrons from organic matter under aerobic conditions, it functions as an oxidizing agent because it oxidizes the organic matter. It is the role of the soil to

$$O_2 + \text{Organic Matter} \rightarrow CO_2 + H_2O$$

provide electron acceptors for the oxidation of organic matter and other compounds. When oxygen is not available, other reducible compounds such as nitrate, Mn (IV), Mn (III), Fe (III), and S (VI) can function as electron acceptors (Bolt and Bruggenwert, 1978). In cases where a soil is flooded or becomes saturated with water and O_2 is not available, the activity of anaerobic microbes increases and soil reduction starts. Once the remaining oxygen has been reduced in flooded soil, the reduction of nitrate and manganese will then follow.

The oxidation-reduction process can be measured by Eh, which represents the ability of the redox ions to donate or accept electrons at equilibrium and standard conditions.

$$Eh = Eh^o - \frac{RT}{nF} \ln \frac{(Red)}{(Ox)\ (H^+)^m}$$

where Eh is the electrode potential in volts, Eh^o is the electrode potential at standard conditions, n is the number of faradays or moles of electrons, F is the Faraday constant, Red is the reduced species and Ox is the oxidized species.

As a result of the release of carbonate in the reduction process, an increase in the pH of acid soils is noted after flooding. As the pH rises to near neutral values, ferrous and manganous hydroxides and carbonates precipitate, thus stabilizing the pH. In calcareous soils, the production of carbonate by reducing iron and manganese oxides is inhibited and, thus, insignificant. However, the production of CO_2 by anaerobic decomposition of organic matter acidifies the soil and thus a decrease of 0.5 to 1 pH unit after flooding is usually noted (Bolt and Bruggenwert, 1978).

Soil Composition: Anions and Cations in the Soil Solution--

On an average in unamended soils, O and Si constitute 49 and 33% by weight of the total mineral composition of soils. Aluminum, Fe, C, Ca, and K are present in the much smaller quantities of 7, 4, 2, 1, and 1%, respectively. All other elements are part of the remaining 3% (Table 1) (Bohn, McNeal and O'Connor, 1979). However, these metals are either basic constituents of soil minerals such as S, adsorbed or associated with soil surfaces, precipitated, occluded or fixed in precipitates, or part of a biological system or organic-complex and, thus, are not abundant in the soil solution. For example,

TABLE 1. TYPICAL CONCENTRATIONS AND RANGES ELEMENTS IN SOILS [a,b]

Element	Typical Value	Range
O (%)	49	-
Si (%)	33	25-35[b]
Al (%)	7	1-30[b]
Fe (%)	4	0.7-55[b]
C (%)	2	-
Ca (%)	1	0.7-50[b]
K (%)	1	0.04-3[b]
Na (%)	0.7	0.07-7[b]
Mg (%)	0.6	0.06-6[b]
N (%)	0.1	0.02-.25[b]
P (%)	0.08	-
S (%)	0.05	0.003-.09[b]
F (%)	0.02	0.003-.03[b]
Cl (%)	0.01	-
B (%)	0.001	0.0002-.01[b]
I (ppm)	5	-
Mo (ppm)	3	0.02-5
Cd (ppm)	0.06	0.01-7
Co (ppm)	8	1-40
Cu (ppm)	20	2-100
Pb (ppm)	10	2-200
Mn (ppm)	850	100-4000
Ni (ppm)	40	10-1000
Zn (ppm)	50	10-300
As (ppm)	5	1-50
Be (ppm)	1	0.2-10
Cr (ppm)	20	5-1000
Se (ppm)	0.5	0.1-2.0
Hg (ppm)	0.05	0.02-0.2

[a] Bohn, McNeal, and O'Connor, 1979
[b] Overcash and Pal, 1979

highly charged ions (P^{+5}, Se^{+6}, Mo^{+6}, As^{+5}) are surrounded by strongly associated oxygen or hydroxyl ligands. These weak acids, H_3PO_4, H_2SeO_2, H_2MoO_4 and H_3AsO_4 tend to form insoluble salts by associating with other cations like Ca^{+2}, Mg^{+2}, Fe^{+3}, Al^{+3} and H^+. As was mentioned in the previous section, Al and Fe form hydroxy polymers that adsorb other ions and are not readily soluble in the soil solution. Whether metals that can promote or hamper agricultural production will precipitate from the soil solution, or will remain very soluble, depends on the metal and its soil solution concentration, soil properties like pH, and environmental conditions.

This section will highlight the inorganic reactions and behaviors of various essential or potentially harmful elements in soils. Elements such as C, N, S, and O that depend on microbial activity and are affected greatly by organic matter content will be discussed in Factors Affecting the Degradation and Sorption of Wastes in Soils.

Anions: Phosphorus, Arsenic, Boron, Molybdenum, Sulfur, Selenium

Phosphorus concentrations in the soil solution are usually low ranging from 0.1 to 1 ppm because P is mostly associated with the solid phase in soils. In acid soils, P reacts with Fe and Al hydroxides to produce adsorbed forms of P that are in equilibrium with the soil solution or are precipitated and thus occluded by the minerals. In calcareous soils, at low concentrations, P is adsorbed onto $CaCO_3$; at high concentrations, Ca phosphate minerals are formed (Mattingly, 1975). The common forms of P in soil are $H_2PO_4^{-2}$ in basic soil solutions.

Arsenic behaves much like P in soils in that adsorption of As increases as iron oxide content increases and that Fe and Al hydrous oxides specifically adsorb As. Arsenate (AsO_4^{-3}), however, can be reduced to arsenite (AsO_2^-) but arsenate is the most common form in soils. Generally, both forms are strongly retained in soils.

In the soil solution boron exists as the weak acid H_3BO_3 that can dissociate in dilute solution to form $B(OH)_4^-$, a monomer. In more concentrated solutions, polymers will form. Boron is primarily adsorbed by Fe and Al hydrous oxides, with maximum adsorption occurring at pH 8 to 9 for Fe forms and pH 7 for Al forms (Ellis and Knezek, 1972). The range for deficiency of several tenths of a ppm or toxicity of a few ppm in soils is narrower than for any other element.

Molybdenum behaves much like arsenic in that it usually exists in soil solutions as the monomer $HMoO_4^-$. Only in solutions with greater than 29 ppm do polymeric forms of Mo exist (Carpeni, 1947). However, maximum adsorption of molybdate occurs around pH 4.2, with the solubility of Mo in the soil solution increasing as the pH increases. Adsorption of molybdate in acid soils is dependent upon the Fe and Al oxide and P status of a soil in that molybdate is specifically adsorbed by Fe oxides but can be replaced by phosphate if present in large quantities.

Sulfur reactions in the soil are greatly dependent on pH and free Fe and Al oxides, with retention increasing as pH decreases. Actual retention of sulfate in soils is also dependent upon other anions in the soil solution in the order phosphate > sulfate > nitrate = chloride (Harward and Reisenauer, 1966).

Selenium behaves similarly to sulfur in the soil solution, existing as selenates (SeO_4^{2}) and selenite (SeO_3^{2}). Selenate, the predominant form in alkaline soils, is quite soluble as $CaSeO_4$ and can move readily in these soils.

Cations: Copper, Zinc, Chromium, Nickel, Cadmium, Lead and Mercury

In soils, copper usually occurs in association with hydrous oxides of Mn and Fe, where specific adsorption is more important than cation exchange. McLaren and Crawford (1973) found that organic matter and free Mn oxides are the dominant soil constituents contributing towards the specific adsorption of Cu. The adsorption maxima showed that manganese oxides > organic matter > iron oxides > clay minerals. Due to this phenomemon, very small amounts of Cu^{+2} are found in the soil solution; most is complexed by organic matter, with availability affected by soil pH.

The sorption of zinc in the soil solution is much like Cu but Zn is not complexed by organic matter to the same extent as is Cu in calcareous soils (Hodgson, Lindsay and Trierweiler, 1966). Sorption on hydrous oxides, reactions with phosphates and carbonates, and formation of hydroxides are the mechanisms for Zn retention on clays in soils (Keeney and Welding, 1977).

The most soluble, mobile, and toxic form of chromium in soils occurs in the hexavalent state as chromate (CrO_4^{2}) or dichromate ($Cr_2O_7^{2}$). When aerobic conditions exist, the hexavalent form is rapidly reduced to trivalent chromium which forms insoluble hydroxides and oxides and is unable to leach. A near neutral soil pH will also prohibit the formation of dichromates.

Nickel availability in soils is governed by Fe and Mn hydrous oxides and by organic chelates which complex Ni less strongly than Cu (CAST, 1976). Nickel differs from Cu and Zn in that it is more available from organic sources than inorganic sources. Acid soils increase the solubility of Ni in the soil solution.

Cadmium is complexed by organic matter, oxides of Fe and Mn, and chlorides (Chaney and Hornick, 1978). In alkaline, low organic matter, sandy soils, precipitation of Cd compounds such as $CdCO_3$ occurs. The soil pH is the most important factor governing Cd solubility and resultant availability, with the solubility increasing as pH decreases.

Lead solubility is determined by the amount of sulfate, phosphate, hydroxides, carbonates, and organic matter present in a soil system. The formation of lead sulfate at very low pH, lead phosphate and hydroxides at intermediate pH, and lead carbonates at a calcareous pH limits the mobility of Pb in soils. As the soil pH and available phosphorus in a soil decrease, soluble Pb increases.

Mercury reacts in soils with chlorides and sulfur to form insoluble HgS, $HgCl_3$ and $HgCl_4^2$. Mercury can be chelated by organic matter as $HgCl_4^2$ or can be absorbed by sesquioxide surfaces (CAST, 1976). Through chemical and microbial degradation, Hg can be volatilized or associated with clay particles and organic matter. Mercury is not very mobile in the soil profile due to its strong sorption reactions with soil constituents.

Soluble salts--The predominant soluble salts in most soils are sodium, calcium, magnesium, chloride and sulfate. Minor amounts of potassium bicarbonate, carbonate, and nitrate may also be present. Usually, sodium and chloride predominate in a saline soil since Ca, Mg, K and SO_4^2 tend to form somewhat insoluble secondary minerals. Bicarbonate is formed from the dissolution of CO_2 in water. The presence of carbonate or bicarbonate is dependent on the pH of the soil solution. Carbonate will predominate only when the pH is greater than 9.5 (U.S. Salinity Laboratory, 1954).

The salinity of a soil can be determined by the electrical conductivity of a saturation extract of a soil, generally expressed in millimhos per centimeter(mmhos/cm). Saturation extracts having electrical conductivities (EC) greater than 4 millimhos per centimeter indicate excess salinity. It would be expected that yields of many crops would be restricted or reduced if grown on these soils. Generally an excess of soluble salts leads to a breakdown of soil aggregation or dispersion of the soils. This dispersion results in a soil texture or structure unfavorable for the necessary physical and chemical processes needed for plant growth.

The addition of wastes or wastewater to soils can aggravate or imbalance non-saline soils. An empirical measure relating the relative and total soluble cation in the saturation extract to the exchangeable cation composition can be made of the sodium imbalance by utilizing the sodium adsorption ratio (SAR).

$$SAR = \frac{(Na^+)}{\sqrt{[(Ca) + (Mg)]/2}}$$

where the concentration of the cations is expressed in milliequivalents per liter (meq/l).

The exchangeable sodium percentage (ESP) is the proportion of the exchange capacity which is occupied by sodium. By utilizing the SAR value, the ESP can be determined with the following equation

$$ESP = \frac{(1.47)\ (SAR) - 1.26}{(0.0147\ (SAR) + 0.99}$$

(U.S. Salinity Laboratory, 1954). An ESP of greater than 15 indicates that the amount of sodium present is sufficient to reduce plant yields.

Saline soils are soils that contain an excess of soluble salts. If the electrical conductivity of the saturation extract is greater than 4 mmhos/cm and the ESP is less than 15, the soil is considered saline. When excessive sodium causes a reduction in plant growth through deflocculation of the soil and reduced availability of Ca and Mg, the ESP will be greater than 15 and the soil is considered sodic. If the ESP is greater than 15 but the salinity is less than 4 mmhos the soil will likely have a pH greater than 8.5 and will be alkaline (Overcash and Pal, 1979). A soil is both saline and sodic or alkaline when the ESP is greater than 15 and the conductivity of the saturation paste is greater than 4 mmhos/cm. Whenever rainfall or irrigation are even slightly high in soluble salts and drainage is inadequate, the inability of the salts to leach through the profile and, thus, concentrate in the surface horizon will create a sodic or saline soil.

Organic Matter - Metal Reactions in Soils --

Organic matter is generally an amorphous organic residual in soils which when present in sufficient amounts has a beneficial effect on the physical and chemical properties of soil because of its high CEC, high specific or reactive surface, and large amounts of exchangeable bases. Fractionation of soil organic matter gives basically three categories: the humin and ulmin fraction, the fulvic fraction, and the crude humin fraction (Kononova, 1967; Schnitzer and Kahn, 1978).

The humin and ulmin fraction is inert due to a lack of surface charge. This fraction is insoluble in both alkali and acid. The fulvic acid fraction is soluble in dilute mineral acid and contains tannins, glucosides, polyuronides, soil N compounds, and organic P and S. Polyuronides are polysaccharides which contain aldehydes and uronic acids that are important cementing agents formed through microbial activity.

These agents are responsible for maintaining soil structure. Polysaccharides, uronic acids and free $CaCO_3$ prevent aggregate breakdown in soils (Greenland, 1965; Greenland, Lindstrom, Quirk, 1962).

The crude humin consists of humic acid and hymatomelanic acids containing functional carboxyl and phenolic hydroxyl groups responsible for exchange and adsorption reactions. Gums and resins are also part of the crude humin fraction which is soluble in dilute alkali but insoluble in dilute acid.

Both the humates and fulvates show a high degree of reactivity due to their acidic functional groups. The reaction of these materials with cations in the soil solution is strongly pH dependent.

Other organic materials which are involved in metal reactions and complexation in soils are plant root exudates and various degradation products which can serve as the base for the humic fraction of the soil. The addition of waste materials which are easily decomposable can become part of an important soil process and result in a substantial increase in beneficial organic materials.

Chelation or complexing of metals -- Metal ions can be bound to organic matter by cation exchange and by chelation. A metal chelate is a complex that is formed when a metal ion is bound to an organic component by more than one bond to form a ring or cyclic structure. The stability of a chelated metal can be greater than the stability of a noncyclic complex if the metal chelate stability is dependent upon several factors such as pH, the stability constants of other metal ions present, temperature, and ionic strength (Keeney and Wildung, 1977).

As is mentioned in the preceeding section on metals, organic matter is a major factor in metal sorption to soils. This has been evidenced by the metal contents of humic and fulvic acid extracts, the correlation between metal sorption and organic matter content, and the increased extractability of metals once organic matter has been oxidized by H_2O_2 (Ellis, 1973; Ellis and Knezek, 1972; Stevenson and Ardakani, 1972).

Hodgson, Lindsay and Trierweiler (1966) showed that 98% of the Cu was in a complexed form suggesting that very little Cu^{+2} was available for adsorption reactions. Early work with organic matter-extracts had indicated that Cu was adsorbed as $CuOH^+$ by carboxyl groups and as Cu^{+2} by phenols (Lewis and Broadbent, 1961). Similarly, the species of Zn that complex with humic acid vary. Randhawa and Broadbent (1965) showed that at pH 7.0, 70% of the Zn was divalent and at pH 3.5, 75%

was monovalent. Schnitzer (1969) showed that stability constants for fulvic acid-metal complexes increase with increasing pH.

at pH 3.5: $Al^{+3} > Cu^{+2} > Ni^{+2} > Pb^{+2} > Co^{+2} > Ca^{+2} > Zn^{+2} > Mn^{+2} > Mg^{+2}$

at pH 5.0: $Cu^{+2} > Pb^{+2} > Fe^{+2} > Na^{+2} > Mn^{+2} > Co^{+2} > Ca^{+2} > Zn^{+2} > Mg^{+2}$

However, widely divergent stability constants, especially for Fe and Zn, have been recorded even at the same pH (Stevenson and Ardakani, 1972). These discrepancies underline the need for the accurate determination of stability constants in soil systems.

Wastes containing large amounts of heavy metals can be land applied due to chelation and immobilization of the metal carboxylic acids, phenols, enols and oxygen-containing functional groups. Analyses of the waste-amended soils has shown that metals such as Cd and Cu are bound by the fulvic acid with retention increasing as pH and organic matter content increase (Chaney and Hornick, 1978).

Summary and Research Needs

Organic material is very important in the soil matrix. Its role in the metal reactions or sorption processes which occur in the soil determines the availability of metals and essential nutrients to plants. However, the exact mechanisms and stabilities of various reactions and organo-metal complexes are still uncertain. More information concerning the reactive nature and properties of organic compounds contained in organic wastes, and the factors governing the competition of clays and organic matter for the binding of metals needs to be obtained by research in order to predict the long term effects of waste additions on soils and cropland. In addition, the nature of trace metal transformations and interactions in microbial degradation processes must be elucidated to prevent waste or trace metal overloadings which will inhibit normal soil processes.

REFERENCES

Bohn, H. L., B. L. McNeal, and G. A. O'Connor. 1979. Soil chemistry. John Wiley and Sons, New York.

Bolt, G. H., and M. G. M. Bruggenwert (ed.). 1978. Soil chemistry - Part A: basic elements. Elsevier Co., New York.

Carpeni, G. 1947. Sur la constitution des solutions aqueuses d'acide molybdique et de molybdates alcalines. IV. Bull. Soc. Chem. 14:501-503.

CAST. 1976. Application of Sewage Sludge to Cropland. Report No. 64, EPA-430/9-76-013. U. S. Environ. Prot. Agen., Wash., D. C.

Chaney, R. L., and S. B. Hornick. 1978. Accumulation and effects of cadmium on crops. pp. 125-140. In Proc. First International Cadmium Conference. Metals Bulletin Ltd., London.

Ellis, B. G. 1973. The soil as a chemical filter. p. 46-70. In W. E. Sopper and L. T. Kardos (ed.) Recycling treated municipal wastewater and sludge through forest and cropland. Penn State Press., University Park, PA.

Ellis, B. G., and B. D. Knezek. 1972. Adsorption reactions of micronutrients in soils. p. 59-68. In J. J. Mortvedt, P. M. Giordano, and W. L. Lindsay (ed.) Micronutrients in agriculture. Soil Sci. Soc. Am., Madison, WI.

Epstein, E. 1975. Effect of sewage sludge on some soil physical properties. J. Environ. Qual. 4:139-142.

Greenland, D. J. 1965. Interactions between clays and organic compounds in soils. Part I. Mechanisms of interaction between clays and defined organic compounds. Soils and Fertilizers 28:415-425.

Greenland, D. J., G. R. Lindstrom, and J. P. Quirk. 1962. Organic materials which stabilize natural soil aggregates. Soil Sci. Soc. Amer. Proc. 26:366-371.

Harward, M. E., and H. M. Reisenauer. 1966. Reactions and movement of inorganic soil sulfur. Soil Sci. 101:326-335.

Hingston, F. J., R. J. Atkinson, A. M. Posner, and J. P. Quirk. 1967. Specific adsorption of anions. Nature 215:1459-1461.

Hingston, F. J., R. J. Atkinson, A. M. Posner, and J. P. Quirk. 1968. Specific adsorption of anions on goethite. Trans. 9th Intl. Cong. Soil Sci. 1:669-678.

Hodgson, J. F., W. L. Lindsay, and J. F. Trierweiler. 1966. Micronutrient cation complexing in soil solution: II. Complexing of zinc and copper in displaced pollution from calcareous soils. Soil Sci. Soc. Amer. Proc. 30:723-726.

Hornick, S. B., J. J. Murray, and R. L. Chaney. 1979. Overview on utilization of composted municipal sludges. In Proc. Natl. Conference on Municipal and Industrial Sludge Composting. Information Trans., Inc., Silver Spring, MD.

Jenne, E. A. 1968. Controls on Mn, Fe, Co. Ni, Cu, and Zn concentrations in soils and water: the significant role of hydrous Mn and Fe oxides. In Trace inorganics in water. Adv. Chem. 73:337-387.

Keeney, D. R., and R. E. Wildung. 1977. Chemical properties of soils. pp. 74-97. In L. F. Elliott and F. J. Stevenson (ed.) Soils for management of organic wastes and waste waters. Soil Sci. Soc. Am., Madison, WI.

Kononova, M. M. 1967. Soil organic matter: Its nature, its role in soil formation and in soil fertility. Pergamon Press, Oxford.

Lewis, T. E., and F. E. Broadbent. 1961. Soil organic matter - metal complexes: 4. Nature and properties of exchange sites. Soil Sci. 91:393-399.

Mattingly, G. E. G. 1975. Labile phosphates in soils. Soil Sci. 119:369-375.

McLaren, R. G., and D. V. Crawford. 1973. Studies on copper. II. The specific adsorption of copper by soils. J. Soil Sci. 24:443-452.

Overcash, M. R. and D. Pal. 1979. Design of land treatment systems for industrial wastes - theory and practice. Ann Arbor Science Publishers, Inc., Ann Arbor, MI. 684 pp.

Randhawa, N. S., and F. E. Broadbent. 1965. Soil organic matter - metal complexes: 5. Reactions of zinc with model compounds and humic acid. Soil Sci. 99:295-300.

Schnitzer, M. 1969. Reaction between fulvic acid, a soil humic compound and inorganic soil constituents. Soil Sci. Soc. Am. Proc. 33:75-81.

Schnitzer, M., and S. U. Khan. 1978. Soil organic matter. Developments in Soil Science 8. Elsevier Scientific Publishing Co., Amsterdam.

Soil Conservation Service, USDA. 1973. Guide for rating limitations of soils for disposal of waste. Interim Guide, Advisory Soils-14. Wash., D. C. 26 pp.

Stevenson, F. J., and M. S. Ardakani. 1972. Organic matter reactions involving micronutrients in soils. pp. 79-114. In J. J. Mortvedt, P. M. Giordano, and W. L. Lindsay (ed.) Micronutrients in agriculture. Soil Sci. Soc. Am., Madison, WI.

Unger, P. W., and B. A. Stewart. 1974. Feedlot waste effects on soil conditions and water evaporation. Soil Sci. Soc. Amer. Proc. 38:954-957.

U.S. Salinity Laboratory. 1954. Saline and alkali soils. Agricultural Handbook No. 60. USGPO, Washington, D. C.

Factors Affecting the Degradation and Inactivation of Waste Constituents In Soils

J.F. Parr, L.J. Sikora, and W.D. Burge

Introduction

The various chemical and physical properties of a soil determine the nature of the environment in which microorganisms are found. The soil environment in turn affects the composition of the microbiological population both qualitatively and quantitatively (Alexander, 1961). The rate of decomposition of an organic waste or crop residue depends primarily on its chemical composition and on those factors which affect the soil environment. Factors having the greatest effect on microbial growth and activity will have the greatest potential for altering the rate of residue decomposition in soil. In a recent review, Parr and Papendick (1978) discussed the various factors which affect the decomposition of crop residues by microorganisms. These same factors, which will be discussed here, can also affect the rate and extent of degradation and inactivation of waste constituents in soil.

On a microscale, the soil environment is heterogeneous and highly dynamic, with many components subject to rapid change and frequent fluctuations. Where wastes are applied to soils, the growth and activity of microorganisms within this complex system are controlled by both soil factors and waste or residue factors, some of which are affected through tillage and other agronomic practices. Soil factors of greatest importance in residue decomposition or degradation/inactivation of chemical wastes include (a) water, (b) temperature, (c) soil pH, (d) aeration or oxygen supply, (e) available nutrients (i.e., N, P, K, S), and (f) soil texture and structure. Significant waste or residue factors include (a) chemical composition of the waste, (b) its physical state (i.e., liquid, slurry, sludge), (c) its carbon:nitrogen ratio, (d) water content and solubility, (e) chemical reactivity and dissolution effects on soil organic matter, (f) volatility, (g) pH, (h) biochemical oxygen demand (BOD), (i) chemical oxygen demand (COD), and (j) nature of the indigenous microflora, if any. At this point it is appropriate to refer to the book entitled "Organic Chemicals in the Soil Environment" (in two volumes) edited by C.A.I. Goring and J. M. Hamaker (1972). It is probably the best reference to date on fundamental aspects of sorption, decomposition, diffusion, volatilization, and mass transfer, factors that govern the fate and behavior of organic chemicals in soil.

Most of these factors do not function independently, i.e., a change in one may effect a change in others. For example, high soil moisture may result in lower soil temperatures, and surface-applied wastes and residues may affect soil moisture and temperature simultaneously. Because of these significant interactions, it is sometimes difficult to isolate effects of certain environmental factors on waste degradation in soils. A brief discussion of just how some of these factors can affect the degradation of soil-applied wastes follows.

Factors Affecting Degradation Processes in Soil

Water Relations--

Both residue decomposition and the biodegradation of waste chemicals require water for microbial growth and for diffusion of nutrients and by-products during the breakdown process. Water requirements for growth and survival of different microorganisms vary considerably and, hence, water availability has a selective effect on the total microflora within a system. Species composition of the soil microflora is regulated largely by water availability which, in turn, is governed essentially by the energy of the water in contact with the soil or waste. Because of the great importance of water in residue decomposition and biodegradation of wastes, a brief discussion of the concept of potential energy of water is relevant.

Water potential is a fundamental concept that can be useful for quantifying the energy status of water in soil, crop residues, and in soils amended with different waste chemicals. The concept provides a convenient way of applying thermodynamic terminology to the water relations of soil-plant-microbial-chemical waste systems. The water relations of both plants and microorganisms are best described in terms of the potential energy of the water (i.e., the water potential) because they must expend work or energy to obtain water from their immediate surroundings. A forthcoming book on the theory and application of this concept (including methods and techniques available for proper measurement) to research in soil microbiology, biochemistry, and the management of chemical wastes on land will soon be published by the Soil Science Society of America (Parr, Gardner, and Elliott, 1981).

In brief, the concept of water potential is based on its relationship with relative humidity. Taking water vapor as an ideal gas (a valid assumption over the range of vapor pressures normally encountered), the water potential is related to the relative humidity (rh) by the equation,

$$\psi = \frac{RT}{V} \log (rh)$$

where R is the ideal gas constant, T is the absolute temperature, and V is the volume of a mole of water. At equilibrium the partial molar free energies of the liquid and vapor phases are equal, so that measurements of the equilibrium rh can be used to estimate the water potential. Since rh is always equal to or less than unity, water potentials normally encountered in soil and residues are zero or negative in value. Water potential is dimensionally equivalent to pressure and is most commonly expressed in bars. The principal method for measuring the equilibrium relative humidity to obtain estimates of the water potential of a particular system is by thermocouple psychrometry (Wiebe et al., 1971).

While the concept of water potential has been widely accepted by soil scientists in recent years, there are those who continue to use such soil moisture constants as saturation capacity, water-holding capacity, field capacity, and permanent wilting point. In studies where water is a variable, these parameters have little value in predicting and interpreting the response of microorganisms in soils and crop residues, or in other natural systems involving organic wastes. Where water relations of microorganisms are investigated, both the water content and water potential should be reported. This would allow for valid comparisons of data obtained from different investigations involving different soils and experimental conditions, and different chemical wastes. Often such comparisons are virtually impossible because inappropriate constants or indices were used to express the soil water regime.

Information on optimal and marginal water potentials for growth, reproduction, and survival of individual species of microorganisms in soil is limited. Generally, with decreasing water potentials, fewer organisms are able to grow and reproduce. Bacterial activity is greatest at high water potentials (wet conditions) but is noticeably decreased at about -3 bars and markedly so at -15 bars (Clark, 1967). Some fungi, however, tend to grow and survive in soils at much lower water potentials (dry conditions). Optimum water potentials for organic matter decomposition are in the 0 to -15 bar range where most heterotrophic microorganisms exhibit maximum growth and activity. Fungal growth is often decreased in wet soils where bacteria flourish, suggesting that bacteria may be antagonistic to fungi at high water potentials, while at low potentials fungi may escape antagonism since bacteria are less active (Cook and Papendick, 1970).

Little is known of the effect of water potential on the ability of different microbial species to degrade or detoxify waste industrial chemicals. Such information is absolutely essential in developing management strategies for efficient and effective land treatment systems for chemical wastes. A thorough investigation of the effect of water potential on the degradative activity of microorganisms in soils amended with specific waste chemicals should be a high priority for future research.

Temperature--

Temperature also has a profound influence on the growth and activity of microorganisms and consequently will markedly affect the decomposition of crop residues and biodegradation of waste chemicals. Most soil microorganisms are categorized as mesophiles, that is, they exhibit maximum growth and activity in the 20 to 35°C temperature range. Soils also contain some microorganisms known as psychrophiles that grow best at temperatures below 20°C, and those that develop only at higher temperatures called thermophiles which exhibit maximum growth rates between 50 and 60°C. Sommers and Biederbeck (1973) reviewed the selective effect of temperature on the total soil microflora, and concluded that temperature increases in the mesophilic range usually result in increased numbers and activities of the microflora. They also reported that when soils were incubated at increasing temperatures from 20 to 60°C, organic matter decomposition was greatly accelerated. At the higher temperatures, actinomycetes became predominant over fungi and bacteria. Conversely, according to Stefaniak (1968), as soil incubation temperatures are decreased below 20°C the growth of actinomycetes decreased to a greater extent than either fungi or bacteria.

Composting research conducted by the United States Department of Agriculture at Beltsville, MD has shown that maximum decomposition rates of cellulosic substrates occur in the thermophilic range. The reason for this is that in composting, the substrate allows an enrichment of thermophilic microorganisms which have their optimum degradative activity between 50 and 60°C. Composting may offer considerable potential for maximizing the biodegradation rate of waste industrial chemicals. It might also provide for the initial detoxification and/or degradation of these wastes prior to their application in land treatment systems. Research is needed to fully assess the potential of composting for detoxification, degradation, and inactivation of organic and inorganic waste chemicals.

In attempting to quantify the relationship between temperature and rate of degradation, simple mathematical relationships have been applied and mesophilic temperatures have been assumed. One concept is based on a Q_{10} relationship which is empirically determined. It quantifies a rate factor at 10-degree intervals and assumes linearity over the mesophilic range. Another mathematical model for describing temperature effects is the Arrhenius equation which is a linear function and has the following relationship:

$$\frac{d \ln k}{dT} = \frac{E}{RT^2}$$

where E = activation energy of the reaction
R = gas constant
T = temperature in Kelvin
k = reaction rate constant

While chemical reactions closely follow the Arrhenius relationship over a wide range of temperatures, biological reactions are somewhat limited in temperature range. Composting, however, is a process in which biological reactions are occurring at above normal temperatures. The combination of rapid chemical reactions coupled with higher-rate biological reactions makes composting a unique chemical degradation system.

Soil pH--

While different types of soil microorganisms have different pH optima for maximum growth, the optimum pH range for rapid decomposition of wastes and residues is 6.5 to 8.5. Bacteria and actinomycetes have pH optima near neutrality and do not compete effectively for nutrients under acidic conditions, which would explain why soil fungi often become dominant in acid soils.

The solubility or availability of macro and micronutrients may be affected by pH and thus inhibit or alter biodegradation of materials. Solubility of possible toxic materials such as copper at different pH's may also influence biodegradation. Of all the macronutrients, the availability of phosphorus is influenced most by the change in pH. The reactivity of minerals in soils such as iron or calcium are influenced by pH and, in turn, influence the availability of phosphorus to microorganisms and plants by forming relatively stable complexes (Sikora and Corey, 1976).

Aeration and Oxygen Supply--

In aerobic soils, active fungi, bacteria, and actinomycetes require oxygen for respiration and oxidative assimilation during residue decomposition or waste chemical degradation. Oxygen to support residue decomposition and waste degradation in soil is supplied through diffusion. If oxygen demand by soil microorganisms and plant roots exceeds that which can be supplied by diffusion, anaerobic conditions can soon occur. Maximum rates of residue decomposition and biological/chemical degradation of waste chemicals would generally be dependent upon an adequate supply of molecular oxygen. While many soil bacteria can grow under anaerobic conditions, though less actively, most fungi and actinomycetes do not grow at all. Decomposition processes under anaerobic conditions are performed principally by bacteria that utilize either an anaerobic respiration or fermentation type of metabolism. End products of anaerobic metabolism are reduced compounds, some of which are toxic to microorganisms and plants.

Available Nutrients--

A number of researchers have shown that the addition of inorganic nutrients such as N, P, and S (Stotzky and Norman, 1961a,b; Tenney and Waksman, 1929) can accelerate the decomposition of organic substrates in soil. Where crop residues of a wide C:N ratio are applied to soil, the rate and extent of their decomposition can be increased by addition of inorganic sources of N. This has also been shown to occur under anaerobic soil conditions (Parr, Smith, and Willis, 1970).

While many wastes may indeed contain substantial amounts of potentially available nutrients such as N, P, K, and S, their rate of release (i.e., mineralization) from organic and inorganic configuration may be too slow to sustain microbial activity. Research is needed to determine the rate and extent to which the degradation of waste organic chemicals can be increased by addition of available nutrients for microbial utilization.

Carbon:Nitrogen Ratio--

Because crop residues contain about the same amount of C (approximately 40% on a dry weight basis), their N contents are often compared on the basis of C:N ratios. Thus, residues with a low N content or wide C:N ratio are often associated with a slow rate of decomposition. On the other hand, residues having a high N content and narrow C:N ratio usually decompose at a more rapid rate. Although the N content or C:N ratio of crop residue can be useful in predicting decomposition rates, they should be used with some caution since the C:N ratio says nothing about the availability of the C or N to microorganisms (Alexander, 1961). Many waste organic chemicals contain substantial amounts of C and N and could be compared on the basis of their C:N ratios for composting prior to their application to land treatment and landfarming systems. However, the validity of such comparisons will depend upon how rapidly and to what extent these waste chemicals degrade (i.e., mineralize) under specific experimental conditions.

The Residue or Waste Microflora--

This is often overlooked, i.e. that many organic wastes, including crop residues, contain indigenous populations of bacteria, fungi, and actinomycetes (Parr and Papendick, 1978; Cook, Boosalis, and Doupnik, 1978). The exact role of these microorganisms in the decomposition of crop residues, or the nature of their interaction with the soil microflora after incorporation, is virtually unknown. The same thing can be said for waste chemicals applied to land. A thorough assessment of the types and numbers of microorganisms in various waste chemicals is needed to establish their potential role in the degradation of chemical constituents.

Decomposition by Macrofauna

The role of the soil macrofauna such as insects, protozoa, earthworms, and slugs in the decomposition of organic materials is significant, but predominantly indirect. Of the total respiration associated with soils amended with organic material, 10 to 20% could be from macrofauna. Because only a few of these soil organisms have the ability to produce their own enzymes for the degradation of substrate, their main degradation feature is mechanical.

The gut of most soil animals contains microorganisms which produce the necessary enzymes for the degradation of a substrate to the extent where animal organism can absorb the nutrients. The remainder of the substrate passes into the soil where microorganisms complete the degradation.

Earthworms play a prominent role in the degradation of organic materials in soil predominantly by their movement in soil. As they feed and move they carry nutrients to deeper soil profiles or portions of the profile which previously did not contain the rich castings or crumbs of the worms. These nutrients stimulate microbial growth and decomposition. Earthworms improve the aeration of soils by their movement whereby worm channels create pathways for gaseous exchange to occur. Evidence of the importance of earthworms in soils is the presence of accumulated, dead vegetation when earthworm populations have been removed or killed by chemicals (Satchell, 1967).

Among the arthropods, the beetles and termites are most correlated with extensive degradation of organic material. Both animals often have a rich microflora in their guts which produce the enzymes which degrade the cellulosic substrates.

The role of that macrofauna play in the decomposition of hazardous wastes is minor compared to microorganisms, but it is still essential. Their role in mechanical disintergration of substrate, movement of nutrients in soils, and conditioning of soils for more efficient decomposition is vital to the overall soil treatment system. Studies involving the effects of land application of hazardous wastes should include studies on soil animals because the overall rate of decomposition can be significantly altered by the reduction in numbers of soil animals from the addition of toxic materials.

Decomposition of Residues and Organic Industrial Wastes and Soil Nitrogen Relationships

The range in total N content for a number of crop residues and organic wastes is shown in Table 2 (Parr and Papendick, 1978). Those materials which contain more than 1.5 to 1.7% N would probably need no additional fertilizer N or soil N to meet the requirements of the microorganisms during decomposition. While there is some variation, depending on the chemical composition of a specific residue or waste, this threshold N level, which corresponds to a C:N ratio of about 25 to 30, has been widely substantiated for many plant materials (Allison, 1955; Bartholomew, 1965; Harmsen and Van Schreven, 1955).

The N content of most mature crop residues, however, is considerably less than the critical level of 1.5 to 1.7% (e.g., wheat straw ranges from 0.2 to 0.6 and corn stover from 0.7 to 1.0% N), and there is immediate concern that application to soil might induce a N deficiency. Large amounts of oxidizable C from such residues creates a microbiological demand for N which can immobilize residue N and available inorganic soil N for extended periods. On the other hand, wastes such as fresh cow manure, poultry manure, sewage sludges, and some paper mill and cannery wastes may contain sufficiently high levels of N so that inorganic N (probably as ammonia) would be released through microbial decomposition soon after application to soil. The addition of supplemental inorganic fertilizer N to low N residues can

TABLE 2. RANGE IN TOTAL NITROGEN CONTENT OF SOME ORGANIC WASTES AND RESIDUES

Waste or residue	Nitrogen content (% of dry material)
Cattle feedlot manure	1.2-2.0
Fresh cow manure, litter free	2.2-2.4
Poultry manure	3.5-5.0
Municipal waste compost	0.8-1.0
Sewage sludge compost	1.0-1.6
Sewage sludges	2.0-6.0
Paper mill sludges	0.2-2.3
Cannery wastes	0.5-1.8
Corn stover	0.7-1.0
Wheat straw	0.2-0.6

accelerate their rate of decomposition. However, the increased rate of decomposition from N occurs soon after application. Subsequently, after 6 to 9 months, there will be little difference in the extent of decomposition of the residue whether supplemental N is added or not.

The Nitrogen Requirement Value--

The nitrogen requirement value (NRV) is the amount of N required by microorganisms to decompose/degrade a particular crop residue or organic chemical waste. According to Allison and Klein (1962), the NRV depends mainly on two factors: (a) the chemical composition of the residue or waste and (b) the rate of decomposition. If any of the aforementioned factors that affect the degradation processes in soil are at less than an optimum level, microbial activity will be lowered accordingly, thus slowing the rate of substrate decomposition and in turn decreasing the NRV. For example, a lowering of soil temperature would decrease the rate of microbial growth and metabolism, which would lower the NRV. Also, oxygen depletion in soils during extended flooding might soon lead to anaerobic conditions that would drastically decrease the rate of residue or waste decomposition by the microflora, which again would decrease the NRV.

Some residues or waste chemicals containing less than 1.5 to 1.7% N may immobilize little or no soil N because of their resistance to degradation, which would markedly decrease the potential demand for N by the microflora. Wastes that are readily biodegradable and containing less than the threshold N level might cause a substantial immobilization of soil and fertilizer N. Waste chemicals that are readily biodegradable, but have an initial and lasting biocidal effect on the soil microflora, would have little effect on the NRV, regardless of their N contents. It is well to remember that since N immobilization and mineralization proceed simultaneously in soils, the NRV is a net value and will vary with time (Allison and Murphy, 1963; Allison, Murphy, and Klein, 1963).

Decomposition of Residues and Organic Industrial Wastes, and Oxygen Relationships

The maximum rate of decomposition of crop residues and organic wastes in soil is usually correlated with a continuous supply of available oxygen. As mentioned earlier, excessive loadings of residues can, under some conditions, lead to depletion of O_2 and extended soil anaerobiosis with possible adverse effects on both soils and plants. The development of more effective systems for managing chemical wastes on land depends on a clear understanding of the metabolic events which can occur and their possible consequences.

Types of Microbial Metabolism which Influence Soil-Oxygen Relationships--

Aerobic respiration--by soil microorganisms involves oxidation-reduction reactions in which molecular O_2 serves as the ultimate electron acceptor, while some organic component of the crop residue functions as the electron donor or energy source. If O_2 is available for active decomposition, the system will be dominated by an array of different organisms including bacteria, actinomycetes, and fungi.

Anaerobic respiration--by soil microorganisms includes biological oxidation-reduction reactions in which inorganic compounds, rather than molecular O_2, serve as the ultimate electron acceptor. The organic waste or residue serves as the electron donor or energy source. Thus, if O_2 is depleted during decomposition of a particular waste the system will be dominated by faculative anaerobic bacteria. These microorganisms can readily utilize molecular O_2 as an electron acceptor in aerobic respiration, but under anaerobiosis they are capable of also utilizing nitrate (NO_3^-), manganic (Mn^{+4}), and ferric (Fe^{+3}) ions as electron acceptors, thereby reducing them to nitrite (NO_2^-), manganous (Mn^{+2}), and ferrous (Fe^{+2}) ions, respectively. Under strict anaerobic conditions, obligate anaerobic bacteria of the genus Desulfovibrio utilize sulfate (SO_4^{-2}) as an electron acceptor, reducing it to sulfide (S^{-2}). These organisms are capable of utilizing organic acids as electron donors in this reduction. Most soil actinomycetes and fungi are obligate aerobes and would not be active under soil anaerobiosis.

Fermentation--includes energy-yielding reactions performed by select groups of obligate and facultative anaerobic bacteria in which organic compounds serve as both electron donors and electron acceptors. Fermentation can occur in soils, particularly when extremely reduced conditions exist. These reactions can lead to an accumulation of incompletely oxidized organic compounds, including organic acids and alcohols. A special example of fermentation is performed by the methane bacteria which are obligate anaerobes capable of fermenting organic acids to yield methane (CH_4).

Succession of Reductive Processes in Soils from Heavy Loadings of Organic Wastes or Residues--
There is a great similarity in the events which occur in poorly drained or waterlogged soils compared with soils that might receive heavy loadings of organic wastes or crop residues (Parr, 1974; Parr and Papendick, 1978). As the level of molecular O_2 steadily decreases during decomposition, the type of microbial metabolism changes successively according to the oxidation-reduction state (i.e. redox potential or Eh[1]), ranging from aerobic respiration in the presence of molecular O_2 to methane fermentation under extreme anaerobiosis. The succession of microbiological events which can occur in waterlogged soils, or poorly drained soils receiving heavy application of crop residues, relative to the soil redox potential is shown in Table 3 according to Takai and Kamura (1966).

Two stages are evident in the type of microbial metabolism that governs the reductive processes. During the first stage, that which occurs early in the incubation period, oxidative decomposition of wastes and residues

---

[1] Eh is an expression of the electron density of a system. As a system becomes increasingly reduced, a corresponding increase in the electron density would occur, resulting in a progressively increased negative potential.

TABLE 3. SUCCESSION OF EVENTS RELATED TO THE REDOX POTENTIAL WHICH CAN OCCUR IN WATERLOGGED SOILS, OR POORLY DRAINED SOILS RECEIVING EXCESSIVE LOADINGS OF ORGANIC WASTES OR CROP RESIDUES

Period of incubation	Stage of reduction	System	Redox potential (Millivolts)	Nature of microbial metabolism	Formation of organic acids
Early	First stage	Disappearance of O_2	+600 to +400	Aerobes	None
		Disappearance of NO_3^-	+500 to +300	Facultative anaerobes	Some accumulation after addition of organic matter
		Formation of Mn^{+2}	+400 to +200		
		Formation of Fe^{+2}	+300 to +100		
Later	Second stage	Formation of S^{-2}	0 to -150	Obligate anaerobes	Rapid accumulation
		Formation of H_2	-150 to -220		Rapid decrease
		Formation of CH_4	-150 to -220		

proceeds through the activity of aerobic and facultative microorganisms. Carbon dioxide and, in some cases, ammonia are evolved with little or no accumulation of incompletely oxidized organic compounds. Soon after waterlogging and/or the application of excessive loading of organic wastes or residues, molecular O_2 disappears and the Eh begins to decline. In fairly rapid succession, nitrates disappear because of microbiological reduction, and formation of manganous (Mn^{+2}) and ferrous (Fe^{+2}) ions follows. During the first stage, Eh drops from +600 to about +100 millivolts (mV).

The second stage in the reductive process may occur somewhat later and is characterized by the reduction of sulfates (SO_4^{-2}) to sulfide (S^{-2}), which can result in the formation of phytotoxic hydrogen sulfide (H_2S). Sulfide formation is followed by the appearance of an array of products of incomplete organic matter decomposition such as organic acids, molecular hydrogen (H_2), and methane. Accumulated organic acids soon disappear due to their utilization as electron donors and as a source of carbon by sulfate-reducing and methane-producing bacteria. The amount of molecular hydrogen produced varies greatly while the amount of S^{-2} and CH_4 continues to increase as the Eh drops to -200 mV.

The possible consequences which might occur from the application of organic wastes or residues to soils at excessive loading rates is illustrated in Figure 1. With acceptable loadings, made in accordance with the soil's properties, O_2 balance is maintained as rapid aerobic decomposition occurs. The end products of decomposition are inorganic C, N, and S compounds. However, with excessive loadings, rapid O_2 depletion occurs, resulting in anaerobiosis, which slows the rate and extent of decomposition. This in turn results in the production of malodorous compounds such as amines, mercaptans, and H_2S which can have pronounced phytotoxic effects. Depending on the extent of overloading the soil may remain in a reduced anaerobic state for some time.

Relationship between waste loading rates, application frequency, and soil anaerobiosis--The relationship between loading rate and application frequency relative to the probable period of soil anaerobiosis is shown in Figure 2, which illustrates the effect of frequent, successive application of wastes at acceptable loading rates, compared with a single slug application at an excessive loading rate, on changes in the soil redox potential. Excessive loading rates of high BOD wastes often lead to a rapid depletion of O_2 and equally rapid decrease in soil Eh to an undesirably

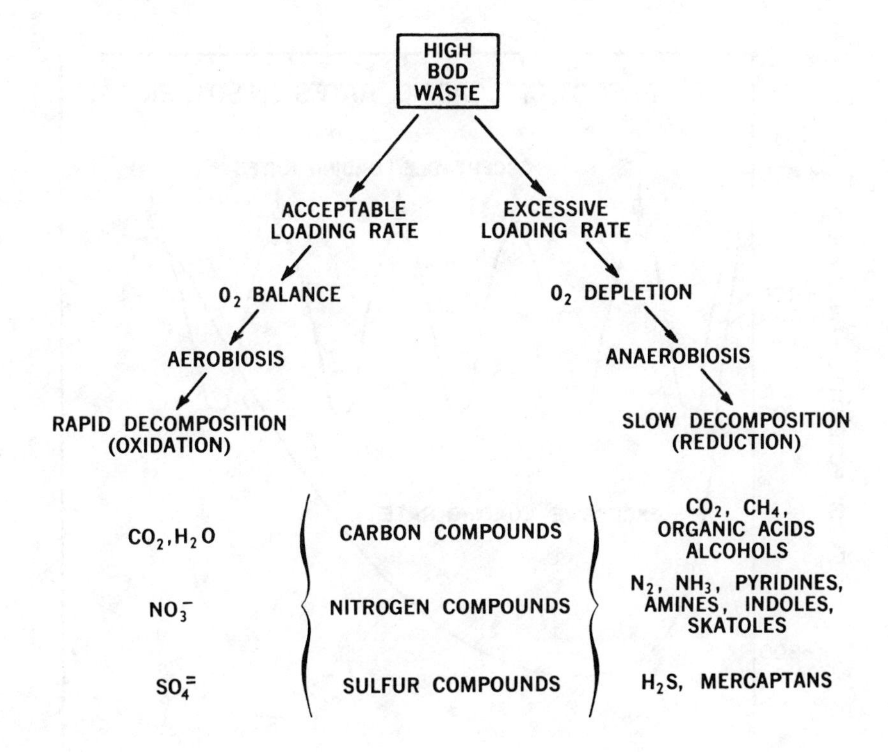

Figure 1. Flow diagram illustrating the possible consequences which can occur from application of organic wastes or crop residues to soil at excessive loading rates compared with lower acceptable loadings.

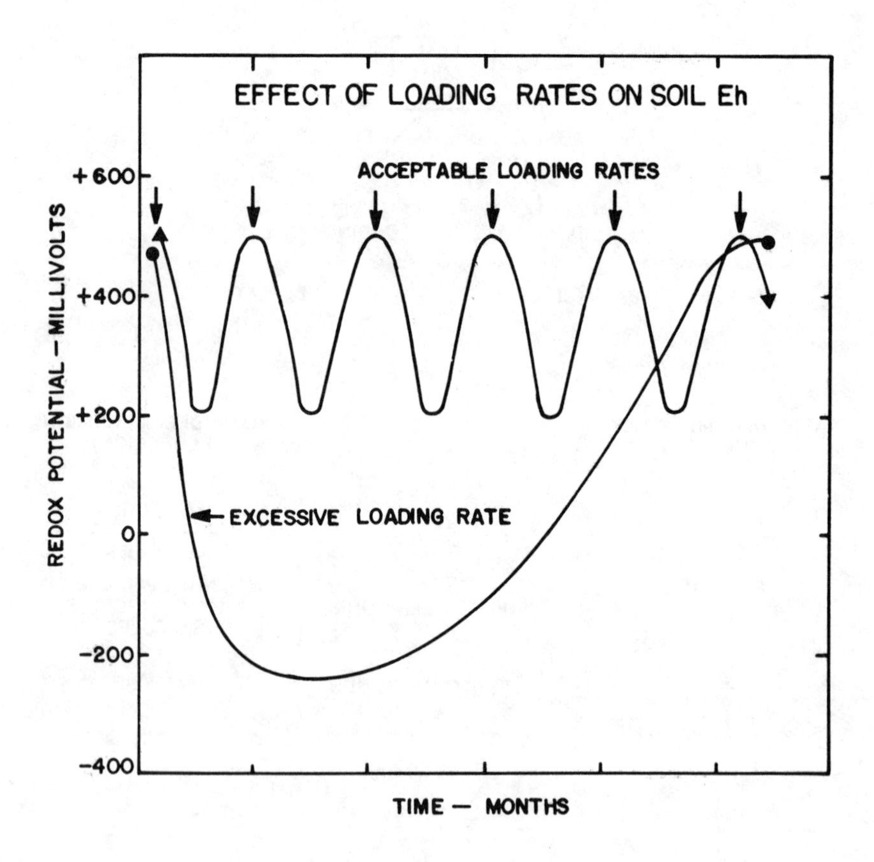

Figure 2. A hypothetical illustration of the relative effects of frequent, successive waste applications at acceptable loading rates compared with a single slug application on changes in the soil redox potential.

reduced state where it might remain for a matter of weeks or months, after which a slow upward trend may occur. On the other hand, a proper balance of loading rate and application frequency can maintain the soil Eh within a more desirable range, where extensive reduction does not occur and where the likelihood of odorous and phytotoxic end products arising is minimal.

While some downward deflection of Eh can be expected even with acceptable loading rates, the system will tend to recover rapidly because it has not been loaded excessively or beyond its limits. Thus, the Eh moves back into the aerobic range. At the beginning of this section it was stated that the maximum rate of residue/waste decomposition in soil is usually dependent upon an adequate level of available oxygen. For the most part this is true; however, it is possible that the degradation sequence of a number of waste industrial chemicals will not be entirely oxidative in nature but may indeed involve some critical reductive steps. For example, DDT is relatively resistant to biodegradation in well-aerated soils. However, it is now known that an important initial step in its degradation is a reductive one which occurs readily under soil anaerobiosis (Guenzi and Beard, 1967). Toxaphene and trifluralin are also known to degrade more rapidly under anaerobic conditions (Parr and Smith, 1973, 1976). Thus, it is conceivable that in the development of land treatment systems for waste organic chemicals it may be necessary to alternate aerobic/anaerobic soil environments to maximize the detoxification, degradation, and inactivation of certain chemicals. Carefully controlled laboratory studies will be needed to identify those chemicals that have reductive steps in their degradation sequences.

Residue and waste chemical decomposition in soil as affected by loading rate and other management factors--A number of researchers have investigated the effect of residue loading rates on decomposition in soil, a subject that has been reviewed in considerable detail by Clark (1967b). Many have reported that the percentage decomposition of residues added to soil was inversely related to the rate of residue added. Although the total cumulative production of CO_2 increases with increased residue loading rates, the percentage of residue carbon evolved as CO_2 progressively decreases. Broadbent and Bartholomew (1949) suggested that smaller percentage decomposition at higher residue loading rates may be attributable to associated physical or biotic factors restricting microbial growth and activity in a given volume of soil. According to Clark (1967b), a satisfactory explanation for this phenomenon has not yet been presented.

In a more recent study, Parnas (1975) points out that although the percentage decomposition decreases as the substrate concentration increases, the absolute rate of decomposition (rate of weight loss of residue, or rate of CO_2 evolution) increases as the substrate loading is increased. This study concluded that measuring decomposition rates in percentage tended to obscure the stimulating effect of increased substrate concentration.

It would appear that there is still a great deal to be learned about the mechanisms, processes, and interactions of factors that affect the relationship between residue loading rates and decomposition of residues, depending on their method of application to soil. Again, the same thing can be said about our present lack of knowledge on the interaction of waste industrial chemicals with soil properties and those factors that affect their decomposition. Such information is a necessary prerequisite for the development of effective and efficient land treatment systems.

In the development of such systems a number of management factors will have to be considered. These include (a) mode or method of application (i.e., surface applied, plowed down, disked in, or injected), (b) rate of application, (c) time of application, and (d) frequency of application. It is essential that we learn how these factors interact with one another and with the soil characteristics because it provides the basis for formulating a waste application scheduling program (WASP) -- one that seeks to maximize the soil's capacity to detoxify and degrade a particular waste chemical and to minimize impairment of the soil environment and contamination of surface and groundwater.

Effects of Organic Chemicals on Soil Microorganisms

Most of our knowledge of the effects of organic chemicals on soil microorganisms comes from studies on the effects of different pesticides (herbicides, insecticides, fungicides, and fumigants) on the types, numbers, and activities of microorganisms relative to various aspects of soil fertility, plant growth and development, and plant pathological relationships. The scientific literature that has accumulated on this subject over the past 20 years is indeed voluminous. A number of reviews (Fletcher, 1960; Bollen, 1961; Kreutzer, 1963, 1965; Alexander, 1969; Helling, Kearney, and Alexander, 1971; Parr, 1974) have characterized microbial responses to these chemicals as none or negligible, direct or indirect, stimulatory or inhibitory, specific or non-specific, and temporary or lasting--all depending upon the pesticides' chemical properties, rate of application, mode of application, and on soil properties, weather and climatic factors, and techniques for measuring the responses.

Despite the volume of published papers on interactions of pesticides and soil microorganisms, meaningful correlations of soil chemical and physical properties with microbiological responses to chemicals of known composition and structure are lacking. The significance of many reported results on effects of pesticides is not known. While considerable data exist on the acute effects of pesticides on the soil microflora, i.e., where soils are exposed to massive doses for brief periods, little is known about sublethal or chronic effects on soil microorganisms exposed repeatedly to recommended rates over longer periods of time. For example, the persistence of some chemicals in soil suggests the possibility of cumulative effects on various aspects of microbial metabolism.

In the land treatment of waste organic chemicals it is possible that certain compounds will exhibit greater toxic and biocidal effects on the soil microflora than others. It is also possible that in some cases partial sterilization could occur, resulting in marked quantitative and qualitative changes in the soil microflora that might require weeks or even months to reestablish a new equilibrium or climax microbial population that can again degrade waste chemicals. To avoid shock loadings and such adverse impacts on the land treatment system, it may be highly advisable to develop a method for rating the relative responses of soil microorganisms to different waste chemicals.

One such method was reported by Wensley (1953) for rating the effectiveness of selected soil fumigants. In his system he defined critical exposure concentration (c) as that concentration of a fumigant necessary to lower the population of a particular microbial group by 90% in a specific period of time. The critical exposure period (t) is that period of confined exposure of a given amount of soil to a specific fumigant concentration necessary to lower the population of a particular microbial group by 90%. The efficiency index (E) is the numerical product of these two values,

$$E = ct,$$

and has proven to be a reliable means for rating soil fumigants. The lower the critical exposure concentration and critical exposure period, the lower the efficiency index and the more lethal the fumigant. The concept and principles applied here could be adopted for evaluating the relative fungicidal and bactericidal effects of soil-applied waste chemicals.

Another concept proposed by Kreutzer (1963, 1965) might be useful in assessing/estimating the real and potential effects of waste chemicals on the soil microflora. He defines the degree of disinfestation (B) of soil by a pesticide (also referred to as the "degree of killing" or "inhibiting power") as a product of the concentration (c) at critical biological sites and the increment time of exposure (dt), according to the equation

$$B = cdt.$$

Thus, B is actually equivalent to the chemical dose or dosage.

A more accurate expression of chemical disinfestation was presented by Hemwall (1962) as

$$B = \int_o^i cdt$$

where disinfestation or biological control (B) is equal to the integral of different pesticide concentrations during the treatment (c) times the increment of exposure time (dt). The actual or effective concentration of a toxicant at a particular biological site is often somewhat lower than the theoretical maximum because of different adsorptive and absorptive interactions of the chemical with organic and inorganic colloids. Accordingly, the Hemwall equation was modified to include a soil factor (K) to account for such differences (Kreutzer, 1965):

$$B = K \int_o^i cdt$$

While this is an oversimplified expression of a highly complex situation, it does put certain factors that influence the effects of pesticides and waste chemicals into proper perspective. The concept that dose is the cumulative product of concentration and time of exposure is also discussed by Goring (1967).

Metallic Ion Effects on Microbiological Process

The application of industrial wastes to soil has created a concern that metals applied in the wastes may be detrimental to soil microbiological processes important in the degradation of organic material and in plant nutrient mineralization. Some metals, depending on their oxidation state, concentration, soil pH, and the soil organic matter content, can cause direct biocidal effects on soil microorganisms or exert inhibitory responses to their growth and metabolic processes for extended periods. If metal loadings are restricted, toxicity can be alleviated by limestone application. If metal loadings are excessive and toxicity occurs at neutral pH, such effects are difficult to correct. Removal of the contaminated soil may be the only remedial alternative. Therefore, careful consideration should be given to the metal content of industrial wastes before they are considered as candidates for land treatment.

The effects of metals on microorganisms have been summarized in reviews such as that by Gadd and Griffith (1978). A summary of studies evaluating the effects of metals on microbial activity in soils and sewage treatment processes is presented in Table 4. In most cases, inhibition of enzyme activity was recorded after a certain threshold concentration of a metal was reached. Some of the metals tested had little or no effect and some were stimulatory (Bollag and Barabasz, 1979; Drucker et al., 1974), but the latter is more likely to be associated with very low concentrations.

Important factors which governed the level of inhibition by metals in soils were (a) the metal ion or complex species tested (Cole, 1977; Bollag and Barabasz, 1979; Wildung, et al., 1976), (b) the amount of organic matter present (Wildung, et al., 1976; Drucker, et al., 1975; Mather and Rayment, 1977); (c) the amount and type of clay present (Babich and Stotzky, 1977); (d) the pH (Babich and Stotzky, 1977) and (e) the organism or enzyme system evaluated (Bollag and Barabasz, 1979; Liang and Tabatabai, 1977; Wildung et al. 1977). Only a few studies, however, evaluated microbial activity in soils that had been contaminated by metals for several years (Tyler, 1974; Mathur and Rayment, 1977). These studies are probably a more valid interpretation of metal effects because the metals were adequately equilibrated with the soils before evaluation.

Silver and Hg appeared to be the most toxic metals affecting microbial processes and Cd, Zn, Cu, Cr, Pb, and Ni also were inhibitory but at much higher concentrations. Many of these metals are used in industrial processes and end up in the solid wastes which then may be land applied. To ensure that land application of metal containing wastes does not cause lasting or permanent impairment of soil microbial activity, research is needed to evaluate (a) the effect of single metals or their combinations (representative of specific industrial wastes) on several microbial processes such as respiration, N mineralization, and phosphatase activity and (b) ways and means of alleviating the toxic effects of metals in soils by addition of organic amendments such as crop residues, municipal wastes, composts, and by pH adjustments.

TABLE 4. SUMMARY OF LITERATURE DEALING WITH BIOLOGICAL EFFECTS OF METALS

Metal	Biological Function Tested	Medium	Response	Reference
Cu, Zn	respiration, urease, acid phosphatase	soil near foundry	inhibition	Tyler (1974)
Hg, Cd	cellulase, xylanase	forest litter extracts	inhibition	Spaulding (1979)
Hg, Cd, Cn Zn, Ni, Pb	respiration	forest litter	inhibition	Spaulding (1979)
Cd, Zn, Cu	denitrification	soil	inhibition	Bollag and Barabasz (1979)
Pb	amylase, -glucosidase	soil	inhibition	Cole, 1977
Ag, Hg, Cd, Ni, Cr, Fe, Al, B, Se	nitrification	soil	inhibition	Liang and Tabatabai (1977)
Cu	acid phosphatase	soil	inhibition	Mathur et al. (1979)
Cu	cellulase, xylanase cellobiose, amylase, inulase, lichenase, lipase, respiration	soil	inhibition	Mathur and Sanderson (1980)

TABLE 4. Continued

Metal	Biological Function Tested	Medium	Response	Reference
Cu	activated sludge	2° waste treatment	increase in solids	Salotto et al. (1964)
As, Co, Mn, Sb, W	respiration	soil	no effect	Drucker et al. (1974)
Fe, Sn, Zn	respiration	soil	stimulation	
Ag, Cd, Cr, Cu, Hg, Ni, Tl	respiration	soil	inhibition	
Hg	dehydrogenase	soil	inhibition	Vanderpost and Corke (1975)
Hg, Cd, Zn, Cu	nodulation by *Rhizobium*	soil	inhibition inhibition	
Cr	carbohydrate util.	soil	inhibition	Drucker et al. (1975)

Models for Describing the Degradation of Hazardous Wastes in Soil

As previously discussed, the decomposition of crop residues and other organic wastes in soil is a complex process influenced by a number of factors (i.e., temperature, moisture, aeration, pH, C:N ratio, and particle size) and their interactions. The process is further complicated in the case of waste industrial chemicals because certain constituents can function as potent biocides and metabolic inhibitors, dissolve soil organic matter, cause drastic shifts in soil pH, and disperse inorganic colloids. All of these can adversely affect the chemical, physical, and microbiological properties of soils and greatly impair their capacity to detoxify and degrade wastes for extended periods.

Mathematical models can greatly enhance the interpretation and usefulness of information acquired by practical experience and experimentation. They may have considerable merit in optimizing the safe, effective, and efficient management of waste industrial chemicals in land treatment systems. Such models must necessarily account for the waste chemical characteristics, the soil characteristics, and interactions of the rate, method, time, and frequency of application. They should provide a basis for predicting feasible and acceptable waste application scheduling programs (WASP) that would (a) sustain and maximize the soil's degradative potential and (b) minimize environmental impact from volatilization, runoff, and leaching. An ideal model here should also predict the rate and extent of mineralization and release of plant nutrients for utilization by crops. Another useful purpose of models is that they can identify certain voids in our knowledge that were not evident before, thereby allowing proper focus and direction of research.

A number of models have recently been proposed for predicting the rate and extent of decomposition of crop residues and soil organic matter, and resulting nutrient transformations (Parnas, 1975; Gilmour, Broadbent, and Beck, 1977; Smith, 1979a,b; Frissel and van Veen, 1978, 1980; Reddy, Khaleel, and Overcash, 1981), the decomposition of animal wastes in soil (Reddy, Khaleel, and Overcash, 1981), and the decomposition and nutrient transformations of sewage sludge applied to soil (Hsieh, Douglas, and Motto, 1981a,b).

To our knowledge there have been no models reported for predicting the detoxification, degradation, or inactivation of hazardous wastes such as waste industrial chemicals in land treatment systems. Thus, there is a need to determine how accurately some of these models can predict organic matter and nutrient transformations of chemical wastes in soil once the biocidal and inhibitory effects of toxic constituents are overcome. It may be that some existing models can be adopted directly, without modification, for modeling specific waste chemicals. In some cases, however, it may be necessary to extensively modify some models to fit specific situations, or develop entirely new ones.

Of the models mentioned here the one that appears to be most flexible and adaptable for hazardous waste chemicals in land treatment systems is that of Smith (1979a,b). The model consists of four submodels that describe the fate of C, N, P, and K as plant and soil organic matter decompose. Michaelis-Menten formalism is used to describe each enzymatically mediated process within the submodels. Smith subjected this model to extensive testing and has reported excellent agreement with a number of established and predictable phenomena including (a) changes in C:N ratio during organic matter decomposition in soil, (b) patterns of mineralization and nitrogen immobilization from addition of organic substrates to soil, and (c) correlation of microbiological activity with CO_2 production as affected by alternate cycles of soil wetting and drying, and temperature fluctuations.

Summary and Research Needs

Most soils have a tremendous capacity to decompose organic chemical wastes. The soil provides (i) a medium for dilution of waste concentrates, (ii) a highly important buffering system, and (iii) a potentially active microflora for rapid and sustained degradative activity. Where wastes are applied at acceptable loading rates, the potential problems associated with waste disposal, including extended anaerobiosis and production of undesirable end products, can be prevented or minimized. However, each soil has certain physical, chemical, and biological limits as to how much of a particular chemical waste it can accommodate at one time, and if loaded excessively there will be certain deleterious effects of rapid O_2 depletion, extended anaerobiosis, chemical reduction, and the accumulation of odorous and/or phytotoxic end products which could impair the soil's fertility and productivity for some time.

Shock loadings of certain chemical wastes and their excessively toxic and biocidal effects on the soil microflora should be avoided, if possible, to prevent serious and lasting impairment of the soil's biodegradative potential. If effective and efficient land treatment/landfarming systems for chemical wastes are to be developed, research will be urgently needed to investigate interactions of the method, rate, time, and frequency of waste applications, and their impact on the soil environment.

A number of metals, including those classified as "heavy metals", are often used in industrial processes and, thus, may be constituents of waste industrial chemicals applied in land treatment systems. Some of these metals can function as biocides or exert inhibitory effects on the growth and metabolic processes of soil microorganisms. Research is needed (a) to characterize the effects of industrial metals and metal combinations on microbial growth and activity, and (b) to develop soil management practices (e.g., the use of composted organic wastes and soil pH adjustment) that can alleviate the toxic effects of metals.

While models have been developed for predicting the rate and extent of crop residues, animal manures, and sewage sludges in soil, and the nutrient transformations therefrom, no such models have been reported for the detoxification and degradation of hazardous and toxic industrial chemical wastes in land treatment systems. The applicability and adaptability of existing models to hazardous waste chemicals in land treatment systems should be thoroughly explored since they may greatly optimize the safe, effective, and efficient management of these wastes.

REFERENCES

Alexander, M. 1969. Microbial degradation and biological effects of pesticides in soil. p. 209-240. In Soil biology: Reviews of research. UNESCO, Paris.

Alexander, M. 1961. Introduction to soil microbiology. John Wiley and Sons, Inc., New York. p. 139-162, 248-271, 425-441.

Allison, F. E. 1955. Does nitrogen applied to crop residues produce more humus? Soil Sci. Soc. Am. Proc. 19:210-211.

Allison, F. E., and C. J. Klein. 1962a. Rates of immobilization and release of nitrogen following additions of carbonaceous materials and nitrogen to soils. Soil Sci. 93:383-386.

Allison, F. E., and R. M. Murphy. 1963. Comparative rates of decomposition in soil of wood and bark particles of several species of pines. Soil Sci. Soc. Am. Proc. 27:309-312.

Allison, F. E., R. M. Murphy, and C. J. Klein. 1963. Nitrogen requirements for the decomposition of various kinds of finely ground woods in soil. Soil Sci. 96:187-190.

Babich, H., and G. Stotzky. 1977. Effect of cadmium on fungi and on interactions between fungi and bacteria in soil: Influence of clay minerals and pH. Appl. and Environ. Microbial. 53:1059-1066.

Bartholomew, W. V. 1965. Mineralization and immobilization of nitrogen in the decomposition of plant and animal residues. p. 285-306. In W. V. Bartholomew and F. E. Clark (eds.) Soil nitrogen. Agronomy Monograph No. 10. Am. Soc. Agron., Madison, WI.

Bollag, J-M., and W. Barabasz. 1979. Effects of heavy metals on the denitrification process in soil. J. Environ. Qual. 8:196-201.

Bollen, W. B. 1961. Interactions between pesticides and soil microorganisms. Ann. Rev. Microbiol. 15:69-92.

Broadbent, F. E., and W. V. Bartholomew. 1949. The effect of quantity of plant material added on the rate of decomposition. Soil. Sci. Soc. Am. Proc. 13:271-274.

Clark, F. E. 1967a. Bacteria in soil. p. 15-49. In A. Burges and F. Raw (ed.) Soil biology. Academic Press, New York and London.

Clark, F. E. 1967b. The growth of bacteria in soil. p. 441-457. In T. R. G. Gray and D. Parkinson (eds.) The ecology of soil bacteria. Liverpool Univ. Press, Liverpool, England.

Cole, M. A. 1977. Lead inhibition of enzyme synthesis in soil. Appl. Environ. Microbiol. 33:262-268.

Cook, R. J., M. G. Boosalis, and B. Doupnik. 1978. Relationship of crop residues to plant disease. p. 147-163. In W. R. Oschwald (ed.) Crop residue management systems. Am. Soc. Agron., Madison, WI.

Cook, R. J., and R. I. Papendick. 1970. Soil water potential as a factor in the ecology of *Fusarium roseum* f. sp. *cerealis* 'Culmorum". Plant Soil 32:131-145.

Drucker, H., T. R. Garland, and R. E. Wildung. 1974. The effects of metals on soil microbial population and metabolism: A ranking system. Agron. Abst. p. 128.

Drucker, H., R. E. Wildung, and T. R. Garland. 1975. Effects of chromium on glucose metabolism in soil--A potential biochemical basis for the effects of chromium on soil microflora. Agron. Abst. p. 127.

Fletcher, W. W. 1960. The effect of herbicides on soil microorganisms. p. 20-62. In E. K. Woodford and G. R. Sagar (ed.) Herbicides and the soil. Blackwell Scientific Publ., Oxford, England.

Frissel, M. J., and J. A. van Veen. 1978. A critique of "computer simulation modeling for nitrogen in irrigated croplands." pp. 145-162. In D. R. Nielsen and J. G. MacDonald (ed.). Nitrogen in the Environment, Vol. I. Academic Press, New York.

Frissel, M. J., and J. A. van Veen. 1980. Simulation model for nitrogen immobilization and mineralization. Chapter 13. In I. K. Iskandar (ed.) Modeling wastewater renovation by land disposal. U.S. Army, Cold Regions Research and Engineering Laboratory, Hanover, New Hampshire. (in press).

Gadd, G. M., and A. J. Griffiths. 1978. Microorganisms and heavy metal toxicity. Microb. Ecol. 4:303-317

Gilmour, C. M., F. E. Broadbent, and S. M. Beck. 1977. Recycling of carbon and nitrogen through land disposal of various wastes. pp. 173-194. In L. F. Elliott and F. J. Stevenson (eds.) Soils For Management of Organic Wastes and Waste Waters. Am. Soc. Agron., Madison, WI. 650 p.

Goring, C. A. I. 1967. Physical aspects of soil in relation to the action of soil fungicides. Ann. Rev. Phytopathol., 5:285-318.

Goring, C.A.I. and J. W. Hamaker. 1972. Organic chemicals in the soil environment (in two volumes). Marcel Dekker, Inc., New York. 968pp.

Guenzi, W. D., and W. E. Beard. 1967. Anaerobic biodegradation of DDT to DDD in soil. Science 156:1116-1117.

Harmsen, G. W., and D. A. Van Schreven. 1955. Mineralization of organic nitrogen in soil. p. 299-398. In A. G. Norman (ed.) Advances in agronomy, Vol. 7 Academic Press, Inc., New York.

Helling, C. S., P. C. Kearney, and M. Alexander. 1971. Behavior of pesticides in soils. Advan. Agron. 23:147-240.

Hsieh, Y. P., L. A. Douglas, and H. L. Motto. 1981a. Modeling sewage sludge decomposition in soil: I. Organic carbon transformation. J. Environ. Qual. (in press).

Hsieh, Y. P., L. A. Douglas, and H. L. Motto. 1981b. Modeling sewage sludge decomposition in soil: II. Nitrogen transformations. J. Environ. Qual. (in press).

Kreutzer, W. A. 1963. Selective toxicity of chemicals to soil microorganisms. Ann. Rev. Phytopathol. 1:101-126.

Kreutzer, W. A. 1965. The reinfestation of treated soil. pp. 495-508. In K. F. Baker and W. C. Snyder (eds.) Ecology of soil-borne plant pathogens: prelude to biological control. Univ. Calif. Press, Berkeley.

Liang, C. N., and M. A. Tabatabai. 1977. Effects of trace elements on nitrogen mineralization in soils. Environ. Pollut. 12:141-147.

Mathur, S. P., H. A. Hamilton, and M. P. Levesque. 1979. The mitigating effect of residual fertilizer copper on the decomposition of an organic soil in Litn. Soil Sci. Soc. Am. J. 43:200-203.

Mathur, S. P., and A. F. Rayment. 1977. Influence of trace element fertilization on the decomposition rate and phosphatase activity of a Mesic Febrisol. Can. J. Soil Sci. 57:397-408.

Mathur, S. P. and R. B. Sanderson. 1980. The partial inactivation of degradative soil enzymes by residual fertilizer copper in histosols. Soil Sci. Soc. Am. J. 44:750-755.

Parnas, H. 1975. Model for decomposition of organic material by microorganisms. Soil Biol. Biochem. 7:161-169.

Parr, J. F. 1974. Chemical and biological considerations for land application of agricultural and municipal wastes. pp. 227-252. In Organic wastes as fertilizers. FAO Soils Bulletin NO. 27, Rome, Italy.

Parr, J. F., W. R. Gardner, and L. F. Elliott. 1981. Water potential relations in soil microbiology. SSSA Special Publ. Soil Sci. Soc. Am. Madison, WI. (in Press).

Parr, J. F., and R. I. Papendick. 1978. Factors affecting the decomposition of crop residues by microorganisms. pp. 101-129. In W. R. Oschwald (ed.) Crop residue management systems. ASA Special Publ. No. 31, Am. Soc. Agron. Madison, WI.

Parr, J. F., and S. Smith. 1973. Degradation of trifluralin under laboratory conditions and soil anaerobiosis. Soil Sci. 115:55-63.

Parr, J. F., and S. Smith. 1976. Degradation of toxaphene in selected anaerobic soil environments. Soil Sci. 121:52-57.

Parr, J. F., S. Smith, and G. H. Willis. 1970. Soil Anaerobiosis: I. Effect of selected environments and energy sources on respiratory activity of soil microorganisms. Soil Sci. 110:37-43.

Reddy, K. R., R. Khaleel, and M. R. Overcash. 1981. Carbon transformations and transport in the land areas receiving organic wastes in relation to nonpoint source pollution: a conceptual model. J. Environ. Qual. (in press).

Salotto, B. V., E. F. Barth, W. E. Tolliver, and M. B. Ettinger. 1964. Organic load and toxicity of copper to the activated - sludge process. pp. 1205-1304. In Proc. of 19th Indust. Waste Conf., Purdue Univ., Lafayette, IN.

Satchell, J. E. 1967. Lumbricidae. In Soil Biology, (ed.) A. Burges and F. Raw. pp. 259-322.

Sikora, L. J. and R. B. Corey. 1976. Fate of nitrogen and phosphorus in soils under septic tank waste disposal fields. Trans. ASAE 19:866-875.

Smith, O. L. 1979a. An analytical model of the decomposition of soil organic matter. Soil Biol. Biochem. 11:585-606.

Smith, O. L. 1979b. Application of a model of the decomposition of soil organic matter. Soil Biol. Biochem. 11:607-618.

Sommers, L. E., and V. O. Biederbeck. 1973. Tillage management principles: Soil microorganisms. pp. 87-108. In Conservation tillage: The proceedings of a national conference. Soil Conservation Society of America, Ankeny, IA.

Spaulding, B. P. 1979. Effects of divalent metal chlorides on respiration and extractable enzymatic activities of Douglas - Fir needle litter. J. Environ. Qual. 8:105-107.

Stefaniak, O. 1968. Occurrence and some properties of aerobic psychrophilic soil bacteria. Plant Soil. 29:193-204.

Stotzky, G., and A. G. Norman. 1961a. Factors limiting microbial activities in soil. I. The level of substrate, nitrogen, and phosphorus. Arch. Mikrobiol. 40:341-369.

Stotzky, G., and A. G. Norman. 1961b. Factors limiting microbial activities in soil. II. The effect of sulfur. Arch. Mikrobiol. 40:370-382.

Takai, Y., and T. Kamura. 1966. The mechanism of reduction in waterlogged paddy soil. Folia Microbiol. (Prague) 11:304-313.

Tenney, F. G., and S. A. Waksman. 1929. Composition of natural organic materials and their decomposition in the soil. IV. Nature and rapidity of decomposition of various complexes in different plant materials under aerobic conditions. Soil Sci. 28:55-84.

Tyler, G. 1974. Heavy metal pollution and soil enzymatic activity. Plant Soil 41:303-311.

Vanderpost, J. and C. T. Corke. 1975. Interactions of metals with microbial populations in soils. pp. 63-78. In Metals in the Biosphere Proc. Symp. spons. by Dept. of Land Resource Sci., Univ. of Guelph, Guelph, Ontario.

Wiebe, H. H., G. S. Campbell, W. H. Garner, S. L. Rawlins, J. W. Cary, and R. W. Brown. 1971. Measurement of plant and soil water status. Utah Agric. Exp. Stn. Bull. 484. 71 pp.

Wildung, R. E., H. Drucker, T. R. Garland, and R. A. Pelroy. 1976. Transformation of trace metals by soil micro-organisms. Agron. Abst. p. 141.

Wildung, R. E., T. R. Garland, and H. Drucker. 1977. Complexation of nickel by metal-resistant soil bacteria and fungi. Agron. Abst. p. 53.

Plant Uptake of Inorganic Waste Constituents

R.L. Chaney

Introduction

Plant uptake of macro- and micronutrients from land treatment sites is important in the management of these sites for several reasons. First, plants need certain nutrients which are essential for growth; cover crop success depends on supply of these essential nutrients. Second, many wastes contain substantial amounts of N and P. When these elements are land-applied, plants can remove much of the applied N and P while they grow, thereby protecting the groundwater. Third, excessive uptake of some essential and non-essential microelements can reduce the yield of plants or even kill them. This interferes with macronutrient removal and promotes soil erosion. Fourth, plant uptake of some elements (Cd, Mo, Se, Co) can cause health effects in the food chain. And fifth, plants may be able to extract excessive levels of some microelements from toxic land treatment sites, producing an ore resource (plant ash) and possibly rehabilitating the site for normal crops.

These points will be reviewed in some detail where they are important to management of land treatment sites. The reader will also need to review the subsections on reactions of waste constituents with soils and on potential effects of waste constituents on the food chain which consider related matters. The latter chapter discusses individual elements in detail while the present chapter presents principles.

Macronutrient Uptake by Plants

Consideration of the primary macronutrients (N, P, and K)

and secondary macronutrients (S, Ca, and Mg) is important in the management of land treatment sites. These elements are normally managed in routine crop production, since optimum supply of each element is required in order to achieve optimum crop production.

In the agronomy of wastes, these elements have added importance. Most wastes, particularly well managed biological sludges, contain substantial amounts of N and P. When sludges are applied to land, the first limiting factor for annual application rates is nearly always N (Overcash and Pal, 1979; Knezek and Miller, 1978; EPA, 1977; Jacobs, 1977; Brown, 1981). Regulations for application of sewage sludge limit annual applications in order to avoid excessive nitrate in groundwater (EPA, 1979). Sludge utilization recommendations are based on supplying the crop's requirement of N or P.

Management problems can develop when wastes contain a high C:N ratio since N is retained by the soil microbes and not made available to crops. When sludges are used to supply P, other sources of N and K are usually needed. Since sludges usually contain a low level of K, crop removal can deplete soil reserves of K when sludge is relied on to supply a crop's N and P. Other macronutrient imbalance situations can lead to grass tetany and other management problems (see section on Potential Effects of Waste Constituents on the Food-Chain). Land treatment site managers should not fear these management requirements, as farmers must consider all these factors in routine soil-crop-livestock management. Managers can rely on soil testing, plant analysis, and general crop production knowledge to avoid all these noted potential problems.

Microelement Uptake by Plants

Introduction--

Historically, both microelement deficiency in plants and toxicity to plants have been important in agriculture. Phytotoxicity was observed on soils naturally enriched in Zn, Cu, and Ni (Dykeman and de Sousa, 1966; Staker, 1942; Staker and Cummings, 1941; Antonovics, Bradshaw, and Turner, 1971; Proctor and Woodell, 1975). Further, excessive use of Cu, Zn, and As pesticides, and Cu fertilizers have caused crop loss in the field (Delas, 1963; Jones and Hatch, 1945; Lee and Craddock, 1969; Lee and Page, 1967; Reuther and Smith, 1954). Industrial waste-applied metals have also caused crop problems (Patterson, 1971).

With that background, intensive research was undertaken during the last decade to learn whether the heavy metals in sewage sludge were likely to cause phytotoxicity problems in agriculture. During this period, an area of waste-soil-plant specialization developed. Appropriate research methodologies were developed and others discarded because they had inherent errors. Many of these concepts have been reviewed, and the reader is referred to these reviews for more detailed information (Allaway, 1968; Antonovics, Bradshaw, and Turner, 1971; Baker and Chesnin, 1976; Bates, 1971; Brown, 1981; Bouwer and Chaney, 1975; Bowen, 1966, 1979; Cannon, 1960; Cataldo and Wildung, 1978; Chaney, 1973, 1975; Chaney and Giordano, 1977; Chaney and Hornick, 1978; Chapman, 1966; Dowdy, Larson, and Epstein, 1976; Foy, Chaney, and White, 1978; Kirkham, 1977; Leeper, 1978; Lisk, 1972; Loneragen, 1975; Olsen, 1972; Page, 1974; Patterson, 1971; Proctor and Woodell, 1975; Ryan, 1977; Sommers, 1980; Walsh, Sumner, and Corey, 1976; Webber, 1972, 1974; Williams, 1975). Major points important to understanding heavy metal uptake, metal translocation from roots to shoots, and metal tolerance by crops are reviewed here within the perspective of managing a site for land treatment of industrial wastes.

The question of whether sewage sludge should be used on cropland became an issue in the early 1970's. Scientific knowledge about heavy metals in agronomic ecosystems rapidly developed for several reasons. Sludge contains much higher levels of metals than do soils, even when the sludge arises from domestic (non-industrial) sources (See Table 5a) (Sommers, 1980; Chaney, 1980). Sludges enriched in metals by industrial discharges can be extremely high in metals. Also, heavy metals reside in surface soils for a prolonged period; the half-life of many metals in perennial grassland is typically considered 1000 years if erosion is managed appropriately (Bowen, 1977).

More recent British surveys have provided more detail on the distribution of metals concentrations in sludges. Table 5b summarizes these results. Sludges high or very high in Co, F, and Mo were found in these surveys (Sterritt and Lester, 1981; Rea, 1979; Davis, 1980).

Based on all this information, recommended management practices were developed for sewage sludge amended privately-owned cropland (EPA, 1977; Knezek and Miller, 1978). The best sludge utilization practice avoids heavy metal phytotoxicity and potential risk to the food chain by 1) monitoring waste composition to avoid polluted wastes, 2) managing annual and cumulative metal loadings to levels acceptable for most crops in moderately acid soils, and 3) maintaining soil pH near 6.5 to 7 to minimize microelement uptake.

TABLE 5a. CONCENTRATION OF SELECTED MICROELEMENTS IN DIGESTED SEWAGE SLUDGES AND BACKGROUND LEVELS IN SOILS (CHANEY, 1980; KIRKHAM, 1979)

	Sludge			Soils	
Element	Reported Range ppm	Typical Median ppm	Maximum Domestic ppm	Normal Range ppm	Typical Median ppm
As	1.1 - 230	10	-	0.1 - 40	6
B	4 - 1000	33	-	2. - 130	10
Cd	1 - 3410	15	25	0.01- 7	0.5
Co	1 - 260	10	-	1 - 40	8
Cu	84 - 17000	800	1000	2 - 100	20
Cr	10 - 99000	500	-	5 - 3000	100
F	2 - 739	86	-	30 - 300	200
Fe, %	0.1 - 15.4	1.7	4	0.5 - 10	2
Hg	0.6 - 56	6	10	0.01- 0.8	0.03
Mn	32 - 9870	260	-	50 - 4000	850
Mo	1.2 - 40	10	-	0.2 - 5	2
Ni	2 - 5300	80	200	5 - 5000	40
Pb	13 - 26000	500	1000	2 - 200	15
Se	1.7 - 17.2	5	-	0.01- 38	0.2
Sn	40 - 700	150	-	2 - 200	10
V	15 - 400	36	-	20 - 500	100
Zn	101 -49000	1700	2500	10 - 300	50

TABLE 5b. COMPOSITION OF SEWAGE SLUDGES IN RECENT BRITISH SURVEYS (STERRITT AND LESTER; 1981: REA, 1979; DAVIS, 1980).

Element	N	Mean	Min	20th %-ile	median	80th %-ile	Max
		----------------	mg/kg	dry sludge	----------------		
Cd	40	24.9	1.5	4.1	16.3	35.1	110
Co	40	10.5	11.3	18.9	30.5	53.2	2490
Cr	40	707.	57.1	115.	367.	774.1	5180
Cu	40	721.	170.	416.	577.	1060.	2080
F	88	366	80.	160.	260.	450.	33500
Mn	40	667.	131.	208.	314.	516.	6120
Mo	40	16.2	0.1	1.0	4.3	16.1	214
Ni	40	290.	10.9	28.1	112.	430.	2020
Pb	40	1550.	27.5	147.	299.	655.	45400
Sn	40	57.5	2.6	130.	14.0	8.7	329
Zn	40	1930	300.	767.	1140.	1890.	9210

Factors Affecting Microelement Uptake --

Microelement properties-- Each element has its unique chemical and physical characteristics in waste-soil-plant systems. If the compounds of an element are essentially insoluble at practical soil pH levels (5.5-8), then that element has a very low concentration in the soil solution and cannot be absorbed at an appreciable rate. If an element is adsorbed or chelated very strongly by the soil, even though it is not precipitated, it has low uptake. If an element is weakly adsorbed, and not precipitated, then the element is subject to plant uptake or leaching through the soil.

Soil properties-- As discussed previously, soils adsorb and/or chelate many microelements. Adsorption occurs on hydrous oxides of Mn and Fe, clays, organic matter, and other soil minerals. Organic matter can chelate microelements. Adsorption, chelation, and dissolution of precipitated mineral forms of an element, are all pH-dependent. Cations are weakly bound at lower pH, strongly bound at high pH. Selenite and molybdate (anions) are more strongly sorbed at low pH than at high pH. Boron forms soluble H_3BO_3 at low pH and greater plant uptake occurs at low pH.

The pH of the soil immediately adjacent to plant roots (the rhizocylinder) is important in plant uptake of metals. Uptake occurs after movement (diffusion) of the metal from the soil particles to the root surface. When roots absorb NH_4^+, the pH of the rhizocylinder soil declines, and when the roots absorb NO_3^-, the pH rises (Barber, 1974; Smiley, 1974). The form of N absorbed by the root has a strong influence on metal uptake (Barber, 1974). Most crop N is absorbed as NO_3-N which raises rhizosphere pH (Nye, 1981). Use of NH_4-fertilizers also causes the pH of the bulk soil to decline since H^+ is generated when NH_4^+ is oxidized to NO_3^- (Jolley and Pierre, 1977). Application of limestone corrects soil acidity. Applying excessive limestone minimizes metal cation uptake, but promotes uptake of anions (Mo, Se).

Soil pH and organic matter are the soil factors most important in plant uptake of microelements. Other factors which influence uptake (soil temperature, soluble salts, added soluble chelators, soil moisture status, and fertility) have been reviewed (Foy, Chaney, and White, 1978; CAST, 1980; Sommers, 1980).

Source properties-- Researchers have noted two types of major errors in experiments conducted to evaluate potential metal uptake into crops (CAST, 1980). First, the source of metals added may strongly affect the result; and Second, the location in which the experiment is conducted may affect the result. The first error is generally called the "salt vs sludge" error. When metals are added as soluble salts, they generally cause greater plant uptake and toxicity than when applied as environmentally relevant forms such as sewage sludge or metal oxides in stack emissions. The wastes should be much nearer to equilibrium with sludge organic matter binding sites, or in sparingly soluble inorganic compounds, or occluded in $CaCO_3$ or other minerals. Sludge organic matter adds metal sorption capacity to the soil (Soon, 1981), and raises the soil C.E.C.; further, sludge adds hydrous oxides of Fe and other elements which can adsorb metals (Garcia-Miragaya and Page, 1978). Usually, the sludge source raises the pH of the sludge-soil mixture, while metal salts lower the pH by displacing adsorbed H+ from the soil. Soluble salts are greatly increased by the sulfate or chloride of the metal salts. Numerous authors have reported results in agreement with the above description (Singh, 1981; Cunningham, Keeney, and Ryan, 1975; Dijkshoorn, Lampe, and Broekhoven, 1981; Dowdy and Ham, 1977.

The second error is generally called the "greenhouse vs field" error. Greenhouse studies offer greater manageability and reproducibility, and lower cost than field studies. However, researchers have found that crop Cd, Zn, and Mn concentrations are increased 1.5 to 5 fold over field studies of the same soil, sludge, and crop. This appears to result from 1) use of NH_4-N fertilizers which lowers soil pH more in pots than field; 2) higher soluble salt levels in greenhouse pots than field due to smaller soil volume for required fertilizer salts; 3) confinement of plant roots to the small volume of treated soil in pots; and 4) abnormal watering of soil required in pots. The smaller the pots, the greater the error. DeVries and Tiller (1978) and deVries (1980), reported larger effects. Another common error in greenhouse pot studies is inadequate supply of required fertilizer nutrients to obtain maximum growth rates (Terman, 1974).

Although pot studies in greenhouse and growth chamber allow the control needed to characterize details of soil-plant interactions, most researchers agree that regulations must be based on field research.

Plant factors-- Crop plants differ widely in uptake of an element, all other factors held constant (Chaney and Giordano, 1977; Sommers, 1980). Growing on the same soil, spinach may contain 10 times more Zn than tall fescue, orchardgrass 15 times more Ni than corn, and chard 5 times more Cu than tall fescue.

Some plant differences are inherent in the uptake by roots (can be observed in nutrient solutions). Other differences in metal uptake are due to soil-plant interactions, and can be observed only in soil (pots in the greenhouse) studies. And still other plant differences can result from differences in root distribution in the soil with depth, and can only be found in field studies.

To date, plant differences are discovered by empirical research. Although specialists can select appropriate crops for specific metal-rich soils, they have a very limited data base to work from. Climate and soil drainage must also be considered in selecting crops for land treatment sites. If properties of the waste other than metals also limit selection of cover crops, crop selection may be difficult. Experimental plantings should be initiated on difficult sites so that management options are available.

Table 6 shows relative metal tolerance of a number of crops grown on a sludge-amended soil. Several crops (orchardgrass, oat, perennial ryegrass) were unexpectedly effective in Ni uptake, and were injured by the absorbed Ni. Among crop plants, metal uptake is closely related to metal tolerance. Crops which are less effective in absorbing a metal from soil are more tolerant of that metal in soil.

Factors Affecting Microelement Translocation--

After a microelement enters the root cells, its translocation to shoots is controlled by metal and plant characteristics. Root cell sap contains high levels of organic acids and amino acids which can chelate many elements. Membrane surfaces and proteins contain functional groups which can chelate some metals. Thus, a metal can be caught in the roots if chelates formed in the root cells sap can not be transported into the xylem. Xylem is the system of non-living tubes in plants in which water and nutrients are translocated from roots to shoots. Most metals reaching the xylem are pumped into it by specialized cells. These cells, and chelates formed in the root cytoplasm, control whether a plant translocates a metal.

Generally, Zn, Cd, Mn, B, Se, and Mo are easily translocated because they are weakly chelated. Copper, Ni, and Co are more strongly chelated; a much smaller portion of the absorbed Cu is translocated to shoots than of Zn. Lead, Cr, and Hg are so strongly held in the root cells that very little is translocated to the shoots of crop plants. Research has characterized chelation of Fe, Ni, Cu, Co, Zn, and Cd in xylem sap, but only Fe citrate has been unequivocally identified (Tiffin, 1967, 1971, 1972, 1977; Foy, Chaney, and White, 1978; White, Chaney, and Decker, 1981; Cataldo, Garland, and Wildung, 1981). Amino acids control translocation of Ni and Cu in crop plants (Tiffin, 1971, 1977; Thompson and Tiffin, 1974; Cataldo, Garland, and Wildung, 1978; Cataldo et al., 1978). Citrate probably chelates Zn and Cd in xylem sap (White, Chaney, and Decker, 1981; Chino and Baba, 1981), although Cataldo et al., (1981) concluded that plant-absorbed Cd appeared in non-citrate complexes.

Many crops form storage or reproductive organs (edible roots or tubers; fruits; seed) which are used as food or feed rather than whole plant shoot. Crops differ widely in botanical type of storage organ formed, and in translocation of microelements into the organ as it forms. The stored fat, protein, and starch come via phloem from foliar photosynthesis. Some species have close control on composition of their storage organs (corn; beans; fruits), while storage organs of other crops readily increase in microelements when the leaves are increased (wheat, oat, soybeans; root crops) (CAST, 1980).

A further source of difference among crops can be expressed as a result of food processing. When many grains are processed into "refined" flour products, the starchy endosperm is separated from the mineral and fiber rich bran. Metals in rice, wheat, and corn refined products are substantially lower than in whole grain products (Hinesly et al., 1979; Chino, 1981; Kitagishi and Obata, 1981). However, oat groats contain the bulk of metals in oat grain (Kirleis, Sommers, and Nelson, 1981), and soybean cotyledons and normal soy protein products are as high in Cd as the whole grain (Braude et al., 1980).

TABLE 6. RELATIVE SENSITIVITY OF CROPS TO SLUDGE-APPLIED HEAVY METALS (CHANEY AND HUNDEMANN, 1980; CHANEY et al., 1978)

Very Sensitive[1]	Sensitive[2]	Tolerant[3]	Very Tolerant[4]
chard	mustard	cauliflower	corn
lettuce	kale	cucumber	sudangrass
redbeet	spinach	zucchini squash	
carrot	broccoli		smooth bromegrass
turnip	radish	flatpea	'Merlin' red fescue
peanut	tomato		
	marigold	oat	
ladino clover		orchardgrass	
alsike clover	zigzag, Red	Japanese bromegrass	
	Kura and		
	crimson clover		
crownvetch		Switchgrass	
'Arc' alfalfa		Red top	
white sweetclover		Buffelgrass	
yellow sweetclover	alfalfa	Tall fescue	
	Korean lespedeza	Red fescue	
Weeping lovegrass	Sericea lespedeza	Kentucky bluegrass	
Lehman lovegrass	Blue lupine		

Table 6. (continued)

Very Sensitive[1]	Sensitive[2]	Tolerant[3]	Very Tolerant[4]
Deertongue	Birdsfoot trefoil		
	Hairy vetch		
	Soybean		
	Snapbean		
	Timothy		
	Colonial bentgrass		
	Perennial ryegrass		
	Creeping bentgrass		

1/ Injured at 10% of a high metal sludge at pH 6.5 and at pH 5.5.

2/ Injured at 10% of a high metal sludge at pH 5.5, but not at pH 6.5.

3/ Injured at 25% high metal sludge at pH 5.5, but not at pH 6.5, and not at 10% sludge at pH 5.5 or 6.5.

4/ Not injured even at 25% sludge, pH 5.5.

Microelement Tolerance by Plants

Plant tolerance of a microelement controls whether the plant grows on element-enriched soils, and tolerance is closely related to concentration of that element in roots or shoots. The concentration of an element in an edible plant tissue is important because it strongly affects whether the food chain could be at risk. Because excessive Cd, Se, Mo, and Co in healthy plants can injure animals, plant tolerance of these elements is otherwise irrelevant (see section on Food-Chain effects).

In managing land treatment sites for industrial wastes, only soil pH and crop species grown are of practical significance. If the site is near the maximum cumulative loading of a phytotoxic metal (Zn, Cu, Ni, Co), improper pH management or localized low pH spots could allow severe phytotoxicity. In this situation, all other soil properties (OM, soluble salts, temperature, P-fertility) are trivial in comparison with small changes in pH. On the other hand, if crops which are very tolerant of the major toxic metals in the site soils are selected as cover crop, soil pH management may not be so critical to maintain the crop.

Crop Plants --

Differences among plant species in tolerance of soil metals depends both upon uptake-translocation of the element by plants, and the tolerance of plant cells to the element. Most crop plants suffer yield reduction at a narrow range of concentrations of metal in leaves (for Zn, Cu, Ni, and Mn).

Boawn (1971), and Boawn and Rasmussen (1971) found that yield reduction due to Zn phytotoxicity occurred at about 500 ppm Zn in leaves of many crops. Only a few leafy vegetables (chard, spinach, etc.) are more tolerant of foliar Zn (Baxter, Chaney, and Kinlaw, 1974). Most crop plants tolerate 25-40 ppm foliar Cu, and about 50-100 ppm foliar Ni or Co.

These same plants which have similar tolerance of foliar metal concentrations vary widely in tolerance of soil metals. The range of soil metal tolerance among crop plants is 10-to-20-fold for Zn, Cu, and Ni.

Unfortunately, little research has been conducted to characterize relative uptake and tolerance of many metals by many crops. Crop tolerance to different metals is independent; thus one plant can tolerate Zn but not Cu, and another can tolerate Cu but not Zn. Greenhouse and field research is badly needed. Experimental designs should reflect methods developed in studying sludge-applied metals

(Foy, Chaney and White, 1978; Chaney et al., 1978). Recent field studies in Britian have shown phytotoxicity to vegetable crops in the field when excessive Zn, Cu, or Ni was sludge-applied; Cr did not injure crops (Marks, Williams, and Chumbley, 1980; Webber, 1980; Williams, 1980).

General crop tolerance is shown in Table 6. Sufficient information is available to manage land treatment sites if soil pH is managed at 6.5 to 7. Legume crops apparently do not tolerate metals at lower soil pH, perhaps because of the Zn-by-Mn interaction (White, Chaney, and Decker, 1979a). Because legume crops are important in vegetation persistence, soil pH and metal loadings should be managed to allow growth of legume crops. More research with legumes is desirable to determine optimum pH, and tolerable metal loadings. Metal interactions also merit study (White, Decker, and Chaney, 1979; White, Chaney, and Decker, 1979a, 1979b; Mitchell, Bingham, and Page, 1978).

Cultivar differences in metal uptake or tolerance have been found for crop plants. Soybean (White, Decker, and Chaney, 1979) and snapbean (Polson and Adams, 1970) tolerance of Zn did not vary widely among cultivars. Corn inbreds varied widely in uptake-translocation of Zn and Cd (Hinesly et al., 1978; CAST, 1980), while northeastern corn hybrids (Bache et al., 1981) and lettuce cultivars (Chaney and Munns, unpublished) differed little in Cd uptake.

Ecotype Variation in Metal Tolerance --

In the 1950's and 1960's, a number of plants were found to be unusually tolerant of heavy metals. Some accumulated high amounts of one metal or another (Cannon, 1960; Antonovics, Bradshaw, and Turner, 1971). Research was initiated to learn how these plants tolerated metals and how they could be used in biogeochemical prospecting for metal ores.

Some species were found to have developed metal-tolerant ecotypes, or local strains, where selection pressure for tolerance was high (mine waste dumps, smelter polluted soils, etc.). The ecotype could be 10 or more times more tolerant of a metal than individual plants of the species growing on normal soils. The fascinating details of mechanism of tolerance, etc., have been reviewed by several authors (Antonovics, Bradshaw, and Turner, 1971; Foy, Chaney, and White, 1978).

Bradshaw (1977), Humphreys and Bradshaw (1977), Johnson, McNeilly, and Putwain (1977), Smith and Bradshaw (1972), McNeilly and Johnson (1978), and Wild and Wiltshire (1971a, 1971b) attempted to use these metal tolerant strains to revegetate metal toxic soils such as mining wastes. A few very tolerant crops are now available. 'Merlin' red fescue is being used to revegetate soils in the U.S. polluted with Zn smelter wastes.

Metal tolerant species and strains include both those which tolerate high amounts of foliar metals (accumulators) and others which have normal metal tolerance of their foliage (excluders or non-translocators). Wild (1970) found both Ni accumulators with foliar Ni over 2000 ppm and Ni-tolerant grasses which had low foliar Ni at the same Ni-rich site. Where available, it would seem wiser to introduce excluder type tolerant species and strains to minimize risk to the food chain.

Hyperaccumulators of Microelements--

The term "hyperaccumulator" was recently introduced to describe a few plant species which have developed extreme uptake-translocation of microelements (over 1000 ppm dry leaves for Ni) without phytotoxicity occuring. These species appear to be endemic to metal-rich natural sites; they developed this capability over millenia.

A few species reach foliar Ni levels over 1% (see below), foliar Cu levels over 1% (Malaisse, et al., 1978), foliar Zn levels up to 5% (Cannon, 1960; Shimwell and Laurie, 1972), and foliar Co levels over 1.5% (Brooks, 1977). These metal tolerant hyperaccumulators appear to offer a unique opportunity to bioconcentrate undesirable metals from land treatment sites while generating an economically valuable metal ore crop.

The Ni hyperaccumulator case-- Information about Ni uptake and translocation will be noted here as an example. Research on Ni hyperaccumulators has progressed rapidly during the last 3 years. Enough information is available to consider different aspects of this example.

Crop plants tolerate only 50-100 ppm foliar Ni. A few crops are much more tolerant of soil Ni (corn, smooth bromegrass) than others which are very sensitive to Ni in acid soils (orchardgrass, peanut, oat). In Ni rich acid soils, nearly every crop plant suffers Ni-induced Fe-deficiency (Roth, Wallihan, and Sharpless, 1971; Hunter and Vergnano, 1952; Halstead, Finn, and MacLean, 1969; Chaney et al., 1978a; Anderson, Meyer, and Mayer, 1973; Patterson, 1971; Mitchell, Bingham, and Page, 1978; Crooke, 1956; Nicholas, Lloyd-Jones, and Fisher, 1957).

Ni-tolerant ecotypes of grasses were discovered on Ni-rich soils in several parts of the world. Only *Alyssum* bertolini was thought to be an accumulator. Then Wild (1970, 1971) reported both types of Ni tolerance: perennial shrubs which accumulated high levels of Ni, and tolerant grasses which did not translocate Ni to shoots.

Lyon et al. (1968, 1970, 1971) reported Ni and Cr levels in some natural vegetation of New Zealand serpentine soil regions. Several had somewhat high Ni levels. Wild (1970, 1971) described Dicoma nicoliffera which hyperaccumulated Ni. Cole (1973) noted an Australian Ni-hyperaccumulator; Severne and Brooks (1972) found the same species with 1% Ni in dry leaves, and 23% Ni in leaf ash; both Ni and Co were accumulated. These results suggested that plants might eventually be used to mine metals from the soil, and research was intensified.

Brooks, Lee, and Jaffre (1974) found other Ni-accumulators. Subsequently, Jaffre et al. (1976) reported on Serbertia acuminata. The common name is "blue sap" after a blue-green latex excreted when the tree is cut. The sap contains 25.7% Ni (dry weight), but is comparatively low in Co. The plant discriminated between Ni and Co. Subsequently Brooks and his colleagues identified the Ni compound present in the sap as Ni(Ni citrate) (Lee et al., 1977, 1978; Kersten et al., 1980). Many Ni-hyperaccumulators appear to use chelation with citrate to protect the foliage from Ni toxicity. Some use malate and malonate rather than citrate. Brooks et al. (1977, 1979) then worked with herbarium specimens of plants from areas with Ni-rich soils to identify other Ni-hyperaccumulators within several genera. Alyssum contained 45 hyperaccumulators out of 168 recognized species. They conducted pot studies with normal and hyperaccumulator species and found normal species suffered phytotoxicity with 50-300 ppm foliar Ni while hyperaccumulators grew well with as much as 19,000 ppm foliar Ni (Brooks et al., 1979; Morrison et al., 1980). If leaves are 1% Ni, leaf ash could be about 25% Ni, a rich ore. If plants can separate Co and Ni, their ash would be even more valuable.

Such extreme hyperaccumulators are available for Ni, Zn, Cu, Co, and Cr. Although far too little is presently known about these plants to plan their use for phyto-extraction of heavy metals from soils, this concept is useful to consider if even this best case for removal might be helpful in managing land treatment sites.

Let us presume that two crops could be grown on a land treatment site rich in Ni. One is the Ni-tolerant corn which would suffer yield reduction at 100 ppm, while the other is a Ni-hyperaccumulator which accumulates 10,000 ppm before yield reduction is clear. The site can be managed to grow each crop at the pH level which just causes visual symptoms of phytotoxicity. The soil contains 1000 ppm Ni in the surface 15 cm. Corn yields 10 dry Mt silage/ha which contains 100 ppm Ni; the whole crop contains 1 kg Ni/ha, 0.05% of the total soil Ni. The hyperaccumulator yields only 5 Mt dry shoots/ha, containing 10,000 ppm Ni. This crop contains 50 kg Ni/ha, and removes 2.5% of the 2000 kg Ni/ha-15 cm surface soil. Upon ashing, the corn ash is only 0.25% Ni, too low to be of economic use, while the hyperaccumulator ash is 25% Ni, a rich ore. Theoretically, by repeatedly cropping the site and removing 50 kg/yr, one could remove half the total soil Ni in 20 years.

This scenario suggests that research might prove useful on working out agronomic systems for metal hyperaccumulators. Until the recent research was conducted by Brooks and colleagues, this discipline appeared to have only academic interest, or possible application in biogeochemical prospecting.

Summary and Research Needs

Wastes may supply enough macronutrients to act as N, P, or K fertilizers and provide optimum macronutrients in forage plants. Alternatively, the waste may provide N at a C:N ratio that causes nitrogen deficiency and low protein levels. Macroelement fertility balance must be managed carefully to obtain both optimum plant growth and forage quality. Growing forage crops is an effective way to remove waste-applied nutrients.

Use of sewage sludge as N and P fertilizer has led to severe K deficiency unless K fertilizer is supplied. Woody wastes cause severe N deficiency unless adequate N is supplied or the wastes are composted with N fertilizer before application. Long term sewage effluent use can deplete soil Mg and Mn and cause greater susceptibility to grass tetany (Mg deficiency of cattle and sheep) or Mn deficiency in sensitive crops.

Some microelements are so insoluble or strongly sorbed in soil (at pH 6-8) that plant shoots do not have significantly increased concentrations of the elements even when soils are greatly enriched (Ti, Cr^{3+}, Si, Pb, F, Zr, Au, Ag, Pd, Pt, Sn, Ga, Hg). Other elements may be increased somewhat in crops, but not enough to adversely affect animals, because the element causes phytotoxicity (As, Zn, Cu, Ni, Be, B, Fe, Mn, Ba, V, Al), the element is tolerated well by animals and/or the maximum increased level in plants is not yet toxic to animals. Under some soil and crop management conditions, foliar levels of Cd, Mo, Se, and/or Co can increase enough to cause adverse health effects.

Plants absorb elements from the soil solution. When chelating or complexing agents (EDTA) or high levels of soluble salts of cloride and sulfate are applied to soils, microelement cations are complexed or chelated, and thus made more soluble. This solubility increases metal diffusion from soil to root and thereby increases plant uptake. Other factors which increase soluble metals (acidifying soils containing microelement cations such as Cd, or Zn, or liming soils containing Mo or Se) cause greater uptake by plants.

Translocation from roots to shoots is usually controlled by chemical properties of plants and elements. Some elements are bound very strongly in the roots after uptake, and essentially are not translocated (Hg, Pb, Cr^{3+}). Other elements are chelated by organic acids and/or amino acids in the root cells, and translocated at low rates as anionic or neutral charged chelates (Cu, Ni, Co, Fe). Some are freely translocated anions (SeO_4^{2}, SeO_3^{2}, MoO_4^{2}). Still other cations are weakly chelated in the roots, and relatively freely translocated to plant shoots (Zn, Cd, Mn).

Microelements can only injure a plant if they are absorbed by the plant. The elements which are relatively easily absorbed and translocated by plants have historically caused phytotoxicity (Zn, Ni, Cu, Mn) and/or toxicity to animals (Se, Mo). For several elements, translocation occurs so freely that the phytotoxic injury occurs directly in the shoots (Zn, Mn, Cd). For some other elements (Cu, Ni, Co, Be), shoot yield is reduced indirectly by metal-induced Fe-deficiency or by decreased root growth limiting nutrient or water uptake from the soil. Research has recently begun to assess microelement toxicity effects on specific biochemical processes within plants. For example, excess Zn and Co interfere with transport of sugars from mesophyll cells into the phloem system.

Research Needs--

Although much has been learned about Zn, Cu, Ni, and Cd in soils and plants from research on use of sewage sludge, most was done under conditions (high pH; tolerant crops; low metal content sludges) which avoided phytotoxicity. It is difficult to extrapolate from sewage sludge to even more poorly characterized industrial wastes. Thus, phytotoxicity of the major toxic metals (Zn, Cu, Ni) present in industrial wastes needs study regarding soil pH and organic matter effects, crop and cultivar differences in uptake tolerance, and residue in plant shoots and edible tissues at 25% yield reduction. Can excess soil Cr^{+3} become phytotoxic in high MnO_2 soils? Other elements present in industrial wastes need studies on nearly all apsects of uptake and tolerance (Co, Be, Sn, V, W). Other elements which appear to be no problem under practical pH conditions may need study to verify that problems should not occur (Ba).

Some hyperaccumulators of heavy metals have been found among plants inhabiting soil naturally high in metals. These plants may be sufficiently enriched in metals that their ash is a metal ore. It may be possible to use these crops to remove excessive metal burdens from land treatment sites while generating a resource. Alternatively, metal-tolerant, metal-excluder ecotypes of some crops may be useful to provide effective, persistent plant ground cover for sites which are toxic for most crop plants. Thus, many areas of plant uptake and tolerance of heavy metals need research to facilitate planning, management, and closure of land treatment sites for industrial wastes.

REFERENCES

Allaway, W. H. 1968. Agronomic controls over the environmental cycling of trace elements. Adv. Agron. 20:235-274.

Anderson, A. J., D. R. Meyer, and F. K. Mayer. 1973. Heavy metal toxicities: Levels of nickel, cobalt, and chromium in the soil and plants associated with visual symptoms and variation in growth of an oat crop. Aust. J. Agric. Res. 24:557-571.

Antonovics, J., A. D. Bradshaw, and R. G. Turner. 1971. Heavy metal tolerance in plants. Adv. Ecol. Res. 7:1-85.

Bache, C. A., W. H. Gutenmann, W. D. Youngs, J. G. Doss, and D. J. Lisk. 1981. Absorption of heavy metals from sludge-amended soil by corn cultivars. Nutr. Rep. Int. 23:499-503.

Baker, D. E., and L. Chesnin. 1976. Chemical monitoring of soils for environmental quality and animal and human health. Adv. Agron. 27:305-374.

Barber, S. A. 1974. Influence of the plant root on ion movement in soil. pp. 525-564. In E. W. Carson (ed.) The Plant Root and its Environment. Univ. Press of Virginia, Charlottesville, VA.

Bates, T. E. 1971. Factors affecting critical nutrient concentrations in plants and their evaluation: A review. Soil Sci. 112:116-130.

Baxter, J. C., R. L. Chaney, and C. S. Kinlaw. 1974. Reversion of Zn and Cd in Sassafras sandy loam as measured by several extractants and by Swiss chard. Agron. Abstr. 1974:23.

Boawn, L. C. 1971. Zinc accumulation characteristics of some leafy vegetables. Commun. Soil Sci. Plant Anal. 2:31-36.

Boawn, L. C. and P. E. Rasmussen. 1971. Crop response to excessive zinc fertilization of alkaline soil. Agron. J. 63:874-876.

Bouwer, H., and R. L. Chaney. 1975. Land disposal of wastewater. Adv. Agron. 26:133-176.

Bowen, H. J. M. 1966. Trace Elements in Biochemistry. Academic Press, New York. 241pp.

Bowen, H. J. M. 1979. Environmental Chemistry of the Elements. Academic Press, New York. 334pp.

Bradsahw, A. D. 1977. The evolution of metal tolerance and its significance for vegetation establishment on metal contaminated sites. In Proc. Intern. Conf. Heavy Metals in the Environment II(2):599-622.

Braude, G. L., A. M. Nash, W. J. Wolf, R. L. Carr, and R. L. Chaney. 1980. Cadmium and lead content of soybean products. J. Food Sci. 45:1187-1189, 1199.

Brooks, R. R. 1977. Copper and cobalt uptake by Haumaniastrum species. Plant Soil. 48:541-544.

Brooks, R. R., J. Lee, R. D. Reeves, and T. Jaffre. 1977. Detection of nickeliferous rocks by analysis of herbarium specimens of indicator plants. J. Geochem. Explor. 7:49-57.

Brooks, R. R., J. Lee, and T. Jaffre. 1974. Some New Zealand and New Caledonian plant accumulators of nickel. J. Ecol. 62:493-499.

Brooks, R. R., R. S. Morrison, R. D. Reeves, T. R. Dudley, and Y. Akman. 1979. Hyperaccumulation of nickel by Alyssum Linneaus (Cruciferae). Proc. Roy. Soc. Lond. B203:387-403.

Brown, K. W., and Associates, Inc. 1980. Hazardous Waste Land Treatment. EPA Draft Report No. SW-874. 974pp.

Cannon, H. L. 1960. Botanical prospecting for ore deposits. Science 132:591-598.

CAST, 1980. Effects of sewage sludge on the cadmium and zinc content of crops. Council for Agricultural Science and Technology Report No. 83. Ames, IA. 77pp.

Cataldo, D. A., T. R. Garland, and R. E. Wildung. 1978. Nickel in plants. I. Uptake kinetics using intact soybean seedlings. Plant Physiol. 62:563-565.

Cataldo, D. A., T. R. Garland, and R. E. Wildung. 1981. Cadmium distribution and chemical fate in soybean plants. Plant Physiol. 68:835-839.

Cataldo, D. A., T. R. Garland, R. E. Wildung, and H. Drucker. 1978. Nickel in plants. II. Distribution and chemical form in soybean plants. Plant Physiol. 62:566-570.

Cataldo, D. A., and R. E. Wildung. 1978. Soil and plant factors influencing the accumulation of heavy metals by plants. Environ. Health Perspect. 27:145-149.

Chaney, R. L. 1973. Crop and food chain effects of toxic elements in sludges and effluents. pp. 129-141. In Recycling Municipal Sludges and Effluents on Land. Nat. Assoc. State Univ. and Land-Grant Coll., Wash., D.C.

Chaney, R. L. 1975. Metals in plants - absorption mechanisms, accumulation, and tolerance. Proc. Symp. Metals in the Biosphere. Dept. Land Resource Science, Univ. Guelph, Guelph, Ontario. pp. 79-99.

Chaney, R. L. 1980. Health risks associated with toxic metals in municipal sludge. pp. 59-83. In G. Bitton et al. (eds.) Sludge -- Health Risks of Land Application. Ann Arbor Science Publishers, Inc., Ann Arbor, MI.

Chaney, R. L., and P. M. Giordano. 1977. Microelements as related to plant deficiencies and toxicities. pp. 234-279. In L. F. Elliot and F. J. Stevenson (eds.) Soils for Management of Organic Wastes and Waste Waters. American Society of Agronomy, Madison, WI.

Chaney, R. L., and S. B. Hornick. 1978. Accumulation and effects of cadmium on crops. pp. 125-140. In Proc. First International Cadmium Conference. Metals Bulletin Ltd., London.

Chaney, R. L., P. T. Hundemann, W. T. Palmer, R. J. Small, M. C. White, and A. M. Decker. 1978. Plant accumulation of heavy metals and phytotoxicity resulting from utilization of sewage sludge and sludge composts on cropland. pp. 86-97. In Proc. Natl. Conf. Composting Municipal Residues and Sludges. Information Transfer, Inc., Rockville, MD.

Chapman, H. D. 1966. Diagnostic criteria for plants and soils. Univ. Calif., Division Agr. Sci., Riverside. 793pp.

Chino, M. 1981. Uptake-transport of toxic matals in rice plants. pp. 81-94. In K. Kitagishi and I.Yamane (eds.) Heavy Metal Pollution of Soils of Japan. Japan Scientific Societies Press, Tokyo.

Chino, M. and A. Baba. 1981. The effects of some environmental factors on the partitioning of zinc and cadmium between roots and tops of rice plants. J. Plant Nutr. 3:203-214.

Cole, M. M. 1973. Geobotanical and biogeochemical investigations in the sclerophyllosis woodland and shrub associations of the Eastern goldfields area of Western Australia, with particular reference to the role of Hybanthus Floribundas (Lindl.) F. Muell. as a nickel indicator and accumulator plant. J. Appl. Ecol. 10:269-320.

Crooke, W. M. 1956. Effect of soil reaction on uptake of nickel from a serpentine soil. Soil Sci. 81:269-276.

Cunningham, J. D., D. R. Keeney, and J. A. Ryan. 1975. Phytotoxicity and uptake of metals added to soils as inorganic salts or in sewage sludge. J. Environ. Qual. 4:460-462.

Davis, R. D. 1980. Uptake of fluoride by ryegrass grown in soil treated with sewage sludge. Environ. Pollut. B1:277-284.

Delas, J. 1963. The toxicity of copper accumulating in soils. Agrochimica 7:258-288.

deVries, M. P. C. 1980. How reliable are results of pot experiments? Commun. Soil Sci. Plant Anal 11:895-902.

deVries, M. P. C., and K. G. Tiller. 1978. Sewage sludge as a soil amendment, with special reference to Cd, Cu, Mn, Ni, Pb, and Zn -- Comparison of results from experiments conducted inside and outside a greenhouse. Environ. Pollut. 16:231-240.

Dijkshoorn, W., J. E. M. Lampe, and L. W. van Broekhoven. 1981. Influence of soil pH on heavy metals in ryegrass from sludge-amended soil. Plant Soil 61:277-284.

Dowdy, R. H., and G. E. Ham. 1977. Soybean growth and elemental content as influenced by soil amendments of sewage sludge and heavy metals: Seedling studies. Agron. J. 69:300-303.

Dowdy, R. H., W. E. Larson, and E. Epstein. 1976. Sewage sludge and effluent use in agriculture. pp. 138-153. In Land Application of Waste Materials. Soil Conservation Society of America, Ankeny, IA.

Dykeman, W. R., and A. S. de Sousa. 1966. Natural mechanisms of copper tolerance in a copper swamp forest. Can. J. Bot. 44:871-878.

E.P.A. 1977. Technical Bulletin. Municipal Sludge Management: Environmental Factors. EPA 430/9-27-004. MCD-28.

E.P.A. 1979. Criteria for classification of solid waste disposal facilities and practices. Federal Register 44(179):53438-53464.

Foy, C. D., R. L. Chaney, and M. C. White. 1978. The physiology of metal toxicity in plants. Ann. Rev. Plant Physiol. 29:511-566.

Garcia-Miragaya, J., and A. L. Page. 1978. Sorption of trace quantities of cadmium by soils with different chemical and mineralogical composition. Water, Air, Soil Pollut. 9:289-299.

Halstead, R. L., B. J. Finn, and A. J. MacLean. 1969. Extractability of nickel added to soils and its concentration in plants. Can. J. Soil. Sci. 49:335-342.

Hinesly, T. D., D. E. Alexander, E. L. Ziegler, and G. L. Barrett. 1978. Zinc and cadmium accumulation by corn inbreds grown on sludge-amended soils. Agron. J. 70:425-428.

Hinesly, T. D., V. Sudarski-Hack, D. E. Alexander, E. L. Ziegler, and G. L. Barrett. 1979. Effect of sewage sludge applications on phosphorus and metal concentrations in fractions of corn and wheat kernels. Cereal Chem. 56:283-287.

Humphreys, M. O., and A. D. Bradshaw. 1977. Genetic potentials for solving problems of soil mineral stress: heavy metal toxicities. pp. 95-105. In M. J. Wright (ed.) Plant Adaptation to Mineral Stress in Problem Soils. Cornell Univ., Ithaca, NY.

Hunter, J. G., and O. Vergnano. 1952. Nickel toxicity in plants. Ann Appl. Biol. 39:279-284.

Jacobs, L. W. (ed.) 1977. Utilizing municipal sewage wastewaters and sludges on land for agricultural production. North Central Regional Extension Publication No. 52. 75pp.

Jaffre, T., R. R. Brooks, J. Lee, and R. D. Reeves. 1976. Sebertia acuminata: a hyperaccumulator of nickel from New Caledonia. Science 193:579-580.

Johnson, M. S., T. McNeilly, and P. D. Putwain. 1977. Revegetation of metalliferous mine spoil contaminated by lead and zinc. Environ. Pollut. 12:261-277.

Jolley, V. D., and W. H. Pierre. 1977. Soil acidity from long-term use of nitrogen fertilizer and its relationship to recovery of the nitrogen. Soil Sci. Soc. Am. J. 41:368-373.

Jones, J. S., and M. B. Hatch. 1945. Spray residues and crop assimilation of arsenic and lead. Soil Sci. 60:277-288.

Kersten, W. J., R. R. Brooks, R. D. Reeves, and T. Jaffre. 1980. Nature of nickel complexes in Psychotria douarrea and other nickel-accumulating plants. Phytochem. 19:1963-1965.

Kirkham, M. B. 1977. Trace elements in sludge on land: Effects on plants, soils, and groundwater. pp. 209-247. In R. C. Loehr (ed.) Land as a Waste Management Alternative. Ann Arbor Sci. Publ. Inc., Ann Arbor, MI.

Kirkham, M. B. 1979. Trace elements. pp. 571-575. In R. W. Fairbridge and C. W. Finkl (eds.) The Encyclopedia of Soil Science. Part 1. Dowden, Hutchinson, and Ross, Inc., Stroudsburg, PA.

Kirleis, A. W., L. E. Sommers, and D. W. Nelson. 1981. Heavy metal content of groats and hulls of oats grown on soil treated with sewage sludges. Cereal Chem. 58:530-533.

Kitagishi, K., and H. Obata. 1981. Accumulation of heavy metals in rice grains. pp. 95-104. In K. Kitagishi and I. Yamane (eds.) Heavy Metal Pollution in Soils of Japan. Japan Scientific Societies Press, Tokyo.

Knezek, B. D., and R. H. Miller (eds.) 1978. Application of sludges and wastewaters on agricultural land: A planning and educational guide. EPA-MCD-35.

Lee, C. R., and G. R. Craddock. 1969. Factors affecting plant growth in high-zinc medium: II. Influence of soil treatments on growth of soybeans on strongly acid soil containing zinc from peach sprays. Agron. J. 61:565-567.

Lee, C. R., and N. R. Page. 1967. Soil factors influencing the growth of cotton following peach orchards. Agron. J. 59:237-240.

Lee, J., R. D. Reeves, R. R. Brooks, and T. Jaffre. 1977. Isolation and identification of a citrato-complex of nickel from nickel-accumulating plants. Phytochem. 16:1503-1505.

Lee, J., R. D. Reeves, R. R. Brooks, and T. Jaffre. 1978. The relation between nickel and citric acid in some nickel-accumulating plants. Phytochem. 17:1033-1035.

Leeper, G. W. 1978. Managing the Heavy Metals on the Land. Marcel Dekker, Inc., New York. 121pp.

Lisk, D. J. 1972. Trace metals in soils, plants, and animals. Adv. Agron. 24:267-325.

Loneragen, J. F. 1975. The availability and absorption of trace elements in soil-plant systems and their relation to movement and concentrations of trace elements in plants. pp. 109-134. In D. J. D. Nicholas and A. R. Egan (eds.) Trace Elements in Soil-Plant-Animals Systems. Academic Press, Inc., New York.

Lyon, G. L., R. R. Brooks, P. J. Peterson, and G. W. Butler. 1968. Trace elements in a New Zealand serpentine flora. Plant Soil 29:225-240.

Lyon, G. L., R. R. Brooks, P. J. Peterson, and G. W. Butler. 1970. Some trace elements in plants from serpentine soils. N. Z. J. Sci. 13:133-139.

Lyon, G. L., P. J. Peterson, R. R. Brooks, and G. W. Butler. 1971. Calcium, magnesium, and trace elements in a New Zealand serpentine flora. J. Ecol. 59:421-429.

Malaisse, F., J. Gregoire, R. R. Brooks, R. S. Morrison, and R. D. Reeves. 1978. Aeolanthus biformifolius De Wild.: A hyperaccumulator of copper from Zaire. Science 199:887-888.

Marks, M. J., J. H. Williams, and C. G. Chumbley. 1980. Field experiments testing the effects of metal contaminated sewage sludges on some vegetable crops. pp. 235-251. In Inorganic Pollution and Agriculture. Min. Agr. Fish. Food Reference Book 326, Her Majesty's Stationary Office, London.

McNeilly, T., and M. S. Johnson. 1978. Mineral nutrition of copper - tolerant browntop on metal-contaminated mine spoil. J. Environ. Qual. 7:483-486.

Mitchell, G. A., F. T. Bingham, and A. L. Page. 1978. Yield and metal composition of lettuce and wheat grown on soils amended with sewage sludge enriched with cadmium, copper, nickel, and zinc. J. Environ. Qual. 7:165-171.

Morrison, R. S., R. R. Brooks, and R. D. Reeves. 1980. Nickel uptake by Alyssum species. Plant Sci. Lett. 17:451-457.

Nicholas, D. J. D., C. P. Lloyd-Jones, and D. J. Fisher. 1957. Some problems associated with determining iron in plants. Plant Soil 8:367-377.

Nye, P. H. 1981. Changes of pH across the rhizosphere induced by roots. Plant Soil 61:7-26.

Olsen, S. R. 1972. Micronutrient interactions. pp. 243-264. In J. J. Mortvedt, P. M. Giordano, and W. L. Lindsay (eds.) Micronutrients in Agriculture. Soil Sci. Soc. Am., Madison, WI.

Overcash, M. R. and D. Pal. 1979. Design of land treatment systems for industrial wastes - theory and practice. Ann Arbor Science Publishers, Inc., Ann Arbor, MI. 684 pp.

Page, A. L. 1974. Fate and effects of trace elements in sewage sludge when applied to agricultural lands. A literature review study. U. S. Environ. Prot. Agency Rept. No. EPA-670/2-74-005. 108 pp.

Patterson, J. B. E. 1971. Metal toxicities arising from industry. In Trace Elements in Soils and Crops. Min. Agr. Fish. Food, Tech. Bull. 21:193-207.

Polson, D. E., and M. W. Adams. 1970. Differential response of navy beans to zinc. I. Differential growth and elemental composition at excessive zinc levels. Agron. J. 62:557-560.

Proctor, J., and S. R. J. Woodell. 1975. The ecology of serpentine soils. Adv. Ecol. Res. 9:255-366.

Rea, R. E. 1979. A rapid method for the determination of fluoride in sewage sludges. Water Pollut. Contr. 78:139-142.

Reuther, W., and P. F. Smith. 1954. Toxic effects of accumulated copper in Florida soils. Soil Sci. Soc. Fla., Proc. 14:17-23.

Roth, J. A., E. F. Wallihan, and R. G. Sharpless. 1971. Uptake by oats and soybeans of copper and nickel added to a peat soil. Soil Sci. 112:338-342.

Ryan, J. A. 1977. Factors affecting plant uptake of heavy metals from land application of residuals. pp. 98-105. In Proc. Nat. Conf. Disposal of Residues on Land. Information Transfer, Inc., Silver Spring, MD.

Severne, B. C., and R. R. Brooks. 1972. A nickel-accumulating plant from Western Australia. Planta 103:91-94.

Shimwell, D. W., and A. E. Laurie. 1972. Lead and zinc contamination of vegetation in the Southern Pennines. Environ. Pollut. 3:291-301.

Singh, S. S. 1981. Uptake of cadmium by lettuce (Lactuca sativa) as influenced by its addition to a soil as inorganic forms or in sewage sludge. Can. J. Soil Sci. 61:19-28.

Smiley, R. W. 1974. Rhizosphere pH as influenced by plants, soils, and nitrogen fertilizers. Soil Sci. Soc. Amer. Proc. 38:795-799.

Smith, R. A. H., and A. D. Bradshaw. 1972. Stabilization of toxic mine wastes by the use of tolerant plant populations. Trans. Inst. Min. Metall. 81:A230-A237.

Sommers, L. E. 1980. Toxic metals in agricultural crops. pp. 105-140. In G. Bitton et al. (eds.) Sludge -- Health Risks of Land Application. Ann Arbor Science Publ., Inc., Ann Arbor, MI.

Soon, Y. K. 1981. Solubility and sorption of cadmium in soils amended with sewage sludge. J. Soil Sci. 32:85-95.

Staker, E. V. 1942. Progress report on the control of zinc toxicity in peat soils. Soil Sci. Soc. Amer. Proc. 7:387-392.

Staker, E. V., and R. W. Cummings. 1941. The influence of zinc on the productivity of certain New York peat soils. Soil Sci. Soc. Amer. Proc. 6:207-214.

Sterritt, R. M., and J. N. Lester. 1981. Concentrations of heavy metals in forty sewage sludges in England. Water, Air, Soil Pollut. 14:125-131.

Terman, G. L. 1974. Amounts of nutrients supplied for crops grown in pot experiments. Commun. Soil Sci. Plant Anal. 5:115-121.

Thompson, J. F., and L. O. Tiffin. 1974. Fractionation of nickel complexing compounds of peanut seeds with ion exchange resins. Plant Phys. 53 (Suppl.): Abstract 125.

Tiffin, L. O. 1967. Translocation of manganese, iron, cobalt, and zinc in tomato. Plant Physiol. 42:1427-1432.

Tiffin, L. O. 1971. Translocation of nickel in xylem exudate of plants. Plant Physiol. 48:273-277.

Tiffin, L. O. 1972. Translocation of micronutrients in plants. pp. 199- 229. In J. J. Mortvedt, P. M. Giordano, and W. L. Lindsay (eds.) Micronutrients in Agriculture. Soil Sci. Soc. Am., Madison, WI.

Tiffin, L. O. 1977. The form and distribution of metals in plants: an overview. pp. 315-334. In Biological Implications of Metals in the Environment. Proc. 15th Annu. Hanford Life Sciences Symp. ERDA-TIC-Conf. No. 750929 (NTIS).

Walsh, L. M., M. E. Sumner, and R. B. Corey. 1976. Consideration of soils for accepting plant nutrients and potentially toxic nonessential elements. pp. 22-47. In Land Application of Waste Materials. Soil Conserv. Soc. Am., Ankeny, IA.

Webber, J. 1972. Effects of toxic metals in sewage on crops. Water Poll. Contr. 71:404-413.

Webber, J. 1974. Sludge handling and disposal practises in England. Sludge Handling and Disposal Seminar, Toronto, Sept. 18-19. Environment Canada and Ontario Min. Environ. pp. B1-B17.

Webber, J. 1980. Metals in sewage sludge applied to the land and their effects on crops. pp. 222-234. In Inorganic Pollution and Agriculture. Min. Agr. Fish. Food Reference Book 326. Her Majesty's Stationary Office, London.

White, M. C., R. L. Chaney, and A. M. Decker. 1979a. Differential cultivar tolerance in soybean to phytotoxic levels of soil Zn. II. Range of soil Zn additions and the uptake and translocation of Zn, Mn, Fe, and P. Agron. J. 71:126-131.

White, M. C., R. L. Chaney, and A. M. Decker. 1979b. Role of roots and shoots of soybean in tolerance to excess soil zinc. Crop Sci. 19:126-128.

White, M. C., R. L. Chaney, and A. M. Decker. 1981. Metal complexation in xylem fluid. III. Electrophoretic evidence. Plant Physiol. 67:311-315.

White, M. C., A. M. Decker, and R. L. Chaney. 1979. Differential cultivar tolerance in soybean to phytotoxic levels of soil Zn. I. Range of cultivar response. Agron. J. 71:121-126.

Wild, H. 1970. Geobotanical anomalies in Rhodesia. 3. The vegetation of nickel-bearing soils. Kirkia 7, Suppl. 1-62.

Wild, H. 1971. The taxonomy, ecology and possible method of evolution of a new metalliferous species of Dicoma cass. (Compositae). Mitt. Bot. Staatssamml. Munchen 10:266-274.

Wild, H., and G. H. Wiltshire. 1971a. The problem of vegetating Rhodesian mine dumps examined. Chamber Mines J. 13(11):26:30.

Wild, H., and G. W. Wiltshire. 1971b. The problem of vegetating Rhodesian mine dumps examined. II. Suggestions for future research and practical trials. Chamber Mines J. 13(12):35-37.

Williams, J. H. 1975. Use of sewage sludge on agricultural land and the effects of metals on crops. Water Pollut. Contr. 74:635-644.

Williams, J. H. 1980. Effect of soil pH on the toxicity of zinc and nickel to vegetable crops. pp. 211-218. In Inorganic Pollution and Agriculture. Min. Agr. Fish. Food Reference Book 326. Her Majesty's Stationary Office, London.

Fate of Toxic Organic Compounds in Land-Applied Wastes

D.D. Kaufman

Although man has made use of chemicals for centuries, it is only within the last three decades that man has become seriously concerned with the environmental fate and behavior of those chemicals he releases into the environment. Despite this comparatively short period of concern a vast amount of knowledge concerning the environmental significance of chemicals has been generated.

An understanding of the processes that operate on organic chemicals in soil is imperative if we are to minimize their undesirable effects and ultimately devise methods for controlling their persistence. Several processes are known to determine the fate of organic chemicals in soil. These include physical, chemical, and biological processes such as leaching, adsorption, desorption, photodecomposition, oxidation, hydrolysis, and adsorption and metabolism by plants and soil microorganisms. Physical, chemical, and biological properties of the organic chemical, the soil, and environmental variables act directly, indirectly, and/or interact in many combinations to influence an organic chemical's reactivity in the environment. One or two of these processes often predominate in effecting loss of a particular chemical or class of chemicals in soil.

Sorption of Organic Chemicals in Soil

Sorption is perhaps the most important single factor affecting the behavior of organic chemicals in the soil environment. Adsorption to soil constituents will affect the rate of volatilization, diffusion, or leaching, as well as the availability of chemicals to microbial or chemical degradation, or uptake by plants or other organisms. The relevance of adsorption to organic chemical behavior in soil has been reviewed by numerous authors (Burchfield, 1960; Kreutzer, 1960; Goring, 1962, 1967; Munnecke, 1967; Bailey and White, 1970).

Sorption Models--

The concentration of an organic chemical in soil, water, or air depends on the physical, chemical, and biological properties of the organic chemical and the soil, and their interactions. Sorption is chemical or physical. Anionic or cationic compounds are chemically sorbed, whereas nonionic compounds are physically sorbed. Hydrogen bonding is intermediate between physical and chemical sorption. Physical sorption generally involves several layers and low binding strengths. Chemical sorption involves high binding strength and, while several layers may be present, only the first layer is chemically bonded to the surface. The Langmiur isotherm equations generally apply to compounds adsorbed in monomolecular layers. *S*-shaped isotherms have been obtained from volatile compounds in soils, and the Brunauer, Emmett, and Teller (BET) equations have been applied to these cases, based on the assumption that the divergency from linearity is due to multimolecular adsorption resulting from action of van der Waals forces (Jurinak and Volman, 1957). Although the soundness of this basic theory has been questioned (Burchfield, 1960), the BET equations have been widely used.

More recently, however, adsorption equilibria have been described by the empirical Freundlich equation:

$$C_S = KC_W{}^{1/n}$$

$$\text{or } \ln C_S = \ln K + \frac{1}{n} \ln C_W$$

$$\text{or } \log x/m = \log K + \frac{1}{n} \log C_e$$

where C_S is the amount (x) of adsorbed chemical per unit amount of adsorbent (m); Cw = Ce is the equilibrium solution concentration of the chemical; and K and n are constants. The Freundlich adsorption coefficient, K, can be obtained by plotting log x/m versus log Ce, yielding a linear curve of slope 1/n. The units of x/m and Ce are often in µg/g and µg/ml, respectively. When C_e is 1 ppm (1 µg/ml), the corresponding value of log x/m is equal to log K. Deviation of the value of n from 1, a common observation, reflects the nonlinearity of the adsorption process. If n were 1, the K would be identical with the partition coefficient.

The adsorption coefficient, K, is commonly reported on a whole soil basis. Hamaker and Thompson (1972) have compiled such data from many investigations; the results for individual chemicals have great variation, reflecting in part the various soils and experimental conditions. Distribution constants ranged from 0, for 3,4-dichloro-*o*-anisic acid, to 14,000, for 1,1,1-trichloro-2,2-bis(*p*-chlorophenyl)ethane (DDT). One method for standardizing adsorption data is based on the high correlation of adsorption with soil organic matter content. Thus, the Freundlich K, divided by the percent soil organic matter content yields Q, a value expressing the grams of chemical adsorbed per unit of organic matter. Some authors prefer the expression K_{OC}, i.e., adsorption based on organic C content, since this quantity is determined without the additional conversion factor (commonly 1.724) needed to derive soil organic matter content.

Sorption Processes--

Several mechanisms or combinations of mechanisms can be postulated for sorption of organic compounds: (1) physical adsorption - adsorption due to van der Waals forces (a summation of dipole-dipole interactions, dipole-induced dipole interactions, and induced dipole-induced dipole interactions, ion dipole interactions), (2) hydrogen bonding, (3) coordination complexes, and (4) chemical adsorption. There is some question of whether to classify hydrogen bonding as physical adsorption or chemical adsorption. Also it is realized that while not all mechanisms occur simultaneously, two or more may occur simultaneously depending on the nature of the functional group and the acidity of the system.

Physical adsorption involves van der Waals forces which result from short range dipole-dipole interactions of several kinds. The dispersion interaction (induced dipole-induced dipole) appears to be the most important factor in determining the van der Waals forces for simple molecules. Due to the complex nature of the soil colloid-cation-water system, it is not feasible to treat the individual interactions which are responsible for van der Waals forces.

Chemical adsorption by soils and soil constituents can occur by at least four different mechanisms: (1) ion exchange, (2) protonation at the silicate or colloidal surface by reaction of the base with the hydronium ion on the exchange site, (3) protonation in the solution phase with subsequent adsorption of the organic molecule via ion exchange, and (4) in systems having water of hydration, protonation by reaction with the dissociated protons from residual water present on the surface or in coordination with the exchangeable cation. Considerable information is available in the literature describing chemical adsorption of organic chemicals in soil. The fact that ion exchange occurs with organic cations is very well documented (Smith, 1934; Gieseking, 1939; Hendricks, 1941; Grim, Allaway, and Cuthbert, 1947; Jordan, 1949; Reay and Barrer, 1957; Cowan and White, 1958; McAtee, 1959; Kuritenko and Mikhalyuk, 1959; Dodd and Satyabrata, 1960; Black and van Winkle, 1961; Sutherland and MacEwan, 1961; McAtee, 1962; Weiss, 1963; Kul'chitskii, 1961; Greenland and Quirk, 1962; Kinter and Diamond, 1963; Rowland and Weiss, 1963; Kesaree, Demirel, and Rosauer, 1960; Brooks, 1964; McAtee and Hachman, 1964; Greenland, 1965; Farqui, Okuda, and Williamson, 1967; Bodenheimer and Heller, 1968). Adsorption due to protonation may occur at or near the surface. Protonation is a very important process, especially in regard to adsorption of organic molecules which are basic in nature.

As indicated above, there is difficulty in classifying hydrogen bonding as physical adsorption or chemical adsorption. Hadzi, Kloflutar, and Oblak (1968) have suggested that there is a parallel between hydrogen bonding and protonation. Protonation may be considered as a full charge transfer from the base (electron donor) to the acid (proton) electron acceptor ---- the hydrogen bond is a partial charge transfer. Low (1961) has suggested that hydrogen bonding may occur at clay surfaces between the surface oxygens and protons of adsorbed molecules, such as water, due to the distortion of the lone-pair electrons of surface oxygen-atoms by the protons of the adsorbed molecules. For those organic compounds possessing a basic chemical character and containing an N-H group, adsorption could occur by formation of a hydrogen bond between the amino group and the oxygen of the clay surface. This would be a prime mechanism for the adsorption of the molecular form of the basic organic compounds.

The importance of functional groups in influencing bonding to the surface of clays has been considered by Brindley and Thompson (1966). They conclude that chain molecules terminating in -OH, in -COOH, and in $-NH_2$ readily form complexes with montmorillonite, whereas similar molecules terminating in -Cl and -Br do not.

Coordination compounds or metal complexes are compounds that contain a central atom or ion, usually a metal surrounded by a cluster of ions or molecules. The complex is formed by the donation of electron pairs by the ligand and the acceptance by the metal (usually a transition metal ion) resulting in the filling or partial filling of inner "d" orbitals (i.e., a coordinate covalent bond). Coordination type bonding may be quite important in determining the fate and behavior of organic molecules in soil. Ashton (1961) deduced from chromatographic evidence that 3-amino-s-triazole (amitrole) formed complexes in solution with nickel, cobalt, and copper ions, but did not react with magnesium, manganese, ferric or ferrous iron. Sund (1956) similarly showed that 3-amino-s-triazole formed "complexes" with the transition metal ions - nickel, cobalt, and iron - as well as with the alkaline earth metal magnesium. Presumably, these were coordination complexes. Aliphatic amines, alcohols, and heterocyclic ring compounds have also been observed to form coordinate complexes.

Factors Affecting Adsorption and Desorption--

Characteristics of both the adsorbent (adsorbing substance) and the adsorbate (substance adsorbed) influence the sorption process. The properties of the adsorbent which influence its behavior in interactions with the adsorbate are primarily related to the area and configuration of the surface, and to the magnitude, distribution, and intensity of the electrical field at the surface (Bailey and White, 1970). Properties which determine the role of the adsorbate in adsorption and desorption of organic chemicals in soils are: (1) chemical character, shape, and configuration; (2) acidity or basicity of the molecule (pKa or pKb); (3) water solubility; (4) charge distribution on the organic cation; (5) polarity; (6) molecular size; and (7) polarizability. There are also a number of extraneous factors which are important to the sorption process: (1) soil pH; (2) surface acidity; (3) temperature; (4) soil organic matter content and type; (5) moisture; and (6) clay content and type (Greenland, 1965; Bailey, White, and Rothberg, 1968).

Soil organic matter and clay--Considerable effort has been expended to determine the relative importance of clay and organic matter in sorption. The task is difficult because both clay and organic matter sorb chemically and physically. Hartley (1964) suggested that clay had "preoccupied" most researcher's minds, and that it was not important with large uncharged molecules. It is now generally accepted that organic matter is most important, except in dry soils where clay content and specific surfaces are the principal factors regulating sorptive capacity.

Soil pH and moisture--Soil pH and moisture are critical factors affecting sorption of some organic chemicals in soil. The pH of soil as well as the charge on clays affects the dissociation of organic chemicals and thus is very important in soil processes. Frissel (1961) concluded that all compounds are adsorbed strongly at low pH; anionic substances are adsorbed negatively at slightly basic conditions; and nonionic compounds are moderately adsorbed.

According to Crafts (1962), cationic compounds are immediately affected by the base-exchange complex in moist soil in a manner similar to the rapid immobilization of ammonia by soil. A similar effect occurs with quaternary nitrogen compounds. Although they may be precipitated as insoluble compounds of soil bases, nonpolar and anionic compounds are not so affected by soil pH. The pH effect is strikingly exemplified by 2-sec-butyl-4,6-dinitrophenol (Harris and Warren, 1964) (Figure 3) which is strongly adsorbed at low pH, but only slightly adsorbed at high pH. A similar effect has been observed in three soils with methyl isothiocyanate (Munnecke and Martin, 1964). Although the slope for the three soils varied, the amount of methyl isothiocyanate released increased linearly as the soil pH increased.

Sorption of some organic chemicals is more complete in dry soils than in wet soils. Sorption of S-ethyl dipropylthiocarbamate (EPTC) on dry montmorillonite was attributed to coordination of cations with the carbonyl group or nitrogen, and bonding of the hydrogen on the methylene groups to the clay surfaces. Although stable complexes were formed in a humid atmosphere, EPTC was completely displaced when the EPTC-clay complex was added to water. Goring (1967) expressed doubt that any nonionic toxicant could compete successfully with water for the surface of clay minerals, which accounts for observations in which toxicants sorbed by dry soils or clays are rapidly released by the addition of water.

Wade (1954, 1955), Call (1957a,b), and Jurinak and Volman (1957) observed that ethylene dibromide was rapidly sorbed by dry soils, but the amount sorbed dropped off sharply as moisture contents increased. Call (1957a) examined ethylene dibromide sorption on 20 different soils held at moisture contents corresponding to field capacity. The data were expressed as sorption coefficients and correlations with surface area, organic matter content, moisture content, and clay content were obtained. Clay content and specific surface area were the principal factors regulating sorptive capacity of dry soils, whereas organic matter was critical at field capacity. Similar results were obtained with cis- and trans-1,3-dichloropropene, but effects of temperature and isomeric form were also noted. The amount adsorbed at 2°C was about three times greater than that adsorbed at 20°C. The trans-isomer was more strongly adsorbed than the cis-isomer.

Temperature--Since organic wastes are polycomponent systems, the influence of temperature on the behavior of these materials upon addition to the soil system may not

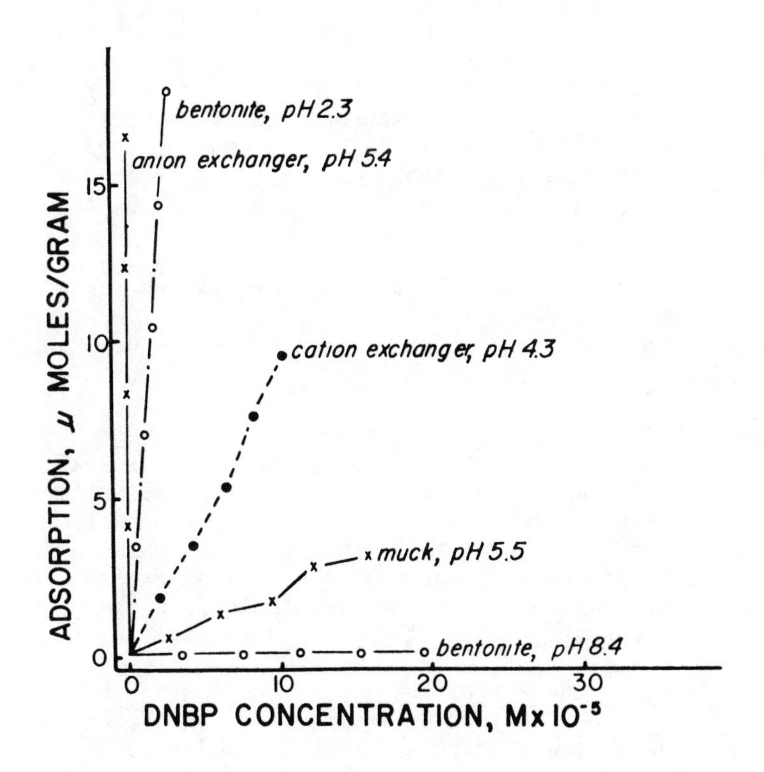

Figure 3. Adsorption of 2-sec-butyl-4,6-dinitrophenol (DNBP)

always be predictable on the basis of generalizations from two or three-component systems.

Adsorption processes are exothermic and desorption processes are endothermic in nature, and an increase in temperature would normally be expected to reduce adsorption and favor the desorption process. This corresponds to a weakening of the attractive forces between the solute and the solid surface (and between adjacent adsorbed solute molecules) with increasing temperature, and corresponding increase in solubility of the solute in the solvent. Exceptions to this generalization have been reported (Harris and Warren, 1964) in which adsorption of 2-chloro-4,6-bis(ethylamino)-s-triazine, 2-chloro-4-(ethylamino)-6-(isopropylamino)-s-triazine and 3-(p-chlorophenyl)-1,1-dimethylurea by bentonite was greater at 0° than at 50°C.

Exchange reactions tend to be temperature independent. For example, 6,7-dihydrodipyridol[1,2-a:2',1'-c]pyrazinediium ion is completely adsorbed at both 0°C and 50°C (Harris and Warren, 1964).

Temperature may influence adsorption through its effects on solubility and vapor pressure. Generally speaking, an increase in temperature leads to decreased adsorption; however, there are exceptions in which the effect of temperature on solubility is such that increased adsorption occurs at higher temperatures; i.e., the increased adsorption of S-ethyl dipropylthiocarbamate at higher temperatures (Freed, Vernetti, and Montgomery, 1962).

Chemical character, shape, and configuration of the adsorbate--Four factors determine the role of the chemical character of the adsorbate in the overall sorption reaction: (1) nature of the functional group; (2) nature of the substituent groups; (3) position of the substituent group with respect to the functional group; and (4) presence and magnitude of unsaturation in the molecule.

The nature of the functional group determines: (1) whether a compound is acidic, basic, or amphoteric in nature; (2) ability to undergo hydrogen bonding; and (3) ability to undergo coordinate covalent bonding. The nature of the substituent on an aromatic molecule, as well as the position of the substituent group, enhances or lessens each of the preceding phenomena. The principal ways in which substituents can affect pKa values are: (1) inductive and field effects; (2) resonance effects; (3) steric factors; (4) tautomerism; (5) solvation effects; (6) internal hydrogen bonding; and (7) stereoisomerism. The nature and position of the substituent group may affect the ability of the molecule to undergo intermolecular hydrogen bonding by promoting intramolecular hydrogen bonding.

The nature and position of the functional groups and the position and length of the alkyl portion of the molecule determines the lyophobic and lyophilic balance of the molecule. The exact nature of the balance will determine the relative affinity the molecules will have for polar and nonpolar adsorbents. Molecular configuration and shape are important in determining if there can be rotation about a double bond which, in turn, may influence the orientation of molecules at the adsorbing surface. The type of molecular orientation may, in turn, determine the contribution of each type of bonding mechanism and thus the overall adsorption potential. More detailed discussions of various aspects of these phenomena are presented by Bailey and White (1970).

Dissociation constant--The pKa of a compound indicates the degree of acidity or basicity that a compound will exhibit and therefore should be very important in determining both the extent of adsorption and the ease of desorption.

Frissel (1961) studied the adsorption of such organic acids as (2,4-dichlorophenoxy)acetic acid and (2,4,5-trichlorophenoxy)acetic acid and noted that negative adsorption occurred until the pH of the adsorbing system approached the pKa for the particular compound. Thereafter, positive adsorption commenced and increased gradually as the pH of the system decreased. Amphoteric antibiotics such as terramycin and aureomycin and the basic antibiotic streptomycin were found to be strongly adsorbed by bentonite or illite while the neutral or acidic antibiotics were adsorbed to a lesser extent (Siminoff and Gottlieb, 1951; Gottlieb and Siminoff, 1952; Martin and Gottlieb, 1952, 1955; and Gottlieb, Siminoff, and Martin, 1952). Pinck, Holton, and Allison (1961) found that the amphoteric antibiotics were adsorbed substantially more by various clay minerals than were the basic antibiotics and that the neutral or acidic antibiotics were hardly adsorbed. Such clay minerals as vermiculite, illite, or kaolinite were incapable of adsorbing either acidic or neutral antibiotics, whereas montmorillonite was capable of adsorbing both types. The release or desorption of such antibiotics was also affected by their degree of acidity or basicity. Amphoteric antibiotics were released from all the clay minerals, the maximum amount being released from kaolinite and the least by montmorillonite. Not one of the four basic antibiotics was released from vermiculite or montmorillonite; one was partially released from illite and two were released to some degree from kaolinite (Pinck, Soulides, and Allison, 1961). The influence of a chemical's pKa on its behavior has also been reviewed by others (Bailey and White, 1970).

Water solubility--The influence of a chemical's water solubility on its adsorbability has been examined by numerous investigators (Bailey, White, and Rothberg, 1968; Harris and Warren, 1964; Ashton, 1961; Frissel, 1961; Wolf et al., 1958; Yuen and Hilton, 1962; Hilton and Yuen, 1963; Talbert and Fletchall, 1965; Hance, 1965; Leopold, Van Schaik, and Neal, 1960; Ward and Upchurch, 1965). In general, while there is a definite relationship between water solubility and adsorptive characteristics within analagous compounds, there appears to be no hard and fast relationship between different chemical classes. In general, however, within an individual class of chemicals a direct relationship exists between water solubility and relative adsorptivity.

Movement of Organic Chemicals in Soil

It is commonly realized that adsorption has a great influence on the nature and extent of leaching and movement of organic chemicals in soil. However, the relationship between adsorption and movement is not fully understood. At least two steps are involved in the leachability of an organic chemical in soil: (1) entrance of the compound into solution; and (2) adsorption of the compound to soil surfaces (Upchurch and Pierce, 1957, 1958). Entrance of the chemical into solution can take place either from the dissolution of the chemical present in particulate form or from the desorption of chemical present on colloidal surfaces. In addition to these factors, the leaching of organic chemicals into soil is influenced by the existing moisture level of the soil at the time of application and the evapo-transpiration ratio. Thus, the factors which appear to affect overall chemical movement most are (1) adsorption, (2) physical properties of the soil, and (3) climatic conditions. The four principal means for chemical transport within soils are: (1) downward flowing water; (2) upward moving water; (3) diffusion in soil water; and (4) diffusion in the air space of the soil.

Leaching--

Organic chemicals are moved into soil by rainfall or irrigation. The movement of an organic chemical in soil is governed by three processes; desorption, diffusion through water, and hydrodynamic dispersion (Adams, 1966; Thomas, 1963). Solubility of the organic chemical has also been considered important (Munnecke, 1967; Goring, 1967; Hartley, 1964). Desorption is proportional to the amount of chemical sorbed (Adams, 1966; Thomas, 1963). Diffusion through the water obeys Fick's law (Munnecke, 1967). Hydrodynamic dispersion (Thomas, 1963) is caused by water percolating downward through the centers of the pores more rapidly than along the sides. The net result of these three processes is increased spread of the organic chemical throughout the soil. Thomas (1963) discussed in detail the distribution patterns for solutes moved through soil by water. Figure 4 shows typical patterns for an organic chemical placed on the soil surface and leached with increasing amounts of water. The amounts of water required to move concentration peaks to selected depths are directly related to the organic matter/water ratios of the organic chemicals being leached (Goring, 1967).

Movement of water in soil is upward as well as downward. If it is neither too easy nor too difficult to leach the organic chemical in question, this alternating movement tends to distribute the chemical evenly throughout the soil profile.

Upward movement of organic chemicals may be a factor, particularly in areas of irrigation and where high evapotranspiration ratios are prevalent and may affect the overall movement and persistence of chemicals in soil. Upward movement of 3,6-dichloro-o-anisic acid and N,N-dimethyl-2,2-diphenylacetamide occurred under subirrigation conditions and also when surface applied water was allowed to evaporate freely from the soil surface (Harris, 1964). Presumably, the upward movement was due to the chemicals being dissolved in upward-flowing capillary water. Thus, weather conditions may also be important in addition to total rainfall in determining the overall movement of chemicals in soil. Other investigators (Ashton, 1961; Jordan, Day, and Clerx, 1963) demonstrated that lateral movement of chemicals results from the lateral movement of capillary water under furrow irrigation. Evidence has also been presented (Comes, Dohmont, and Alley, 1961) for the upward movement of 7-oxabicyclo[2.2.1]heptane-2,3-dicarboxylic acid in subirrigated soil. The upward movement of phenols in soil has also been reported (Phillips, 1964).

The soil thin-layer chromatographic (soil TLC) technique has proved useful in studies of relative mobility of organic chemicals (Helling, 1971a,b; Helling and Turner, 1968; Stipes and Oderwald, 1970). Water moves in the soil layer by unsaturated flow (Helling, 1971a), thus resembling typical water flow under field conditions. The soil TLC technique has received wide acceptance as an indicator of the ability of an organic chemical to move through soil. Thirty-three fungicides and 50 herbicides were evaluated in terms of leaching and diffusion characteristics with fungal and algal bioassay methods developed for use with soil TLC (Helling, Dennison, and Kaufman, 1974; Helling, Kaufman, and Dieter,

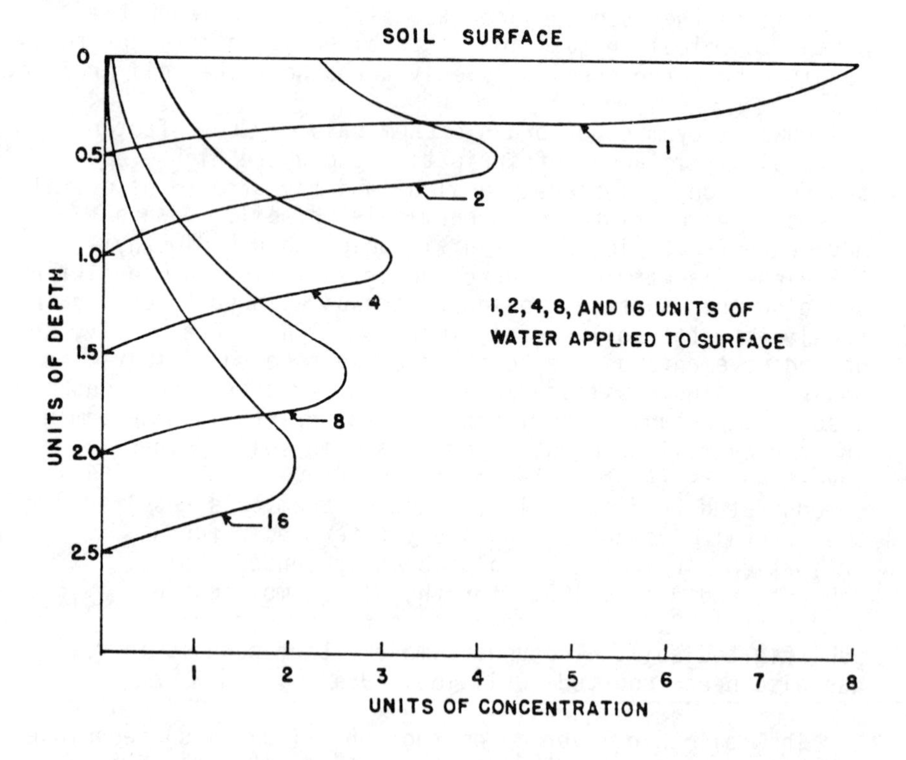

Figure 4. Relation between amounts of water applied and concentration of toxicant at various depths, assuming application of toxicant to soil surface.

1971). The relative mobility in soil TLC of the compounds evaluated was quite similar to that reported by other more cumbersome methods.

One must consider the inherent persistence of each chemical in soil in evaluating whether any mobile chemical might pollute groundwater. The half-life of many organic chemicals in soil is sufficiently short to make it highly unlikely that the chemical would ever reach the water table under ordinary field leaching conditions.

Diffusion and Volatilization--

Diffusion and volatilization are important factors affecting the distribution and persistence of some organic chemicals in soil. The significance of these processes becomes most apparent when organic chemicals, such as pesticides or other toxicants, are found in air or on soils at sites far from those originally treated. Although the majority of organic chemicals used in pest control or considered for land disposal have vapor pressures that are low by the standards of practical organic chemistry, the ability of these chemicals to diffuse and to volatilize enables their movement from the application site. Compounds with high vapor pressure move in field soils best if the gas is not allowed to escape into the air. For example, the vapor pressure of methyl bromide is 1,380 mm Hg at 20°C as opposed to 20 mm Hg for trichloronitromethane (Youngson, Baker, and Goring, 1962). It is useless to apply methyl bromide without covering the soil. Tarping the soil frequently enhances the pest control effectiveness of compounds with lower vapor pressures such as trichloronitromethane and methyl isothiocyanate.

Goring (1967) suggested that Fick's first law of diffusion (Crank, 1965) was probably applicable to all diffusion processes in soil, including movement of organic chemicals through air, water, and organic matter. Simply stated, the rate of movement of a chemical is directly proportional to the concentration of the toxicant and its diffusion coefficient. The diffusion coefficient is a quantitative way of expressing the difficulty diffusing molecules have in moving through different transfer media such as air, water, or organic matter.

Goring (1967) stressed the significance of water/air ratios and organic matter/water ratios. Organic chemicals with water/air ratios under 10,000 should diffuse primarily through air, whereas those with ratios over 30,000 should diffuse primarily through water. Methylene bromide and carbon disulfide have water/air ratios under 10 and move so rapidly that they are lost within hours to days when introduced into soil. Compounds such as trichloronitromethane, ethylene dibromide, 1,3-dichloropropene, and methyl isothiocyanate with ratios of 10 to 100 disappear within a few weeks. 1-Chloro-2-nitropropane, 1,2-dibromo-3-chloropropane (DBCP), and 2-chloro-6(trichloromethyl)pyridine with ratios of 100-2,000 could remain in the soil for months if not decomposed by other mechanisms.

It appears that percolating water is the principal means of movement of relatively nonvolatile chemicals, and that diffusion in soil water is important only for transport over very small distances. Calculations by Hartley (1976) indicate that several years would be required for as little as one percent of the concentration of surface-applied chemicals to migrate by diffusion to a depth of only two feet in a moist soil. Mathematical treatment for the diffusion of (2,4-dichlorophenoxy)acetic acid in saturated soils revealed that the diffusion coefficient for (2,4-dichlorophenoxy)-acetic acid in nine soils was inversely related to soil texture.

The partition coefficient suggested by Goring (1967) is important in the process of volatilization. Usually the amount of an organic chemical applied is much less than that required to saturate the soil-water phase. Dry soil surfaces may adsorb these chemicals strongly, but at normal soil moisture content, volatilization of a chemical takes place from a dilute aqueous solution at the exposed soil surface. An estimate of potential volatility can be obtained from the ratio of water solubility to vapor pressure, since this indicates what portion of the chemical is in the vapor phase. This ratio is useful as a guide, but is affected by adsorption, which decreases the amount of material present in the vapor phase.

The diffusion rate of a gas is influenced by molecular weight, temperature, presence of codiffusing gases, continuity of air spaces, and distribution of the chemical between air, water, and solid phases of soil. This latter distribution is influenced by temperature, moisture, air space, clay, and organic matter (Munnecke, 1967). Diffusion is more important than movement in water in affecting soil penetration by methyl bromide and carbon disulfide. Both air diffusion and water flow are important with water soluble compounds such as sodium methyldithiocarbamate and tetrahydro-3,5-dimethyl-2*H*-1,3,5-thiadiazine-2-thione, whose toxic activity depends upon volatile breakdown products (Munnecke, 1967). In either case, however, soil moisture content affects diffusion greatly because the physical blocking of free air space by water hinders movement.

There are many methods for applying volatile compounds to soil and these have been reviewed extensively (Goring, 1962, 1967; Peachey and Chapman, 1966). The basic pattern of diffusion, once incorporated or injected, is illustrated in Figure 5. The area under each curve represents the concentration at that location at that distance from the injection point.

Application of volatile chemicals to soil surfaces generally results in their rapid dissipation. Losses from soil surfaces have even been reported for chemicals such as s-triazines (Kearney, Sheets, and Smith, 1964) and phenylureas (Hill et al., 1955) which have very low vapor pressures. Thus, incorporation into the soil is required for most effective use of some fumigants, or for disposal of organic wastes containing compounds which have relatively high vapor pressures.

Runoff--

The movement of organic chemicals by overland flow is equally as important as their movement downward in the soil profile, especially as it affects water pollution. Occurrence of overland flow has been extensively documented, particularly in the field of pesticide science. Two questions of importance in this regard are: (1) Is the chemical transported mainly in the liquid phase, or (2) is it initially adsorbed on particulate matter and then transported "piggy back" to the water course with eroding soil? The need for such information is two-fold: (1) chemicals which are toxicants and which may be transported in the liquid phase may cause considerably more damage as a pollutant in water than those chemicals which are transported in an adsorbed phase and are not likely to desorb and cause problems of toxicological nature; (2) incorporation at the application site may be extremely important in order to prevent serious runoff of wastes considered potentially hazardous.

Plant Uptake--

It has long been known that organic chemicals move into plants. It was not until after the introduction of synthetic organic chemicals to control insects, plant diseases, and weeds that man became concerned about the ingestion of such chemicals with the food we eat.

Most research on plant uptake of organic chemicals has been conducted since the early 1960's, primarily because the advent of gas-liquid chromatography (GLC) increased analytical sensitivity from 10 to 1,000 fold over previous chemical and bioassay techniques (Koblitsky and Chisholm, 1949; Mills, Onley, and Gaither, 1963; Guiffrida, Ives, and Bostwick, 1966). Modern GLC analyses have also provided

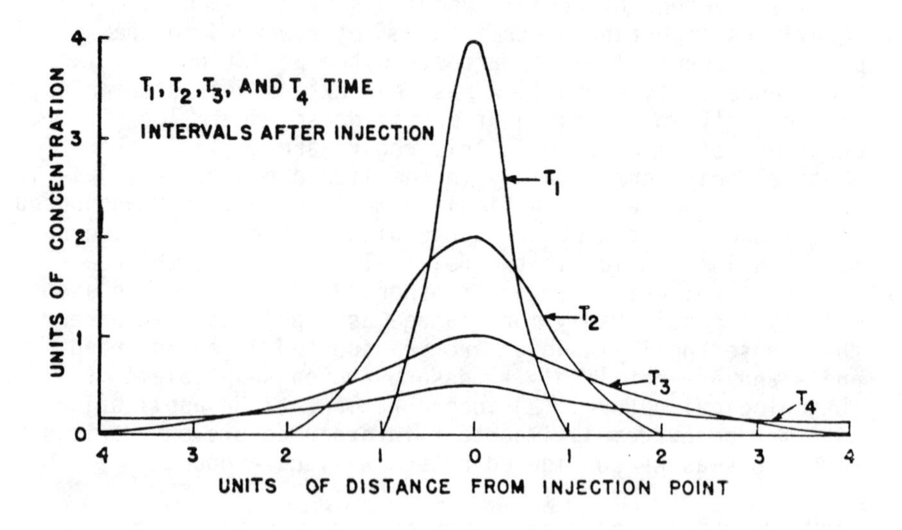

Figure 5. Relation between distance from injection point and concentration of fumigant in soil at various time intervals after injection.

residue identification, whereas previous chemical and bioassay techniques often did not. A few chemical assays are quite specific and provide sensitivities near 0.1 ppm, which is a factor of only about a 10 fold difference from GLC sensitivity with some organic chemicals (Schechter and Hornstein, 1952; Hornstein, 1955). A few bioassays have provided sensitivities nearly equal to many present GLC assays, but do not provide identification of the toxicant (Fleming et al., 1962).

Concerns for the uptake of organic compounds from soil or the atmosphere include: (1) are such compounds phytotoxic; (2) do they enter plants through the root system, stems, or leaves; (3) are they translocated to edible plant parts; (4) are they metabolized, activated, or inactivated in the plant; (5) are they metabolized to more toxic compounds; (6) are they potentially harmful to the consumer of the plant. Numerous reviews have attempted to address the significance of plant uptake of organic chemicals from soil. Foy, Coats, and Jones (1971) extensively reviewed plant uptake of organic chemicals, in particular, herbicides and growth regulators. Their review included reference to plant species, plant parts, organic chemicals, direction of chemical translocation, and the method used to trace translocation. Crafts (1961) also reviewed plant sorption of herbicides and growth regulators, whereas Nash (1974) reviewed plant uptake of insecticides, fungicides, and fumigants from soils. Other aspects of the literature which have been reviewed include: the chlorinated hydrocarbon insecticides (Marth, 1965; Wheatley, 1965; Newsom, 1967; Caro, 1969; Casida and Lykken, 1969; Edwards, 1970; and Young, 1971); the organophosphate insecticides (Spencer, 1965; Finlayson and MacCarthy, 1965; Metcalf, 1967); the antibiotics, fungicides and bactericides (Pramer, 1959, 1961; Brian, 1967); the insecticides, herbicides, and fungicides (Pimental, 1971); and Pb (Warren and Delevault, 1962). In examining such reviews the reader should be aware that all too often only research that reports positive results (i.e., plant uptake) are published, whereas negative results (no uptake) are frequently left unpublished, if not included with positive results.

Of all the synthetic organic chemicals manufactured only a very small number have been examined in plant uptake investigations. Although all compounds considered for soil disposal should be investigated, some information may be lacking simply because the results were negative. Analysis of the available reviews on plant uptake of organic chemicals suggests that plant roots are not very discriminating toward small organic molecules with a molecular weight of less than 500, except on the basis of polarity. Nonpolar molecules tend to adsorb to root surfaces rather than pass through the epidermis. This situation appears to be the case for chlorinated hydrocarbons. Polar molecules readily reach and enter into root tissue, and are translocated throughout the plant.

The concentration of a contaminant in a plant is primarily dependent upon two factors: (1) the residual concentration of the contaminant in the soil; and (2) the plant species and variety. Generally speaking, residue concentrations in plants do not increase after a few days or weeks because: (1) plant growth and metabolism of the contaminant occurs more rapidly than plant sorption of the contaminant; and (2) the degradative mechanisms in soil reduce the concentration of contaminant available for sorption. In some cases, however, the total amount of contaminant absorbed by a single plant over a growing season may increase with time if the contaminant is persistent, e.g. the chlorinated hydrocarbon insecticides.

The most important single factor affecting plant uptake of residual chemicals is the water solubility of the chemical. The route or mechanism of plant contamination may also be dependent upon the solubility of the chemical. The more polar or water soluble residues enter the plant through the roots and are readily translocated. Recent investigations with several chlorinated hydrocarbons (DDT by Beall and Nash, 1971; chlorobiphenyls by Fries and Marrow, 1981) indicate that contamination of upper plant parts by these molecules occurs by volatilization of the chemical from the soil surface and its subsequent adsorption to plant stems and leaves. Additional work needs to be done in this area of research.

The most important soil factor influencing plant uptake of organic chemicals by root sorption is the soil organic matter content. Adsorption to soil organic matter may limit availability of organic chemicals for plant uptake, particularly those compounds of low polarity or water solubility.

Other factors such as soil pH, clay, and microbial activity influence plant sorption of organic chemicals and become more important with increased polarity or water solubility of the chemical.

Degradation of Organic Chemicals in Soil

An important part of the soil-organic chemical interaction involves the rate and mechanism of decomposition of the organic chemical. Decomposition of organic chemicals added to soil may occur by chemical reactions with soil constituents, as a result of biochemical interactions with soil microorganisms, or as a result of photochemical degradation on exposed soil surfaces. Whether chemical, photochemical, or biochemical processes are most important in decomposition of a specific organic chemical is probably most dependent upon the chemical characteristics of the organic compound in question, but may also be dependent upon specific soil characteristics, as well as the location of the chemical within the soil profile. Very astute investigations are needed to accurately determine whether degradation is actually biological or chemical. Occasionally, due to apparent similarities, chemical degradative mechanisms may be misinterpreted as biological mechanisms by the more casual investigator.

Photodegradation--

The action of sunlight may chemically alter and degrade organic chemicals in the environment. Since environmental photochemistry is complicated by the possible interaction of the reacting molecule with many environmental components, laboratory experiments have been necessary to obtain comparative data for determining the effect of factors that modify products and rates of reaction.

There are two fundamental laws of photochemistry: (1) the Grotthus-Draper law which states that only the light absorbed by a molecule is responsible for reaction; and (2) the Stark-Einstein law which states that the absorption of light by a molecule is a one-quantum process, so that the sum of the primary quantum yields must be unity (Calvert and Pitts, 1966). Determination of quantum yields requires elaborate experimental techniques. Major attention has been directed recently to atmospheric photochemistry, where consideration can be given to reactions of organic molecules at low concentrations in the vapor phase (Hecht and Seinfeld, 1972).

"Sensitized" processes, in which light is absorbed by a "sensitizer" molecule which adopts an excited electronic configuration, are believed important in the environment. This excited electronic state possesses a relatively long-life-time during which the molecule can transfer energy to a molecule of another species which does not absorb light in the same region as the sensitizer molecule. Thus, a molecule which does not absorb light in the visible region of the spectrum may undergo reaction in the presence of a sensitizer molecule absorbing at longer wavelengths (Plimmer and Klingebiel, 1971; Plimmer et al. 1967; Rosen and Siewierski, 1970; Foote, 1968).

Solvent, concentration, and aeration are important factors affecting rates of photolysis. Temperature seldom influences rates of photochemical reaction. Water is the predominant solvent for environmental studies of photolysis. Organic solvents are frequently limited by their ability to transmit light, or by the fact that they themselves may participate in the chemical reactions. Many complex polymeric substances may be formed in the presence of high concentrations. First-order rate laws apply to decomposition rates by photolysis in dilute aqueous solutions. The presence or absence of oxygen also influences the course and products of reaction. Photooxidation reactions predominate in the presence of oxygen rather than nitrogen (Plimmer, Klingebiel, and Hummer, 1970).

Except when chemicals enter into aquatic systems, their primary points of exposure are either in the vapor phase or on leaf or soil surfaces. Photochemical studies have been performed on thin layers of pure material deposited on glass plates in attempts to simulate environmental surface photochemistry. Silica-gel coated glass plates or thin layers of soil have been used. Few studies have been reported on soil surfaces but the rates of photolysis appear to be slower than on silica-gel (Plimmer, 1974). Changes in absorption spectra occur, however, with organic chemicals deposited on surfaces (Plimmer, 1972). While such spectral changes are a general phenomenon and are dependent upon the nature of the compound (Leermakers et al., 1966), care must be taken in extrapolating photochemical information from one system to another. The importance of photochemical reactions to the degradation of organic wastes disposed of by land application will depend largely upon the mode of application and soil incorporation. If chemicals are applied to land by spray irrigation, they will be subjected to some photolytic action during the time they are in the spray water or on exposed plant or soil surfaces. Similar effects will probably exist with any surface application method. If, however, they are incorporated into soil, photolytic degradation will be nonexistent, since sunlight does not penetrate soil surface.

Chemical Degradation--

Soil factors known to affect organic chemical degradation include temperature, aeration, microbial population, pH, organic matter, clay, cation exchange capacity, and moisture. Soil chemical and physical processes, and microbial populations may be similarly affected by these factors. The influence of environmental factors on physical processes such as leaching, diffusion, volatilization, or adsorption may parallel those effects observed on microbial populations. In a milieu such as soil, the nature, number, rate, and complexity of chemical interactions can be expected to increase or decrease under many of the same conditions causing increased or decreased microbial activity. Increasing moisture increased tetrahydro-3,5-dimethyl-2H-1,3,5-thiadiazine-2-thione decomposition (Call, 1957a), but sodium methyldithiocarbamate decomposition increases only with decreased moisture (Turner, Corden, and Young, 1962). The rate of decomposition of both chemicals increases with increased temperature (Munnecke and Martin, 1964; Turner, Corden, and Young, 1962). Solubility, sorption by organic matter, and catalysis at clay surfaces are factors believed to affect these interactions (Kotter, Willenbrink, and Junkmann, 1961; Gray, 1962). Microbial decomposition and activity are generally enhanced by increasing temperature and moisture (Goring, 1967; Hill et al., 1955; Aldrich, 1953; Burschell and Freed, 1960; Edwards, 1964, 1966; Lichtenstein and Schulz, 1960, 1964; Schuldt, Burchfield, and Bluestone, 1957). The activity of some microorganisms, however, may be enhanced by decreased soil moisture contents.

Decomposition of ethylene dibromide (Hanson and Nex, 1953), 1,3-dichloropropene (Ashley and Leigh, 1963), sodium methyldithiocarbamate (Burchfield, 1960; Gray, 1962), and methyl isothiocyanate (Kotter, Willenbrink, and Junkmann, 1961) are accelerated by added organic matter. Organic matter content is generally considered a good indicator of microbiological activity, and decomposition also seems to be increased with increased organic matter (Goring, 1967). Soil pH is known to affect microbial populations. Sodium methyldithiocarbamate decomposition is accelerated by metallic cations such as Cu and Fe (Ashley and Leigh, 1963).

Chemical reactions most likely to occur are decomposition in water or hydrolysis (Goring, 1967; Burchfield, 1960; Turner and Corden, 1963; Munnecke and Ferguson, 1953; Munnecke, 1958, 1961; Turner, Corden, and Young, 1962; Gray, 1962; Ashley, and Leigh, 1963; Castro and Belser, 1966; Spanis, Munnecke, and Solberg, 1962; Torgeson, Yoder, and Johnson, 1957), and nucleophilic substitution by active groups of organic matter (Burchfield, 1960; Goring, 1962; Moje, 1959, 1960; Kotter, Willenbrink, and Junkmann, 1961; Gray, 1962; Hanson and Nex, 1953; Burchfield and Schuldt, 1958; Burchfield and Storrs, 1956; Moje, Martin, and Baines, 1957). The role of free radicals (high-energy hydroxyl groups) in decomposition of organic chemicals in soil has been recently examined (Kaufman et al., 1968; Plimmer et al., 1967). The dithiocarbamate pesticides tend to hydrolyze or decompose readily in water (Gray, 1962; Menzie, 1969). It has been suggested (Burchfield, 1960) that 2,3-dichloro-1,4-naphthoquinone, tetrachloro-p-benzoquinone, 2,4-dichloro-6-(o-chloroanilino)-s-triazine, pentachloronitrobenzene, N-[(trichloromethyl)thio]-4-cyclohexene-1,2-dicarboximide, nitrohalobenzenes, and compounds with allylic halogens participated in substitution reactions. Others have also been described (Moje, 1960). Since the concentration of toxicant is usually low relative to the substrate with which it is reacting, first-order kinetics are believed applicable whether the reaction occurring is hydrolysis, or uni- or bimolecular substitution (Goring, 1967; Burchfield, 1960; Moje, 1960).

Numerous other types of chemical reactions are known to occur in soil. One class of non-biological degradation which enters the grey zone approaching the living is that due to enzymatic action. Soil organic matter includes a wide variety of enzymes not necessarily associated with live organisms. These may arise from both living and dead microorganisms, root exudates, and soil animals, and are stabilized to a remarkable degree by adsorption to buffered particulate matter (Skujins, 1967). In this form, the enzymes may survive much longer than they normally would in solution and may even become resistant to heat and pH which would otherwise rapidly deactivate them. In fact, high energy (gamma ray) irradiation sufficient to kill soil microorganisms left extracellular enzymatic activity intact.

Getzin and Rosefield (1968) observed a non-living, heat-labile fraction which catalyzed the hydrolysis of the *O,O*-dimethyl phosphorodithioate *S* ester of diethyl mercaptosuccinate at rates comparable to soil, and lost this capacity upon autoclaving but not upon gamma irradiation. Similar results were obtained by Burge (1972, 1973) with 3',4'-dichloropropionanilide. The intermediacy of 2,4-dihydroxy-7-methoxy-1,4-benzoxazin-3-one in the catalyzed conversion of 2-chloro-4,6-bis(ethylamino)-*s*-triazine to 2-hydroxy-4,6-bis(ethylamino)-*s*-triazine (Roth and Knusli, 1961; Hamilton and Moreland, 1962) has been repeatedly confirmed, and the degradative capacity of the organic fraction of soil may be due in part to this and similar chemical agents.

The use of purely chemical degradative reactions in soil would require a large scale operation with primarily environmentally available reagents in soil. Fortunately, some of these are extremely powerful (if we can learn to harness them) and they are of a molecular size which ensures both pervasiveness and large molar proportions. The possibility of coupling these chemical reactions with microbial activity should also be considered.

The use of oxygen is one mechanism; its free-radical character and solubility in water make it especially intriguing. Oxidation is exceptionally important where light is available to energize the reactions; although there already exists from the petrochemical industry a considerable volume of information about catalyzed autooxidations which proceed in the absence of light, little effort appears to have been directed toward soil applications. Soil bacteria and fungi and their cell-free extracts oxidize even "stable" chlorinated hydrocarbons such as 1,1,1-trichloro-2,2-bis-(p-chlorophenyl)ethane (Engst and Kujawa, 1967; Wedemeyer, 1967). The oxidative decontamination of 1,1,1-trichloro-2,2-bis(p-chlorophenyl)ethane thus is already reduced to chemical terms if the mechanisms involved can be clarified. Pretreatment of biorefractory chemical wastewater by oxidation with hydrogen peroxide or sodium hypochlorite was found to significantly increase the biodegradability of the waste.

Water is another available reagent. Hydration and hydrolysis frequently can be catalyzed effectively by surfaces (Armstrong and Chesters, 1968), metallic ions (Mortland and Raman, 1967), or by even slightly elevated pH. Although the buffering capacity of many soils is very substantial, the local alteration of pH presents a possibility and the alteration of particulate surfaces a certainty. A crude cell extract, from a mixed bacterial culture growing on O,O-diethyl O-(p-nitrophenyl)-phosphorothioate, hydrolyzed this compound (21°C) at a rate 2,450 times faster than 0.1 N sodium hydroxide at 40°C. Eight of 12 other commonly used organophosphate chemicals were also enzymatically hydrolyzed by this enzyme preparation. Other agents which could catalyze or take part in hydrolytic reactions also might be added in dilute aqueous solution or, as with the traditional "lime", in stable, solid form.

Several more powerful nucleophiles, widely available in common use, could be used for the decontamination of soil. For example, the application of agricultural ammonia to a field soil which contained high levels of the soil sterilant sodium methyldithiocarbamate could result in rapid reaction to form harmless 1,1-dimethylthiourea. Sulfides and mercaptides have been demonstrated to be useful for the decontamination of seeds containing N-[(trichloromethyl)thio]-4-cyclohexene-1,2-dicarboximide and related chemicals (Kohn, 1969).

Reducing power is also available from existing soil organic matter if it can be released, and even catalytic reduction is not inconceivable. The application of large increments of organic matter and the flooding of soil has been repeatedly demonstrated to enhance the degradation of some more complex chlorinated hydrocarbons (Farmer et al., 1974; Sethunathan, 1973).

Biodegradation--

Kinetics of microbial degradation--Microbial degradation accounts for much of the loss of organic chemicals from soil (Kaufman, 1974, 1977a; Goring, 1967; Woodcock, 1964; Kaufman and Kearney, 1970). Whether or not a chemical is adsorbed, absorbed, activated, inactivated, persistent, short-lived, mobile, stationary, or eventually constitutes a residue problem, may depend upon its transformation by soil microorganisms. The kinetics of decomposition typical for enrichment cultures were established in early studies with (2,4-dichlorophenoxy)acetic acid (2,4-D) (Audus, 1960). Such results encouraged isolation of soil microorganisms specifically responsible for decomposing select organic chemicals. While many microorganisms capable of degrading specific xenobiotics have been isolated and characterized, some chemicals failed to support microbial growth or enrichment during the degradation process. Other chemicals appeared persistent or otherwise resistant to microbial degradation. Hence the concepts of molecular recalcitrance (Alexander, 1965), microbial fallibility (Alexander, 1965; Alexander and Lustigman, 1966), and cometabolism (Horvath, 1971; Horvath and Alexander, 1970) were introduced. While some microorganisms thrive and become enriched on easily degraded compounds, some chemicals are initially unsuitable as energy sources for microorganisms. The more persistent chemicals are less efficiently degraded by the myriad enzymatic reactions generated by soil microorganisms.

First-order kinetics apply where the concentration of the chemical being degraded is low relative to the biological activity in the soil (Hill et al., 1955; Burschell and Freed, 1960; Schuldt, Burchfield, and Bluestone, 1957; Hamaker, 1966; Sheets, 1964; Sheets and Harris, 1965). Michaelis-Menten kinetics seem to apply when the chemical concentration increases and the rate of decomposition changes from being proportional to being independent of concentration (Hamaker, 1966; Hamaker, Youngson, and Goring, 1967). The dissipation of most chemicals from soil, for example, follows a first order reaction (Figure 6 A), meaning that at any one time, the rate of pesticide loss is proportional to its concentration in the soil (Kearney, Nash, and Isensee, 1969). Although the loss generally assumes the shape of a first order reaction curve, several mechanisms may be acting on a soil residue at any given time. Volatilization, photodecomposition, mechanical removal by water or wind erosion, leaching, cultivation, adsorption, microbial metabolism, and chemical decomposition may contribute to a chemical's dissipation or persistence. Periodic application of degradable chemicals yields the type of curves illustrated in Figure 6 B. Maximum and minimum residue levels remain parallel to the base line and would not exhibit any progressive accumulation under field conditions when the number of chemical units lost in a given time equals the number of units applied. Thus, Figure 6 B illustrates a situation in which 4 units are applied periodically and 20% residue remains at the next application. The minimum residue remains constant at 1 unit after about the fifth application. If the chemical were degraded to only 50% before the next application, the minimum residue would reach 4 units after 11 applications. Similar curves have been developed by Sheets (1964) and Foster et al. (1956).

Periodic application of heavy metal or organometal compounds to soil should result in the situation shown in Figure 6 C. This assumes that a portion of the chemical molecule is not lost, but accumulates in direct proportion to its application. This probably is the case with the organic and inorganic arsenical pesticides or any other organometallic compound. The significance of this phenomenon is discusssed more fully elsewhere (Kearney, Nash, and Isensee, 1969).

The dissipation of certain biodegradable chemicals from soil has been typified by Figure 6 D. A lag phase may occur after the initial application in which relatively little chemical is lost. The lag phase is followed by a period of rapid disappearance as a result of microbial metabolism, for example, the loss of 2,2-dichloropropionic acid (Kaufman, 1964), (2,4-dichlorophenoxy)acetic acid (Audus, 1949, 1950, 1951, 1960), [(4-chloro-*o*-tolyl)oxy]acetic acid (Audus, 1960), isopropyl *m*-chlorocarbanilate (Kaufman and Kearney, 1965), and others. This type of degradation, however, has been observed largely under isolated culture or soil perfusion conditions with high concentrations of the compound, where the influence of other factors may not be limiting. There is limited evidence to support the existence of such a sigmoid pattern in soil under field conditions where other factors are not limiting in their effect. This is not to imply that biodegradation is not a significant factor in the dissipation of many organic chemicals in soil. Indeed, microbial degradation is known to play a signficant role in the dissipation of most organic chemicals in soil. Rather, it is to indicate that other processes such as volatilization, adsorption, leaching, etc. may limit the availability of the chemical for biodegradation. The 1,1'-dimethyl-4,4'-bipyridinium ion is a classic example of this. Although this chemical is readily biodegradable in solution cultures (Funderburk, 1969; Funderburk and Bozarth, 1967), it is not degradable when adsorbed to clay minerals (Weber and Coble, 1968). The dissipation of 2,2-dichloropropionic acid and isopropyl *m*-chlorocarbanilate (Kaufman, unpublished data) from potted soil under greenhouse conditions would appear to follow a sigmoid type pattern. In field conditions, however, such a phenomenon might be expected to occur only where volatilization, leaching, adsorption, or other critical factors were not limiting.

Readily biodegradable chemicals are generally degraded more rapidly, and without the initial lag phase (Figure 6 E), in subsequent applications to soil. This phenomenon has been observed with numerous organic chemicals including (2,4-dichlorophenoxy)acetic acid (Audus, 1960), 2,2-dichloropropionic acid (Kaufman, 1964), isopropyl m-chlorocarbanilate (Kaufman and Kearney, 1965), and others in soil perfusion or other enrichment type systems. Hurle and Rademacher (1970) compared the dissipation of 4,6-dinitro-o-cresol and (2,4-dichlorophenoxy)acetic acid in soil treated for the first time and soil from field plots treated annually over a period of 12 years. 2,4-Dichlorophenoxy)acetic acid dissipation was more rapid in previously treated soil than in soil treated for the first time. Pretreatment had no effect on the rate of 2,4-dinitro-o-cresol dissipation from soil. Similar promotions have been obtained by others in investigations with (2,4-dichlorophenoxy)acetic acid (Newman, Thomas, and Walker, 1952; Newman and Thomas, 1949; Audus, 1949, 1951, 1960; Aly and Faust, 1964; Fryer and Kirkland, 1970), (2,4,5-trichlorophenoxy)acetic acid (Newman, Thomas, and Walker, 1952), [(4-chloro-o-tolyl)oxy]acetic acid (Kirkland and Fryer, 1966; Kirkland, 1967; Fryer and Kirkland, 1970; Audus, 1960), 7-oxabicyclo[2.2.1]heptane-2,3-dicarboxylic acid (Jensen, 1964; Horowitz, 1966), and 2,2-dichloropropionic acid (Leasure, 1964; Kaufman, 1964), but not with 2-chloro-4,6-bis(ethylamino)-s-triazine or 3-(3,4-dichlorophenyl)-1-methoxy-1-methylurea (Fryer and Kirkland, 1970). Newman and Thomas (1949) also observed that pretreatment of soil with other structurally related organic compounds shortened the persistence of (2,4-dichlorophenoxy)acetic acid applied to these soils. The time for [(4-chloro-o-tolyl)oxy]acetic acid applications to reach the limit of detection was reduced from three weeks after three previous annual applications to four days after 10 previous applications (Fryer and Kirkland, 1970). Jensen (1964) and Horowitz (1966) found 7-oxabicyclo[2.2.1]heptane-2,3-dicarboxylic acid to be less persistent in previously treated field soils.

The actual significance of this phenomenon under field conditions would appear to depend upon several factors: (1) the rate and frequency of applications; (2) the time elapsed between applications; (3) the cropping system; (4) the survival of an enriched population; (5) the complexity of the metabolic reactions involved; (6) the physical-chemical behavior of the chemical in soil; and (7) possible

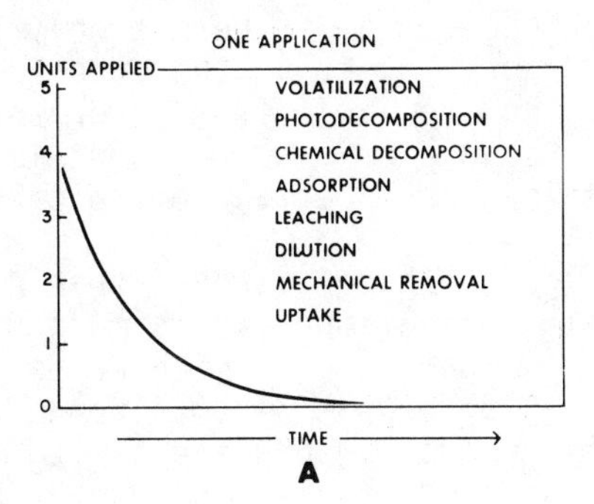

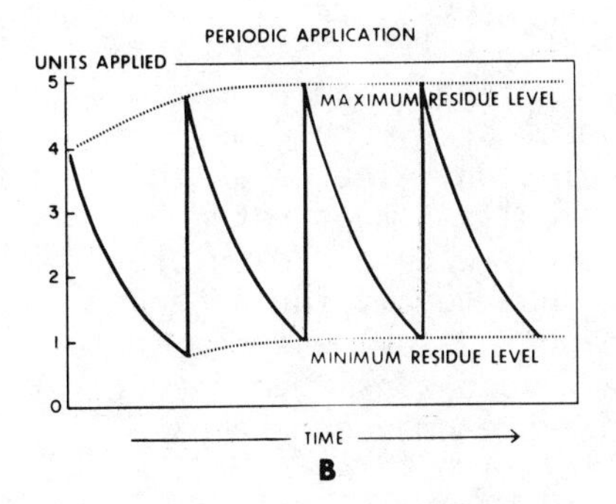

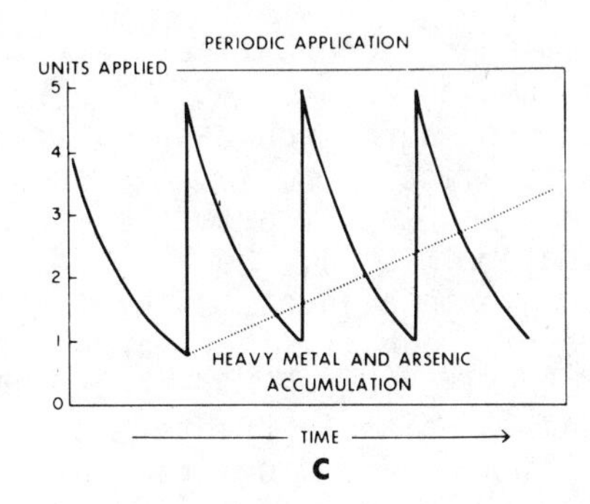

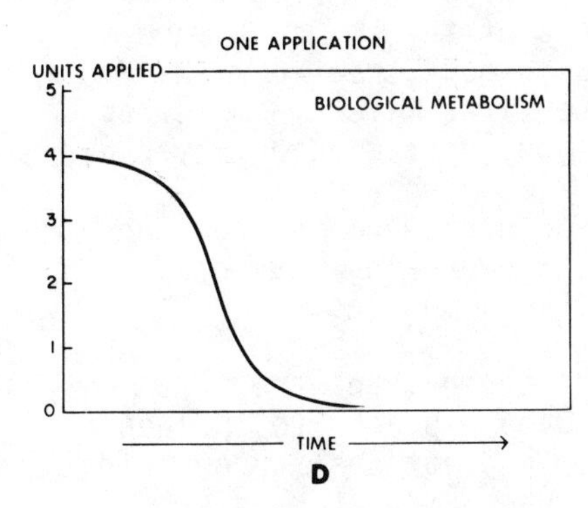

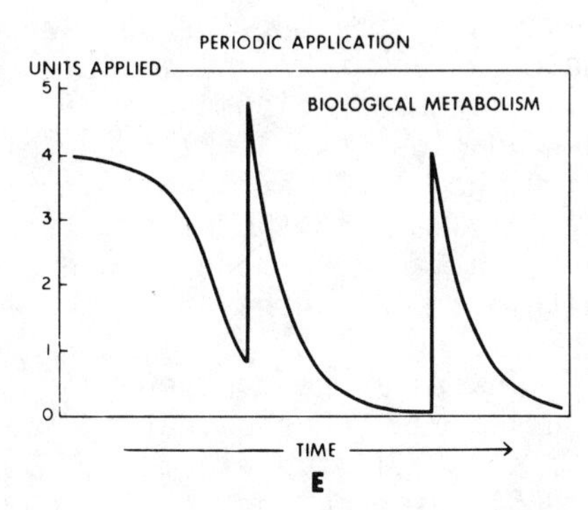

Figure 6. Loss of chemicals from soil after one or periodic applications.

interactions of one chemical with another. Certain of these factors, of course, are quite closely interrelated and subject to multiple interactions. Unfortunately, the available data come from experiments which are either inconclusive in themselves, or are sufficiently narrow in scope, subject, or experimental methodology as to permit only limited conclusions to be drawn. However, there would appear to be sufficient information available from laboratory, greenhouse, and field experiments to permit some generalizations as to the possible significance of certain of these effects.

Several investigators have observed that dissipation of subsequent chemical applications occurs more rapidly in soils initially treated at higher application rates (Audus, 1949; Newman and Thomas, 1949; Newman, Thomas, and Walker, 1952). This observation is generally consistent with the fact that microbial populations tend to respond both quantitatively and qualitatively in accordance with the supply of readily available substrate. Thus, one might expect, barring other limiting factors (i.e., antagonism, toxicity, lysis), that higher concentrations of a readily available and digestible carbon source would permit development of a more dense and metabolically active microbial population. The effect of frequent soil treatments was dramatically demonstrated by Leasure (1964). He treated potted soil with 2,2-dichloropropionic acid at 50 lb/acre (56 kg/ha) weekly for a period of six weeks. After six weeks the potted soil was seeded to four different 2,2-dichloropropionic acid-sensitive species and immediately treated again with a pre-emergence application of 2,2-dichloropropionic acid at 50 lb/acre. All seedlings grew normally in pots which had received a total of 350 lb/acre (390 kg/ha) 2,2-dichloropropionic acid within the seven-week period. Drastic effects of the toxicant were noted on all four species planted in pots treated with 2,2-dichloropropionic acid for the first time. A more rapid degradation of [(4-chloro-o-tolyl)oxy]acetic acid (Fryer and Kirkland, 1970) and 7-oxabicyclo[2.2.1]heptane-2,3-dicarboxylic acid (Horowitz, 1966) occurred when succeeding applications are made within a single growing season, or within a few months after the initial application.

The rate and frequency of application, and the time elapsed between applications could be expected to influence the survival of enriched microbial populations. Brownbridge (1956) observed that (2,4-dichlorophenoxy)acetic acid enriched populations in perfused soil remained viable for at least one year when stored in a moist condition. Microbial populations enriched by (2,4-dichlorophenoxy)acetic acid (Newman and Thomas, 1949), [(4-chloro-o-tolyl)oxy]acetic acid (Fryer and Kirkland, 1970), and 7-oxabicyclo[2.2.1]heptane-2,3-dicarboxylic acid (Jensen, 1964) are known to survive for periods of at least one year under field conditions. More recently, Kaufman (personal communication, 1980), has obtained data indicating that an enriched population can survive under cropped field conditions for a period of two or more years.

Numerous factors are known to affect the survival of specific microbial populations in soil. In the case of soil-borne plant pathogens, an alternate host, crop debris, or resistant structure (i.e., chlamydospore, sclerotia, etc.) are among the critical factors enabling such organisms to survive from one growing season to the next. While resistant spore-like structures are indeed a possibility, little is known about such survival mechanisms of chemically-enriched populations. In the absence of such structural features, however, enriched populations must seemingly rely on either trace amounts of the original enriching substrate, alternate substrates which enable the population to retain its numbers and/or enzyme potential, or the existence of a quiescent state free of competition, antagonism, or lysis, etc. which would assure their survival. Under natural conditions this latter state would be abnormal. Under conditions where the chemical is adsorbed, one might expect that enriched populations might survive on the slowly desorbed materials. Nearly all of the chemicals for which enriched populations have survived under field conditions, however, are subject to only very limited adsorption. Thus, one could reasonably expect complete dissipation of the substrate within a relatively short period of time. The likelihood of a continuous supply of these materials being in soil to sustain the enriched population for long periods of time, therefore, would appear negligible.

The presence of alternate substrates capable of sustaining or aiding in the development of an enriched population has received only cursory examination. Newman and Thomas (1949) examined 13 organic compounds, of which 9 showed some indication of enhancing the subsequent development of (2,4-dichlorophenoxy)acetic acid-enriched populations. Blanckenberg and Loos (personal communication, 1976) examined 65 different aromatic compounds. Many of the compounds tested could occur naturally in soil and had structural features which might permit them to act as alternate substrates or inducers of (2,4-dichlorophenoxy)acetic acid-degrading microbial enzyme systems. Their work demonstrated the use of such compounds as alternative growth substrates for (2,4-dichlorophenoxy)acetic acid-degrading bacteria. Sandmann (1974) subsequently examined the metabolism of 17 phenolic and other aromatic growth substrates by resting cell suspensions of (2,4-dichlorophenoxy)acetic acid-grown *Arthrobacter*. Only 11 of these compounds were degraded rapidly. Of these, only 2, catechol and homogentisic acid, were considered as potential alternative substrates for the (2,4-dichlorophenoxy)acetic acid-degrading enzyme systems. The possible role of these compounds in the maintenance of (2,4-dichlorophenoxy)acetic acid-degrading enzymes in enriched populations under natural conditions needs further investigation.

Cropping systems are also known to have a profound influence on microbial populations. It is widely recognized that the biochemical diversity and reactivity of microbial populations in cropped soils far exceeds that in fallow soils. The ability of enriched populations to survive under pressure of the increased environmental stresses, and competitive or antagonistic interactions associated with various cropping systems is not known. Fryer and Kirkland (1970) compared the persistence of [(4-chloro-o-tolyl)-oxy]acetic acid in previously treated (five annual applications) and untreated cropped soils and concluded that while differences in [(4-chloro-o-tolyl)oxy]acetic acid persistence were indeed statistically significant, they were clearly quite small. Considerably more information is needed to understand fully the survival mechanisms and abilities of enriched populations.

The chemical and physical behavior of a chemical in soil may also preclude its biodegradation, and/or the development of an enriched population. Structural characteristics may cause a chemical to be recalcitrant to biodegradation. Structural characteristics affecting biodegradability have been elucidated for several classes of organic chemicals. In general, however, there are few organic chemicals which are not biodegradable at least to some extent. Adsorption to soil constituents is probably the most significant factor limiting biodegradation of some chemicals in soil. As mentioned earlier, the 1,1'-dimethyl-4,4'-bipyridinium ion, paraquat, is readily degradable in solution (Funderburk, 1969; Funderburk and Bozarth, 1967) but not degradable when adsorbed to clay minerals (Weber and Coble, 1968). Nearly all of the organic chemicals for which enriched soil populations have been detected in vivo are not strongly adsorbed, i.e., [(4-chloro-o-tolyl)oxy]acetic acid, (2,4-dichlorophenoxy)acetic acid, 2,2-dichloropropionic acid, 7-oxabicyclo[2.2.1]heptane-2,3-dicarboxylic acid, and probably trichloroacetic acid. Thus, availability to the microbial population of a substrate in sufficient quantities or at a sufficient rate may be a determining factor in the development of an enriched population. Other factors, however, may be involved. 4,6-Dinitro-o-cresol is readily degraded by isolated soil microorganisms (Moore, 1949; Gundersen and Jensen, 1955; Tewfik and Evans, 1966; Fewson and Nicholas,

1961). Its degradation in soil occurs at a slow rate only, and was not affected by previous treatments (Hurle and Rademacher, 1970). The biodegradability of 4,6-dinitro-*o*-cresol in soil, however, may be a function of soil pH. Adsorption of 4,6-dinitro-*o*-cresol in soil is pH dependent, with greatest adsorption (99+%) occuring on illite and montmorillonite clays at pH 4.6 (Harris and Warren, 1964), whereas no adsorption occurs at pH 7.3. The availability of 4,6-dinitro-*o*-cresol for biodegradation may thus be limited in acidic soils, but unlimited in akaline soils.

Chemicals which do not appear to have induced populations may be degraded by constitutive enzyme systems, or soil microbes which do not proliferate in response to chemical treatment. Chemicals such as 2,3,6-trichlorobenzoic acid (Horvath, 1971) may be degraded by what is described as "cometabolism", and not serve as a carbon source initially. Reduction of the nitro-groups of 4,6-dinitro-*o*-cresol and di- and trinitroanilines is the principal degradative mechanism of these compounds. While isolated soil microorganisms are known to reduce the nitro groups readily, it is conceivable that this process could also occur cometabolically in soil by constitutive nitrate-reducing organisms. More recent information (Willis, Wander, and Southwick, 1974) suggests that at least one or more nitro-groups may be reduced chemically under conditions of low redox potential.

The major reactions which involve a loss of toxicity may not be reactions which are conducive to microbial proliferation. The major degradation mechanism for the chloro-*s*-triazine herbicides in soil is chemical hydrolysis to yield the corresponding hydroxy-*s*-triazine (Harris, 1967). While the dealkylation of *s*-triazines is a microbiological process, this does not always lead to reduced phytotoxicity (Kaufman and Blake, 1970; Kaufman and Plimmer, 1972), nor is it a particularly rapid metabolic process in soil. Similarly, dealkylation of alkyl-phenylureas is a prolonged process in soil, presumably because of adsorption. It is interesting to note that those chemicals microbially degraded by dealkylation are generally more persistent than those degraded by ester hydrolysis or dehalogenation. Further, the preponderance of soil microorganisms which have been identified as dealkylating these chemicals are fungi, whereas bacteria are more commonly associated with chemicals degraded by hydrolytic or dehalogenation mechanisms.

Kinetics of population development and biodegradation--The kinetics involved in the development of soil microbial populations capable of degrading organic chemicals are not fully understood. Available information indicates that it basically follows the well-known sigmoid growth patterns of isolated microbial cultures initially exposed to fresh, suitable substrates under favorable environmental conditions. The population response in a microbial culture which reproduces by binary fission follows a more or less definite growth curve, in which various phases are recognized. An initial lag phase occurs when microbes are initially exposed to fresh medium. This period is characterized as a period of adjustment during which there is little or no growth of the microbes. Toward the end of the lag phase is a period in which the microorganisms begin to increase in numbers. Thomas and Grainger (1952) refer to this as the "phase of physiological youth." By this time the microbial population has made the necessary adjustments enabling it to utilize the available substrate and begin to proliferate. The period when microorganisms proliferate at a maximum rate, increasing by geometric progression as a result of reproduction by fission, is known as the growth or logarithmic phase. It is during this phase that the most rapid utilization of substrate occurs. As substrate availability or space becomes limiting, or toxic growth substances accumulate, the microorganisms cease to multiply at a geometrical ratio and the rate of cell division decreases. Finally, the number of dividing cells equals the number of dying cells and the population enters a stationary or resting phase. Several responses may be observed at this point. If the microbial population is resupplied with additional fresh substrate, it will continue to proliferate with either a shorter lag phase, or no lag phase at all. In the absence of additional substrates, the population will enter a death or decline phase. The rate of decline depends upon the microbial species. In some species, senescence and death may be rapid and logarithmic, whereas in other species the decline may be prolonged for weeks or months. The formation of spores or other resistant resting bodies may occur as the population attains the stationary phase. These structures may ultimately prolong the death or decline phase of that species.

While this series of growth responses is somewhat oversimplified, the microbial degradation of numerous organic chemicals in soil appears to be consistent with the basic sequence of events. In other words, following the introduction of some organic chemicals to soil microbial populations, there is a lag phase during which no degradation occurs. Depending on the chemical-physical characteristics of the soil and the organic chemical in question, a small amount of the chemical may be removed initially by adsorption on soil colloids. The lag period is followed by a period of rapid dissipation of the chemical. The length of the lag period is primarily influenced by the chemical nature of the organic chemical, but may also be influenced by other factors such as concentration and soil type. Subsequent additions of the chemical dissipate more rapidly than the first. This series of events has been observed with numerous organic chemicals in soil perfusion (Audus, 1949, 1951, 1960; Kaufman, 1964; Kaufman and Kearney, 1965; Kaufman and Blake, 1973) and in soil under both glass house (Kaufman, unpublished data) and field conditions (Newman, Thomas, and Walker, 1952; Newman and Thomas, 1949; Aly and Faust, 1964; Fryer and Kirkland, 1970) with a wide variety of chemicals.

The physiological and biochemical changes which occur in the microbial population during the lag phase in the presence of xenobiotics are not fully understood. Two major possibilities have been proposed (Audus, 1960): (1) chance mutation; or (2) inducible enzyme formation. The preponderance of available information supports the formation of inducible enzymes as the major change which has occurred in chemically-enriched populations. Audus (1960) suggested that the chance mutation theory is supported by the random manner in which a wide variety of microbial genera respond to decompose certain organic chemicals. The tendency of multiple samples of a single soil to develop similarly induced microbial populations repeatedly when independently enriched, however, conflicts with this theory. Also, one might reasonably expect the length of the lag phase to vary widely even when multiple samples of a single soil are treated identically. The length of the lag phase, however, is seldom very divergent. Kaufman (unpublished data) has repeatedly perfused as many as 36 samples of a single soil with an individual chemical and observed that detectable levels of degradation occur on precisely the same day in all samples. Loos (1974), however, has demonstrated that the reproducibility of the lag phase for different samples of a soil may not apply as the sample size becomes smaller. While it is possible that longer lag periods in some samples may be due to late mutations, the dilution effects on an inducible population, and the possibly different lag periods among various species may account for some of these differences.

When samples of a soil are enriched independently they tend to produce active populations of the same organisms (Audus, 1960). This phenomenon can be reconciled with the mutation hypothesis only by assuming that those particular microbial species are more prone than others to undergo the mutation(s) necessary for activity. In actuality it would seem reasonable to assume that either or both possibilities may occur.

The occurrence of inducible enzyme formation during this lag phase of microbial populations exposed to xenobiotics has been demonstrated indirectly by numerous investigators (Kearney and Kaufman, 1965; Kearney, Kaufman, and Beall,

1964; Loos, Bollag and Alexander, 1967; Bollag, Helling, and Alexander, 1968; Tiedje and Alexander, 1969; Tiedje et al., 1969; Engelhardt, Wallnofer and Plapp, 1973; Blake and Kaufman, 1973, 1975). In some instances, numerous enzymes associated with the secondary metabolism of chemical degradation products have also been demonstrated. In nearly all cases these enzymes have been shown to be inducible in the sense that they are not present in cells of the same organism cultured in the absence of the specific chemical in question.

The proliferation of inducible microbial populations is an important feature of the enrichment process. In the absence of competition for the energy and mineral constituents of the chemical substrate, induced microorganisms increase and multiply in number. Burge (1969) has confirmed that incubation of soils with 2,2-dichloropropionic acid results in large increases of 2,2-dichloropropionic acid-decomposing organisms where 2,2-dichloropropionic acid is degraded rapidly. The ability of enriched soils to degrade repeated applications of xenobiotic chemicals more rapidly with no, or a considerably shortened, lag phase is indirect evidence that larger, more active populations of adapted organisms exist in the previously treated soil.

Soil microorganisms and enzymes involved in biodegradation-- The ultimate proof of the involvement of soil microorganisms in the dissipation of a xenobiotic from soil is the isolation of the responsible organism, and the demonstration of its ability to inactivate the xenobiotic in culture solution. Isolation of xenobiotic-degrading soil microorganisms from enriched soil systems has generally been accomplished with standard dilution procedures. Selective media prepared for these isolations have usually incorporated the xenobiotic as a primary energy or nutrient source. In theory, this approach encourages and permits the isolation of all those organisms capable of metabolizing the xenobiotic. In fact, however, it isolates only those microorganisms which are capable of utilizing the xenobiotic as a primary or supplemental source of nutrients and proliferating at the expense of the xenobiotic.

Soil microorganisms capable of cometabolically degrading a given xenobiotic may not necessarily follow the normal enrichment sequence of events, i.e., proliferation at the expense of the xenobiotic substrate. Unique techniques may thus be necessary to isolate microorganisms capable of cometabolic degradation of xenobiotics. Unfortunately, there is limited information available suggesting what alternate substrates are suitable for isolation of microbial populations capable of metabolizing xenobiotics by cometabolic processes. Presumably, such techniques could utilize as a primary substrate a chemical analog which would permit enrichment of microbial populations having the necessary cometabolic requirements to degrade the xenobiotic. This, in effect, was the procedure used in isolating microbial cultures capable of cometabolizing a number of chloro- and aminobenzoates (Horvath and Alexander, 1970a,b). Evidence is accumulating which indicates that this method of biodegradation may be a particularly important phenomenon in the dissipation of the more refractory xenobiotics. Considerably more research is needed to understand the importance of this process, however.

Since the initial isolation of a (2,4-dichlorophenoxy)-acetic acid-degrading bacterium (Audus, 1951), a large number of organic chemical degrading microorganisms have been isolated and described. Extensive lists of soil microorganisms capable of metabolizing these and other xenobiotic chemicals have been published (Audus, 1960, 1964; Loos, 1969; Kaufman and Kearney, 1970). Classes of microorganisms involved include the algae, actinomycetes, bacteria, fungi, and numerous microfauna. Although many microorganisms appear limited to a specific group of chemicals, others have demonstrated a wide diversification of substrates which they are capable of metabolizing.

A number of interesting points become apparent upon examination of the literature and various lists of soil microorganisms involved in xenobiotic metabolism. Bacteria have been the predominant organisms isolated from enrichment experiments in which the soil perfusion technique has been employed. Soil fungi capable of degrading xenobiotics have been more frequently isolated from enrichment experiments which have used shake-culture techniques. While the exact reason for this is not known, it would seem quite logical that the cultural techniques employed would ultimately affect those microorganisms isolated. A second point of interest is that the bacteria are predominantly involved with those chemicals which have a higher degree of water solubility and are not strongly adsorbed. In contrast, fungi appear to be predominantly involved in metabolizing those xenobiotics of lower water solubility and greater adsorptivity. The exact reasons for this are not known. The binary fission type reproductive methods of bacteria enable them to compete more successfully for readily available substrates, whereas the mycelial type growth characteristics of fungi perhaps enables them to encapsulate and penetrate soil particles to which xenobiotics may be adsorbed. Soil fungi are generally believed to play a more important role in the formation, metabolism, and interactions of soil organic matter complexes. Such abilities may enable them to cope better with the various bonding mechanisms involved with adsorbed materials. Such observations with regard to xenobiotics, however, are largely empirical and may not necessarily be reflective of the actual significance of either group of organisms in metabolism of these compounds in soil. In isolated culture systems, either group of microorganisms appears capable of metabolizing either group of compounds.

Once effective xenobiotic degrading microorganisms have been isolated from the soil environment, the physiological and biochemical reactions can be studied in detail at the cellular and enzymic levels. The principal biochemical reactions associated with the microbial metabolism of xenobiotics include acylation, alkylation, dealkylation, dehalogenation, amide or ester hydrolysis, oxidation, reduction, aromatic ring hydroxylation, ring cleavage, and condensate or conjugate formation. Many of the enzymes, the reactions, the metabolic pathways, and the intermediate metabolic products have been investigated and described in detail. Although many of these reactions have been demonstrated as occurring in soils, most of the metabolic product information has been obtained through pure culture investigations. Metabolic reactions involved with xenobiotic metabolism will be discussed in more detail in subsequent sections.

There is a growing body of literature describing the enzymes involved in microbial metabolism of xenobiotics. Information is also available concerning microbial enzymes involved in detoxification of pesticides. In most cases, only those enzymes responsible for the initial degradative reaction are known and have been characterized. In the metabolism of (2,4-dichlorophenoxy)acetic acid, however, the specific role of enzymes has been indicated for nearly all reactions involved from the parent molecule to its final natural product, succinic acid. Only a single enzyme appears to be responsible for the initial dehalogenation of 2,2-dichloropropionic acid, and the ultimate formation of pyruvate (Kearney, Kaufman, and Beall, 1964). Similarly, a single enzyme was also responsible for the dehalogenation of trichloroacetic acid (Kearney et al., 1969). The formation of condensates such as the azobenzenes appears to be the action of relatively non-specific peroxidases (Bartha, 1968). Burge (1973), however, noted that not all soil peroxidases had the ability to form 3,3',4,4'-tetrachloroazobenzene from 3',4'-dichloropropionanilide or 3,4-dichloroaniline in soil.

Very little information is known about the actual inductive mechanisms leading to the formation of these induced enzymes. Consequently their true uniqueness can not be fully appreciated at this time. Nearly all of them appear to be "induced" in the sense that they are not present in microbial cells not exposed to the xenobiotic in question. A few enzymes, however, may be induced by other chemicals which themselves are not substrates of the enzyme induced. Enzyme induction by compounds that are not substrates of the enzymes induced is well known in bacteria (Pardee, 1962). Steenson and Walker (1958) investigated the induction of (2,4-dichlorophenoxy)acetic acid and [(4-chloro-o-tolyl)oxy]acetic acid-degrading enzymes in Flavobacterium peregrenum and observed that [(4-chloro-o-tolyl)oxy]acetic acid was an inducer, but not a substrate of the (2,4-dichlorophenoxy)-acetic acid-degrading enzyme induced. Similarly, Blake and Kaufman (1975) observed that the acylamidase of Fusarium oxysporum induced by p-chlorophenyl N-methylcarbamate could not act on p-chlorophenyl N-methylcarbamate, but was highly active in hydrolyzing 3',4'-dichloropropionanilide. This latter phenomenon was of particular interest since p-chlorophenyl N-methylcarbamate also acted as a potent inhibitor of the hydrolytic activity of the acylamidase it induced. The precedence for such a phenomenon in microorganisms has been previously reported by Jacob and Monod (1961). They observed with Escherichia coli that while β-galactosidase and galactoside permease enzymes were induced by four β-D-thiogalactosides, two β-D-galactosides, β-D-glucoside, and galactose, only the β-D-galactosides were substrates for the β-galactosidase enzyme induced.

The work of Sandmann (1974), and Newman and Thomas (1949) is also of particular interest in this regard. A study of the potential ability of naturally occurring compounds analagous to phenoxyalkanoates either to induce or to sustain active phenoxyalkanoate-degrading populations, or provide the basic chemical template from which the induced enzyme is ultimately modified, may provide insights into induction mechanisms. Considerably more information is needed, however, before meaningful conclusions can be reached.

Several types of xenobiotic-degrading enzymes have been isolated from soil microorganisms. Each type is particularly interesting for several reasons. Naturally occurring compounds with carbon-halogen (Br, Cl, F, I) bonds are not particularly common. Synthetic compounds containing carbon-halogen bonds are in widespread use, and their persistence in the environment has become a matter of considerable concern. Although many of the halogenated compounds are chemically or physically transformed in the environment, there remains a concern that the organic halogen may not be returned to an inorganic form and the remainder of the molecule mineralized by appropriate organisms. The role of microorganisms in dehalogenating many of these compounds is well known and in several cases the enzymes responsible for cleaving the carbon-halogen bond have been characterized. The enzymology of carbon-halogen bonds has been reviewed by Goldman (1972). Enzymes capable of removing chloro-, bromo-, iodo-, and fluoro-groups have been isolated and characterized. In some instances these enzymes appear to be truly unique, whereas others are apparently constitutive and their ability to dehalogenate a given organic chemical is coincidental. The enzymes involved in the dehalogenation of 2,2-dichloropropionate (Kearney, Kaufman, and Beall, 1964) and trichloroacetate (Kearney et al., 1969) appear unique, and were not detected in microorganisms which were not adapted to metabolism of these compounds. The 2,2-dichloropropionate enzyme was also capable of dehalogenating dichloroacetate and 2,2-dichlorobutyrate, but was unable to dehalogenate trichloroacetate. Substrate specificity studies were not reported for the trichloroacetate-induced enzyme.

Microbial enzymes capable of breaking metal-carbon bonds have also been described. Tonomura and Kanzaki (1969a,b) reported the isolation of an enzyme from soil microorganisms capable of metabolizing phenylmercuric acetate. The enzyme(s) were capable of cleaving the carbon-mercury bond and required both a sulfhydryl compound and NADH. Organoarsenical compounds are known to be degraded by soil microorganisms (Von Endt, Kearney, and Kaufman, 1968), but the enzymes responsible were not isolated or characterized. Whether or not those enzymes capable of metal-carbon bond cleavage are analagous to those which methylate elements such as Hg or As is not known.

Several acylamidase type enzymes have been isolated from soil microorganisms and characterized. All appear to have unique as well as similar characteristics. Nearly all appear to be of the EC 3.5.1a arylacylamide amidohydrolase type. The pH optima of these enzymes are on the alkaline side of neutrality. Considerable substrate variability exists. All appear to be more active on acylanilide substrates than on the inducing substrate. Those enzymes induced by acylanilides are active only on acylanilides, whereas those induced by the methoxy-substituted phenylureas are also active on acylanilides and phenylcarbamates. The methyl and phenylcarbamate-induced enzymes are active on phenylcarbamates and acylanilides but not phenylureas. With two exceptions, all of the acylamidases can be obtained in cell-free extracts. The 4-chloro-2-butynyl-*m*-chlorocarbanilate and isopropyl carbanilate hydrolyzing enzymes from *Penicillium* sp. and *Arthrobacter* sp., respectively, appear to be closely associated with the cell envelope (Clark and Wright, 1970a,b; Wright and Forey, 1972). The contrast in 3',4'-dichloropropionanilide-induced amidases in *Fusarium solani*

(Lanzilotta and Pramer, 1970a,b) and F. oxysporum (Blake and Kaufman, 1975) is particularly interesting in view of the close phylogenetic relationships of those two organisms. In addition to differences in substrate specificity, there are considerable differences in sensitivity to inhibitors. The acylamidase from F. solani is insensitive to O,O-diethyl O-(p-nitrophenyl) phosphorothioate and 1-naphthyl methylcarbamate, but is competitively inhibited by 2-chloro-N-isopropyl-acetanilide. In contrast, the F. oxysporum acylamidase is insensitive to 2-chloro-N-isopropylacetanilide, but strongly inhibited by O,O-diethyl O-(p-nitrophenyl)-phosphorothioate, 1-naphthol methylcarbamate, and p-chlorophenyl methylcarbamate, as well as 3-(3,4-dichlorophenyl)-1,1-dimethylurea and isopropyl m-chlorocarbanilate.

There is some evidence supporting the possibility that more than one acylamidase may be induced from a single organism. Jacob and Monod (1961) tested the inductive effects of methyl-β-D-thiogalactoside alone and in combination with phenyl-β-D-thiogalactoside (both inducers, but not substrates), but the inductive effect was not additive. They concluded the presence of a single enzyme, capable of being induced by a variety of compounds. Blake and Kaufman (1975) examined the inductive effects of p-chlorophenyl methylcarbamate and 3',4'-dichloropropionanilide alone and in combination. The inductive effect of 3',4'-dichloropropionanilide and p-chlorophenyl methylcarbamate combined at the concentrations tested was more than additive and increased the enzyme activity over that of either 3',4'-dichloropropionanilide or p-chlorophenyl methylcarbamate alone. Differences in substrate specificity, pH optima, and molecular weight were also noted. The experimental parameters examined failed to demonstrate unequivocally whether any of the amidases induced were identical, but rather indicated the probable presence of different amidases. Iwainsky and Sehrt (1970) reported that at least three amidases were induced in Mycobacterium smegmatis by aliphatic amides.

There is evidence for the existence in soil of extracellular enzymes capable of degrading xenobiotics or xenobiotic degradation products added to soil. Getzin and Rosefield (1968) demonstrated the presence of an active, extractable organic entity which behaved like an enzyme, and was capable of hydrolyzing the O,O-dimethyl phosphorodithioate S ester of diethyl mercaptosuccinate. The active entity was heat-labile and persisted in irradiated sterile soil. It was not active in hydrolyzing other organophosphates, nor was the active entity present in all soils examined. Kaufman, Blake and Miller (1971) obtained evidence suggesting the existence of a 3',4'-dichloropropionanilide hydrolyzing system in soil. They observed that 78% of the 3',4'-dichloropropionanilide added to a fresh, previously untreated soil was hydrolyzed to 3,4-dichloroaniline in 4 hours. Only 4.3 % of the 3',4'-dichloropropionanilide added to soil treated with p-chlorophenyl methylcarbamate was hydrolyzed to 3,4-dichloroaniline in the equivalent time period. Burge (1972) subsequently isolated directly from soil a heat-labile, filterable 3',4'-dichloropropionanilide-hydrolyzing factor which was subject to inhibition by p-chlorophenyl methylcarbamate. The presence of peroxidases in soil capable of forming chloroazobenzenes from chloroanilines in soil has been demonstrated (Bartha, Linke, and Pramer, 1968). The presence of peroxidases in soil should be of great interest to researchers investigating soil xenobiotic transformations. Presumably, they could be involved in a vast array of soil metabolic reactions affecting xenobiotic residues. Despite the versatility of peroxidase enzymes, however, Burge (1973) noted that not all soil peroxidases had the ability to form 3,3',4,4'-tetrachloroazobenzene from 3',4'-dichloropropionanilide or 3,4-dichloroaniline in soil.

Chemical Structure: Biodegradability--Ideally, chemicals considered for land treatment should be nontoxic, remain stationary at the site of application, and degrade rapidly once they are applied. Selection of such chemicals necessitates an understanding of each chemical's chemical-physical characteristics and the structure-activity relationships underlying toxicity, mobility, and degradability. Although we have identified large classes of toxic chemicals such as the phenoxyalkanoates, carbamates, toluidines, organophosphates, s-triazines, and phenylureas, the structure-activity relationships relative to toxicity, mobility, and degradability have not been adequately investigated. Kaufman and Plimmer (1972) compared the structural features necessary for toxicity to target organisms with those permitting degradation in the environment for several large classes of pesticides. As would be expected, differences existed among the various chemical classes. In some chemical classes those structural features contributing to toxicity were coincident with those necessary for degradability, whereas in other chemical classes they were diametrical. In all chemical classes, however, the relationships were mediated by substituent type, number, and position.

The structure-toxicity and structure-degradability relationships of certain halogenated aliphatic acids were quite similar, i.e., the most phytotoxic structures were also the most readily degradable. Correlations between structure-toxicity and structure-degradability could also be adduced for the phenoxyalkanoates. Meta-substitution (para to a free *ortho* position) confers resistance to biodegradation and eliminates phytotoxic activity. (2,4,5-trichlorophenoxy)-acetic acid is moderately resistant to biodegradation but is highly active, having the two necessary free positions for activity. In contrast, (2,3,4-trichlorophenoxy)acetic acid is both inactive toxicologically and highly resistant to biodegradation. Halogenation in the *para* position increases both phytotoxicity and biodegradability of phenoxyacetates. Increasing the length of the side chain affects both phytotoxicity and biodegradability.

Carbamate chemicals fall into several categories, i.e., methylcarbamates, phenylcarbamates, and thio- and dithio-carbamates. Insufficient information is available to suggest very strong relationships for either methylcarbamates or the thio- and dithiocarbamates. Structure-activity and structure-biodegradability information are available for the phenylcarbamates. Templeman and Sexton (1945), Shaw and Swanson (1953), Moreland and Hill (1959), and Good (1961) have made comprehensive investigations of structure-activity relationships on a number of phenylcarbamates. Kaufman (1966b, 1971) investigated the biodegradation of phenyl-carbamates and *s*-triazines with mixed populations and pure cultures of soil microorganisms, whereas Kearney (1967) examined the effect of phenylcarbamate structure on enzymic hydrolysis. The results of these investigations suggest that both similarities and differences exist in structure-activity-degradability relationships of this group.

Although extensive structure-activity-degradability studies have not been made (or published) for phenylureas, certain generalizations can be made. In general, phytotoxicity increases with increased structural complexity, i.e. 3-phenyl-1,1-dimethylurea < 3-(*p*-chlorophenyl)-1,1-dimethyl-urea < 3-(3,4-dichlorophenyl)-1-methylurea < 3-(3,4-dichloro-phenyl)-1,1-dimethylurea < 3-(3,4-dichlorophenyl)-1-butyl-1-methylurea. They are all degraded initially by dealkylation of the side chains in isolated microbial culture systems, and their resistance to degradation appears to increase with increased complexity. Thus the structural features providing increased phytotoxicity are diametric to those assuring degradation. While their individual behavior in soil depends on differences in solubility and adsorptivity to soil particles, soil persistence of the phenylureas increases with ring substitution and complexity of the alkyl groups. Whether this is due to resistance to biodegradation or adsorptivity and possible inaccessibility to soil organisms is not fully known.

Some generalizations can be made concerning the effect of various chemical substituents, linkages, or metabolic reactions. Kearney and Plimmer (1970) suggested that several linkages could be classified as readily susceptible to degradation:

(1) $R-NH-CO_2R'$

(2) R-NH-COR'

(3) $\genfrac{}{}{0pt}{}{-O}{-O}\!\!>$ P-S-R

(4) $\genfrac{}{}{0pt}{}{-O}{-O}\!\!>$ P-O-R

(5) R-CHCl-COO-

(6) $R-CCl_2-COO-$

Aliphatic acids, anilides, carbamates, and phosphates are generally degraded within a short time after soil application. The rate at which linkages are hydrolyzed will depend upon the nature of R and R'. For example, replacement of the alkoxy group of a phenylcarbamate with an alkylamine group results in a phenylurea, which is generally more persistent than the phenylcarbamate. Other linkages might also be included under specific conditions. The ether linkage of the phenoxyalkanoates is generally susceptible to microbial attack, but in the methoxy-*s*-triazines, 3-[4-(*p*-chlorophenoxy)-phenyl]-1,1-dimethylurea, 3,6-dichloro-*o*-anisic acid and 3,5,6-trichloro-*o*-anisic acid the ether linkage appears relatively stable in soil.

Kaufman and Plimmer (1972) categorized organic chemical classes according to the initial degradative reactions (Table 7). Chemicals whose initial degradative reaction is ester hydrolysis are relatively short-lived in soil, whereas chemicals which initially undergo dealkylation generally tend to be somewhat more persistent. Chemicals which are initially dehalogenated are variable in their persistence, apparently depending on the complexity of the basic molecule. Halogenated aliphatic acids are readily degraded and thus not very persistent, whereas halogenated benzoic acids and *s*-triazines are intermediate in their persistence. Such observations, however, must also be interpreted in the light of other physicochemical characteristics of the molecule and its environment which may preclude the chemical's availability for degradation.

Considerable new insight into our understanding of chemical structure-activity relationships has been provided by the recent investigations of Hansch and his associates (Hansch, 1969; Hansch et al., 1963; Hansch and Fujita, 1964; Hansch, Kiehs, and Lawrence, 1965; Hansch and Deutsch,

TABLE 7. RELATIVE PERSISTENCE AND INITIAL DEGRADATIVE REACTIONS OF NINE MAJOR ORGANIC CHEMICAL CLASSES

Chemical class	Persistence	Initial degradative process
Carbamates	2-8 weeks	Ester hydrolysis
Aliphatic acids	3-10 weeks	Dehalogenation
Nitriles	4 months	Reduction
Phenoxyalkanoates	1-5 months	Dealkylation, ring hydroxylation or oxidation
Toluidine	6 months	Dealkylation (aerobic) or reduction (anaerobic)
Amides	2-10 months	Dealkylation
Benzoic acids	3-12 months	Dehalogenation or decarboxylation
Ureas	4-10 months	Dealkylation
Triazines	3-18 months	Dealkylation or dehalogenation

1966a,b; Hansch, Deutsch, and Smith, 1965). Explanations of results in earlier investigations have generally been limited to consideration of steric and electronic effects. By using substituent constants and regression analysis, they have demonstrated that the substituent effects can be factored into three groups: electronic, steric, and hydrophobic. They have developed an equation using experimentally based variables for correlating the effect of a given substituent on the biological activity of a parent compound. The value of their equations has been tested on numerous types of biologically active compounds including antibiotics, pesticides, growth regulators, and carcinogens.

In general, their correlations have proven exceptionally good. They have successfully demonstrated the feasibility of predicting biological activity of chemicals within a given class by basing their approach on extra-thermodynamic considerations. The application of their approach, however, has been limited to already established or existing classes of chemicals showing bioactivity, i.e., this approach has not been used in the development of new, previously untried chemicals. Nor has it been applied to structure-biodegradability analysis. The real value of the Hansch approach would appear to be in the selective development of new biocides. The selection of specific unique processes within the target organism and the design of biocides for the process should be a feasible objective with such an approach. Incorporated into such an approach should be an awareness of "structure-activity" relationship relative to the biodegradability of the organic chemical.

Biochemical mechanisms of breakdown--Organic chemicals are subject to alteration by a number of biochemical reactions. In soil, these reactions are catalyzed by enzymes from a large array of microorganisms. In general, metabolites arising from these microbial reactions are usually nontoxic, polar molecules that exhibit little ability to accumulate in food chains. Studies on the biotransformation of organic chemicals in soil and microbial systems have been in progress for several decades; and a number of comprehensive review articles are available (Alexander, 1967; Audus, 1960, 1964; Helling, Kearney, and Alexander, 1971; Kaufman, 1970, 1974; Kearney and Kaufman, 1969; Kearney, Nash, and Isensee, 1969; Kaufman and Kearney, 1970; Menzie, 1969). The reader is referred to these articles for in-depth reviews of the numerous reactions and pathways elucidated for a large number of organic chemicals. The following sections deal with the major reactions of a large number of organic chemicals, wastes, and their metabolites in soils or in microbial systems.

Degradation of Combinations of Chemicals--Combinations of wastes in soil may result from either the successive application of individual wastes, with the concomitant cumulation of their residues, or the intentional application of combined wastes. Problems involving the degradation, persistence, or toxicity of organic chemicals may arise when several wastes or their residues are present in the soil together.

Numerous types of chemical interactions have been observed. Interactions in terms of plant response have been noted for some mixtures of organic chemicals (Colby and Feeny, 1967; Bowling and Hudgines, 1966; Hacskaylo, Walker, and Pires, 1964; Nash, 1967, 1968; Ranney, 1964; Swanson and Swanson, 1968; Nash and Harris, 1969; Arnold and Apple, 1957; Duffield, 1952; Stanier, 1947). While most of these interactions resulted in increased phytotoxicity, a few involved reduced toxicity. Chemical interactions among pesticides are known to occur (Miller and Lukens, 1966), but these are generally recognized during developmental stages of the pesticides and appropriate precautions are observed. Only a few interactions in terms of soil persistence have been reported (Audus, 1951, 1952, 1960, 1964; Brownbridge, 1956; Dawson, 1969; Kaufman, 1966a, 1977b; Kaufman, Blake, and Miller, 1971; Kaufman et al., 1970; Ferris and Lichtenstein, 1980; Kaufman and Miller, 1969, 1970; Kaufman and Sheets, 1965). For more detailed review of these interactions the reader is referred to the cited literature. The purpose of this discussion is to briefly present some of the possible interactions one might observe during the degradation of waste combinations.

Interactions leading to decreased persistence may result from chemical reactions or enhanced biodegradation of the combined wastes. The deactivation of sodium methyldithiocarbamate by certain halogenated hydrocarbons is an example of chemical reaction. Sodium methyldithiocarbamate is degraded when combined with 1,3-dichloropropene, 1,2-dibromo-3-chloropropene, ethylene dibromide, or related polyhalogenated alkenes (Miller and Lukens, 1966). It was suggested that sodium methyldithiocarbamate degradation occurred through esterification of the carbamic acid by the halogenated alkene. Sodium methyldithiocarbamate is a volatile soil fumigant and interactions of this type which further limit its residual life in soil are, therefore, deleterious. Such interactions within chemical wastes, however, may be advantageous, as long as they do not create new problems.

The ability of soil microorganisms to at least partially metabolize a homologous series of molecules, once adapted to metabolizing a single member of the series, is well documented in metabolism research, and is in accordance with the principles of adaptation outlined by Stanier (1947). Audus (1951, 1952, 1960, 1964) reported that soil microbial populations pre-induced with (2,4-dichlorophenoxy)acetic acid could rapidly degrade [(4-chloro-*o*-tolyl)oxy]acetic acid and *vice versa*. He also observed that a [(4-chloro-*o*-tolyl)oxy]-acetic acid-induced microbial population could decompose (2,4,5-trichlorophenoxy)acetic acid which a (2,4-dichlorophenoxy)acetic acid enriched population could not. Brownbridge (1956) observed that [(4-chloro-*o*-tolyl)oxy]-acetic acid-induced microbial populations could rapidly degrade all phenoxyacetates supplied to it. Similar observations have been reported by others (Loos, 1969; Steenson and Walker, 1958).

Practical application of such phenomena may be possible with some combinations of wastes. For example, incorporation of small amounts of [(4-chloro-*o*-tolyl)oxy]acetic acid or a similar inducer into (2,4,5-trichlorophenoxy)acetic acid solutions may facilitate a more rapid degradation of the more persistent (2,4,5-trichlorophenoxy)acetic acid. Some caution should be expressed, however, since the microbial degradation of most phenoxyacetates proceeds by cleavage of the ether linkage to acetate and the corresponding phenol. The accumulation of the toxic 2,4,5-trichlorophenol may present an undesirable problem unless the enhanced degradation of (2,4,5-trichlorophenoxy)acetic acid is accompanied by further metabolism of the phenol. Consideration of the use of inducer molecules to enhance microbial degradation must, therefore be accompanied by an understanding of both the mechanisms involved and the ecological acceptability of the residues formed.

In analagous experiments, Kaufman (unpublished research) examined the degradation of 3,6-dichloro-o-anisic acid in perfused soils treated only with 3,6-dichloro-o-anisic acid, or previously enriched with (2,4-dichlorophenoxy)acetic acid. Degradation was measured by chloride ion liberation. 3,6-Dichloro-o-anisic acid applied to previously untreated soil was quite persistent and no chloride liberation was observed. In soils previously perfused with (2,4-dichlorophenoxy)acetic acid, however, rapid liberation of Cl^- from 3,6-dichloro-o-anisic acid was observed following the primary treatment with 3,6-dichloro-o-anisic acid. Subsequent treatments with this acid, however, were degraded more slowly, presumably due to a loss of "effectiveness" by the soil population.

Cooxidation, or cometabolism, is a process which might be used for the degradation of persistent chemicals applied in combination. Cooxidation is the phenomenon whereby a microorganism may metabolize a compound without being able to utilize the energy derived from the chemical reaction to sustain growth (Foster, 1962). Miyazaki, Boush, and Matsumura (1969) observed that the rate of isopropyl 4,4'-dichlorobenzilate and ethyl 4,4'-dichlorobenzilate degradation was greatly increased when citrate was incorporated into microbial growth media as a carbon supplement. A basic similarity exists between the reductive decarboxylation process of citric acid and 4,4'-dichlorobenzilic acid, a degradation product of both isopropyl 4,4'-dichlorobenzilate and ethyl 4,4'-dichlorobenzilate. The reductive decarboxylation process of the chlorobenzilates was apparently stimulated by citrate. Exploitation of such phenomena may prove successful and ecologically significant in controlling residual accumulations of certain hazardous wastes.

Increased persistence of one or more chemicals applied to soil in combinations has been observed (Kaufman, 1966a; Kaufman and Miller, 1969, 1970; Kaufman and Sheets, 1965; Kaufman et al., 1970). All of the interactions presently known involve the parent organic molecule, i.e., the interaction of one intact molecule with another. The diversity and multiplicity of organic chemical residues and their degradation products increases the probability that similar interactions may also occur at various other stages of microbial degradation.

Increased persistence of chemicals may result from several types of interactions: (1) chemical and physical interactions of the organic molecules occurring in combination may preclude their normal degradation in soil and thus increase their persistence in soil; (2) the biocidal properties of the chemicals to soil microorganisms may preclude their biodegradation; (3) direct inhibition of the adaptive enzymes of effective soil microorganisms; and (4) inhibition of the proliferation processes of effective microorganisms. Increased residue persistence resulting from these types of interactions may prolong the time required between waste applications, or may prevent the combining of certain waste streams.

An apparent increased chemical persistence may also be explained by reactions of organisms sensitive to the chemical combinations. The sensitivity of organisms to a given chemical may be greater in the presence of a second chemical. Thus, the organism may be affected for a longer period of time due to the enhanced activity of lower chemical concentrations during the time when the soil chemical concentration is actually decreasing.

Examples of several of the aforementioned reactions have been described in the literature. Wheat seedling bioassays and chemical analyses of soils treated with various 2,2-dichloropropionic acid and 3-amino-s-triazole combinations substantiated the fact that 2,2-dichloropropionate persisted longer in soils when applied in combination with 3-amino-s-triazole than when 2,2-dichloropropionic acid was applied alone (Kaufman, 1966a). While the precise mechanism by which 3-amino-s-triazole inhibits dalapon degradation is not known, the results of enrichment investigations indicated 3-amino-s-triazole did not interfere with the lag period or adaptation of effective microorganisms, but did inhibit their proliferation (Kaufman, 1966a). 2,2-Dichloroproprionate had very little effect on 3-amino-s-triazole behavior in soil.

1-Naphthyl methylcarbamate, 1,4,5,6,7,8,8-heptachloro-3a,4,7,7a-tetrahydro-4,7,methanoindene, and pentachloronitrobenzene increased the persistence of isopropyl m-chlorocarbanilate in soil when applied either alone or in combination with other chemicals (Kaufman, 1972). Walker (1970) also observed an increased soil persistence of isopropyl m-chlorocarbanilate in the presence of pentachloronitrobenzene. Combinations of 1,1,1-trichloro-2,2-bis-(p-chlorophenyl)ethane and N-[(trichloromethyl)thio]-4-cyclohexene-1,2-dicarboximide also increased soil persistence of isopropyl m-chlorocarbanilate although neither chemical alone affected isopropyl m-chlorocarbanilate persistence. 1,1,1-trichloro-2,2-bis-(p-chlorophenyl)-ethane has been reported to inhibit certain esterases in soil microorganisms (Franzke, Kujawa, and Engst, 1970). The microbial enzymes responsible for isopropyl m-chlorocarbanilate hydrolysis are esteratic or amidase type enzymes (Kearney, 1965; Kearney and Kaufman, 1965). Apparently, however, the combined effect of 1,1,1-trichloro-2,2-bis-(p-chlorophenyl)ethane and N-[(trichloro-methyl)-thio]-4-cyclohexene-1,2-dicarboximide was necessary to effectively inhibit isopropyl m-chlorocarbanilate degradation.

The interaction of methylcarbamates such as 1-naphthyl methylcarbamate and p-chlorophenyl methylcarbamate with phenylamides has been extensively investigated (Kaufman, Blake, and Miller, 1971; Kaufman, 1977a; Kaufman and Miller, 1970). Soil microorganisms are active in the biodegradation of phenylamide chemicals in soil. The degradation of both phenylcarbamates and acylanilides involves an initial hydrolysis to the corresponding phenyl and acid or alcohol moieties. Esterase and amidase type enzymes catalyzing these hydrolyses have been isolated and characterized (Blake and Kaufman, 1973, 1975; Franzke, Kujawa, and Engst, 1970; Kearney, 1965). Detailed enzyme investigations have demonstrated that methylcarbamates lacking steric hinderance of the carbamate linkage are competitive inhibitors of many of these enzymes (Kaufman et al., 1970). Certain of these enzymes are also sensitive to inhibition by organophosphates such as O,O-diethyl-O-(2-isopropyl-6-methyl-5-pyrimidinyl)-phosphorothioate and O,O-diethyl-O-(p-nitrophenyl)-phosphorothioate.

Simultaneous application of p-chlorophenyl methylcarbamate with 3',4'-dichloropropionanilide to soil retarded degradation of the 3',4'-dichloropropionanilide in soil and greatly reduced the formation of 3,3',4,4'-tetrachloroazobenzene (TCAB) (Kaufman, Blake, and Miller, 1971). Thus the presence of some combinations in soil may affect not only the rate of degradation but the pathway of degradation and the products formed.

Interactions affecting biodegradation of a chemical may ultimately affect the mobility of the chemical or its residues in soil. Talbert, Runyan, and Baker (1970) observed an interesting interaction between 3-amino-2,5-dichlorobenzioc acid and 2-sec-butyl-4,6-dintrophenol. 3-Amino-2,5-dichlorobenzioc acid leaches in soil when applied as the ammonium salt. In autoclaved soil, however, the methyl ester remains near the soil surface and does not leach. In nonsterile soil the methyl ester is readily hydrolyzed microbiologically to the free acid which then leaches readily. In the presence of 2-sec-butyl-4,6-dintrophenol, however, hydrolysis of the methyl ester of 3-amino-2,5-dichlorobenzioc acid is inhibited and no leaching occurs. Thus, inhibition of microbial degradation may ultimately affect mobility in soil.

The formation of conjugates and condensation products of organic chemicals has been well documented. The significance or fate of these complexes in our environment is unknown. The occurrence of multiple waste residues in soil provides opportunity for the formation of complex or hybrid residues. Such residues may result from the biochemical interaction of the residues of a single chemical, or the multiple interactions of residues from combinations of chemicals. Our knowledge of such interactions occurring in soil is very limited at present. Another area of major interest should be the effect of antibiotics present either as industrial wastes or in various human and livestock wastes. These chemicals are used extensively to control diseases and pests, but once excreted from the human or animal system are only partially, if at all, understood with regard to their degradation and fate in the environment, or their influence on various components of the environment.

Disposal of sewage sludge and industrial wastes is a pivotal question in their processing. "Recycling" wastes to agricultural land is considered feasible and in some cases even desirable. One of the most actively investigated areas of environmental research has been the decomposition of agricultural, industrial, and municipal wastes in various types of water processing systems. The oxidation of individual organic chemicals in activated sludge has been examined in everything from simple bench-type systems to full scale treatment plants and field experiments. The results of these investigations are obvious in that we now commonly use waste treatment facilities which are dependent upon microbiological processes. Despite concern in recent years over the environmental impact of heavy metals contained in nearly all major wastes, there are few studies on the effects of these pollutants on microorganisms in their natural environment. Although metal tolerances between different groups of organisms have been observed (Bewley, 1979, 1980; Allen et al, 1974; Babich and Stotzky, 1974, 1978; Gingell, Campbell, and Martin, 1976; Little, 1973; Little and Martin, 1972) only a few studies (Doyle, Kaufman, and Burt, 1978) have attempted to characterize the mechanism by which heavy metals affect metabolism of residues in soils.

Doyle (1979) and Doyle, Kaufman, and Burt (1978) examined the degradation of 14 ^{14}C-labeled organic chemicals in sewage sludge and dairy manure-amended soil in laboratory soil incubation studies. The dairy manure and sewage sludge were applied to the soil at rates of 1, 50, and 100 metric (t/ha). The organic chemicals examined were selected as representative of the major microbial degradative pathways in soil, i.e., oxidation, reduction, dehalogenation, hydrolysis, dealkylation, etc. Altered rates of microbial degradation were observed in sludge and/or manure-amended soil for a number of structurally unrelated organic chemicals. ^{14}C-Product distribution varied with soil amendments. Manure amendments increased the rate of degradation of 10 of the compounds tested. The breakdown of several compounds was positively correlated with the increased total microbial activity of manure-amended soil. Manure did not signficantly reduce the degradation of any chemical examined. Sewage sludge, however, enhanced the breakdown of two compounds, and decreased the rate of degradation of nine compounds. The breakdown of chemicals metabolized by microbially mediated dealkylation [2-chloro-4-(ethylamino)-6-(isopropylamino)-*s*-triazine and 3-(3,4-dichlorophenyl)-1,1-dimethylurea] or hydrolytic reactions [2-chloro-*N*-isopropylacetanilide; 3-(3,4-dichlorophenyl)-1-methoxy-1-methylurea; *S*-ethyl dipropylthiocarbamate; and isopropyl *m*-chlorocarbanilate] was inhibited by sewage sludge but enhanced by manure. Sewage sludge treatments also altered pentachloronitrobenzene metabolism by inducing a shift from the reductive to the oxidative mechanisms of breakdown.

In additional studies, dairy manure and sewage sludge were incorporated into soil at rates of 0 and 50 t/ha and then incubated 0, 30, 60, 120, and 240 days before application of ^{14}C-pentachloronitrobenzene and ^{14}C-2-chloro-4-(ethylamino)-6-(isopropylamino)-*s*-triazine. The accelerated rate of microbial degradation in manure treated soils decreased as the length of the preincubation of amended soils increased. The sewage sludge effects did not diminish as the preincubation period increased, suggesting that a stable component of the sludge was involved in the observed effects.

The degradation of 2-chloro-4-(ethylamino)-6-(isopropylamino)-*s*-triazine was further examined in soil amended with: (1) sewage sludges from different locations; (2) composted sewage sludge; (3) fractionated sewage sludge; (4) sewage effluent; and the chloride salts of Ni, Cu, Cd, Zn, and Pb. Collectively, the data indicated that the high concentration of heavy metals in sludge reduced the microbial dealkylation of 2-chloro-4-(ethylamino)-6-(isopropylamino)-*s*-triazine, but did not significantly change the rate of its chemical breakdown through hydrolysis. Inorganic metal salts were not effective in reducing 2-chloro-4-(ethylamino)-6-(isopropylamino)-*s*-triazine dealkylation unless incubated in the soil for 120 days prior to chemical application. This suggested that the type of metal complex or bonding was related to the inhibitory effects. Various combinations of metals and higher metal concentrations were more effective in inhibiting the dealkylation process. In field studies dairy manure, sewage sludge, and composted sewage sludge produced similar results to those observed in laboratory investigations.

Summary and Research Needs

Fate of toxic organic compounds (TO's) applied to land treatment sites depends largely on adsorption of the TO by soil and/or its biodegradation in the soil. Adsorption depends on water solubility, functional groups of the TO, soil pH and organic matter. The more strongly a TO is sorbed, the less likely it is to volatilize, be absorbed by plants, or leach to groundwater. Many TO's are degraded so rapidly that they do not appreciably contaminate ground water under field conditions.

Volatility can be estimated from the ratio of water solubility to vapor pressure, since this relationship indicates the portion of TO in the vapor phase over a dilute soil solution of the TO. This ratio is also affected by adsorption, which reduces the level of TO in solution and vapor phase. Wastes containing volatilizable TO's will need to be mixed into soil, applied by subsurface injection, or pretreated to reduce the volatile TO's.

Plants can absorb organic chemicals from the soil. Plants roots do not appear to discriminate in absorption of polar compounds under 500 molecular weight. Some nonpolar compounds are adsorbed onto root or leaf surfaces rather than being absorbed and translocated (DDT, PCB's). Uptake depends strongly on the water solubility of a compound. Adsorption in the soil (and hence organic matter level on the soil) reduces plant uptake. Much research will be needed to clarify how to manage land treatment sites to avoid plant contamination with toxic organic compounds.

Degradation of organic compounds in soil may result from chemical, photochemical, or biological processes. Degradability of a compound depends on its chemical structure, some being rapidly degraded and others relatively recalcitrant to degradation. Biodegradation can occur in microbial cells, in the soil solution by chemical mechanisms, or by extracellular enzymes sorbed to soil particles. The kinetics of biodegradation are zero-order if the concentration is high relative to microbes which can degrade it, or first order if concentration is not high enough to saturate the ability of microbes.

Often, soil microbes capable of degrading a compound proliferate in soil, and the effective population may remain several years after the last treatment. One can isolate pure cultures capable of degrading many of these chemicals. However, little study has been conducted on degradation of mixtures of many compounds as would occur in wastes. It is likely that maintaining a supply of biodegradable organic matter in site soils would allow a higher population of diverse microbes capable of degrading many kinds of toxic organic compounds. Compounds which are strongly adsorbed by soil are not as prone to be leached, and as such may not be available for degradation, or may be more readily biodegraded. Microbes may utilize a particular chemical as an energy source, or may cometabolize the compound with other normal metabolic processes. It is important to consider both types of biodegradation. Some cells are able to degrade a single group of chemicals while others may degrade diverse types of compounds. Although the kinds of organisms and even types of enzymes involved in key steps in biodegradation are known for some pesticides and organic compounds occurring in wastes, little is known about most compounds in organic chemical wastes and by-products.

Chemical structure vs. biodegradability relationships are complex. Models for biological activity have recently been developed which consider the electronic, steric, and hydrophobic properties of substituents. This approach has not yet been applied to biodegradability.

Research Needs--

The degradation, metabolism, persistence, and movement of many hazardous wastes in soil are not known. Research is needed to characterize the degradative mechanism, the pathway, products formed, persistence, and leachability of these wastes and their degradation products. The influence of environmental factors, and other chemicals (both inorganic and organic) either within the same waste or mixed wastes must be determined, to allow safe mixing and disposal of hazardous wastes. Attention must be given to how much, how often, and in what sequences wastes can be applied to a single land site and have effective degradation of all wastes occur. Methods of managing a soil microbial population for purposes of waste disposal must be developed. Methods of chemically and/or physically managing the waste in soil must also be developed. Pretreatment processes for recalcitrant wastes which enable their more rapid degradation should be developed.

Methods for characterizing the microbiological suitability of a soil for use as a waste management site should be developed. Factors influencing chemical-structure: biodegradability should be determined so that ultimately this feature can be programmed into the determination of whether a waste is suitable for land farming. Factors affecting or regulating plant uptake of a hazardous waste should be examined.

REFERENCES

Adams, R. S., Jr. 1966. Soxhlet extraction of simazine from soils. Soil Sci. Soc. Amer. Proc. 30:689-693.

Aldrich, R. J. 1953. Residues in soil. J. Agr. Food Chem. 1:257-260.

Alexander, M. 1967. The breakdown of pesticides in soils. pp. 331-342. In N. C. Brady (ed.) Agriculture and the Quality of our Environment. Am. Assoc. Adv. Sci., Washington, D.C.

Alexander, M. 1965. Persistence and biological reactions of pesticides in soils. Soil Sci. Soc. Amer. Proc. 29:1-7.

Alexander, M., and B. K. Lustigman. 1966. Effect of chemical structure on microbial degradation of substituted benzene. J. Agr. Food Chem. 14:410-413.

Allen, G., G. Nickless, B. Wibberley, and J. Pick. 1974. Heavy metal particle characterization. Nature (London) 252:571-572.

Aly, O. M., and S. D. Faust. 1964. Studies on the fate of 2,4-D and ester derivatives in natural surface water. J. Agric. Food Chem. 12:541-546.

Armstrong, D. E., and G. Chesters. 1968. Adsorption catalyzed chemical hydrolysis of atrazine. Environ. Sci. and Technol. 2:683-689.

Arnold, E. W., and J. W. Apple. 1957. The compatibility of insecticides and fungicides used for the treatment of corn seed. J. Econ. Entomol. 50:43-45.

Ashley, M. G., and B. L. Leigh. 1963. The action of metham-sodium in soil. 1. Development of an analytical method for the determination of methyl isothiocyanate residues in soil. J. Sci. Food Agric. 14:148-153.

Ashton, F. M. 1961. Movement of herbicides in soil with simulated furrow irrigation. Weeds 9:612.

Audus, L. J. 1949. Biological detoxification of 2,4-dichlorophenoxyacetic acid in soil. Plant and Soil 2:31-36.

Audus, L. J. 1950. Biological detoxification of 2,4-dichlorophenoxyacetic acid in soil: Isolation of an effective organism. Nature (London) 166:356.

Audus, L. J. 1951. The biological detoxification of hormone herbicides in soil. Plant Soil. 3:170-192.

Audus, L. J. 1952. The decomposition of 2,4-dichlorophenoxyacetic acid and 2-methyl-4-chlorophenoxyacetic acid in soil. J. Sci. Food Agric. 3:268-274.

Audus, L. J. 1960. Microbiological breakdown of herbicides in soils. pp. 1-18. In E. K. Woodford and G. R. Sagar (eds.), Herbicides and the Soil. Blackwell, Oxford.

Audus, L. J. 1964. The herbicide behavior. pp. 163-206. In The Physiology and Biochemistry of Herbicides. Academic Press, New York and London.

Babich, H., and G. Stotzky. 1974. Air pollution and microbial ecology. Crit. Rev. Environ. Control 4:353-421.

Babich, H., and G. Stotzky. 1978. Effects of cadmium on the biota: Influence of environmental factors. Adv. Appl. Microbiol. 23:55-117.

Bailey, G. W., and J. L. White. 1970. Factors influencing the adsorption, desorption and movement of pesticides in soil. Residue Rev. 32:29-92.

Bailey, G. W., J. L. White, and T. Rothberg. 1968. Adsorption of organic herbicides by montmorillonite: role of pH and chemical character of adsorbate. Soil Sci. Soc. Amer. Proc. 32:222.

Bartha, R. 1968. Biochemical transformations of anilide herbicides in soils. J. Agric. Food Chem. 16:602-604.

Bartha, R., H. A. B. Linke, and D. Pramer. 1968. Pesticide transformation: Production of chloroazobenzenes from chloroanilines. Science 161:582-583.

Beall, M. L., Jr., and R. G. Nash. 1971. Organochlorine insecticide residues in soybean plant tops: Root vs. vapor sorption. Agron. J. 63:460-464.

Bewley, R. J. F. 1979. The effects of zinc, lead, and cadmium pollution on the leaf surface microflora of *Lolium perenne* L. J. Gen. Microbiol. 110:247-254.

Bewley, R. J. F. 1980. Effects of heavy metal pollution on oak leaf microorganisms. Appl. Environ. Microbiol. 40:1053-1059.

Black, J. L., and M. Van Winkle. 1961. Sorption of n-butyl ammonium and n-dodecyl ammonium acetate by sodium montmorillonite and sodium vermiculite. J. Chem. Eng. Data 6:557.

Blake, J., and D. D. Kaufman. 1973. Isolation of acylanilide-hydrolyzing microbial enzymes. Abstr. 165th Mtg. Amer. Chem. Soc., Div. Pestic. Chem. No. 7.

Blake, J., and D. D. Kaufman. 1975. Characterization of acylanilide hydrolyzing enzyme(s) from Fusarium oxysporum Schlecht. Pestic. Biochem. Physiol. 5:305-313.

Bodenheimer, W., and L. Heller. 1968. Sorption of methylene blue by montmorillonite saturated with different cations. Israel J. Chem. 6:307.

Bollag, J. M., C. S. Helling, and M. Alexander. 1968. 2,4-D metabolism. Enzymatic hydroxylation of chlorinated phenols. J. Agric. Food Chem. 16:826-828.

Bowling, C. C., and H. R. Hudgines. 1966. The effect of insecticides on the selectivity of propanil on rice. Weeds 14:94-95.

Brian, P. W. 1967. Uptake and transport of systemic fungicides and bactericides. Agrochimica 11:203-228.

Brindley, G. W., and T. D. Thompson. 1966. Clay organic studies. XI. Complexes of benzene, pyridine, piperidine and 1,3-substituted propanes with a synthetic Ca-fluorhectorite. Clay Minerals 6:345.

Brooks, C. S. 1964. Mechanism of methylene blue dye adsorption on siliceous minerals. Kolloid Zh. 199:31.

Brownbridge, N. 1956. Studies on the breakdown of 2,4-dichlorophenoxyacetic acid and some related compounds by soil micro-organics. Ph.D. Thesis, London Univ.

Burchfield, H. P. 1960. Performance of fungicides on plants and in soil - physical, chemical and biological considerations. pp.477-520. In J. G. Horsfall and A. E. Dimond (eds.) Plant Pathology, Vol. 3. Academic Press, New York.

Burchfield, H. P., and P. H. Schuldt. 1958. Pyridine-alkali reactions in the analysis of pesticides containing active halogen atoms. J. Agr. Food Chem. 6:106-111.

Burchfield, H. P., and E. E. Storrs. 1956. Chemical structures and dissociation constants of amino acids, peptides, and proteins in relation to their reaction rates with 2,4-dichloro-6-(o-chloroanilino)-s-triazine. Contrib. Boyce Thompson Inst. 18:395-418.

Burge, W. D. 1969. Populations of dalapon-decomposing bacteria in soil as influenced by additions of dalapon or other carbon sources. Appl. Microbiol. 17:545-550.

Burge, W. D. 1972. Microbial populations hydrolyzing propanil and accumulation of 3,4-dichloroaniline and 3,3',4,4'-tetrachloroazobenzene in soils. Soil Biol. Biochem. 4:379-386.

Burge, W. D. 1973. Transformation of propanil-derived 3,4-dichloroaniline in soil to 3,3',4,4'-tetrachloroazobenzene as related to soil peroxidase activity. Soil Sci. Soc. Amer. Proc. 37:392-395.

Burschel, P., and V. H. Freed. 1960. The decomposition of herbicides in soils. Weeds 7:151-161.

Call, F. 1957a. The mechanism of sorption of ethylene dibromide on moist soils. J. Sci. Food Agric. 8:630-639.

Call, F. 1957b. Sorption of ethylene dibromide on soils at field capacity. J. Sci. Food Agric. 8:137-142.

Calvert, J. C., and J. N. Pitts. 1966. Photochemistry. Wiley, New York. 899pp.

Caro, J. H. 1969. Accumulation by plants of organochloride insecticides from soil. Phytopathology 59:1191-1197.

Casida, J. E., and L. Lykken. 1969. Metabolism of organic pesticide chemicals in higher plants. Ann. Rev. Plant Physiol. 20:607-636.

Castro, C. E., and N. O. Belser. 1966. Hydrolysis of cis- and trans-1,3-dichloropropene in wet soil. J. Agric. Food Chem. 14:69-70.

Clark, C. G., and S. J. L. Wright. 1970a. Detoxication of isopropyl N-phenylcarbamate (IPC) and isopropyl N-3-chlorophenylcarbamate (CIPC) in soil, and the isolation of IPC metabolizing bacteria. Soil Biol. Biochem. 2:19-26.

Clark, C. G., and S. J. L. Wright. 1970b. Degradation of the herbicide isopropyl N-phenylcarbamate by Arthrobacter and Achromobacter spp. from soil. Soil Biol. Biochem. 2:217-226.

Colby, S. R., and R. W. Feeny. 1967. Herbicidal interactions of potassium azide with calcium cyanamid. Weeds 15:163-167.

Comes, R. D., D. W. Dohmont, and H. P. Alley. 1961. Movement and persistence of endothal (3,6-endoxohexahydrophthalic acid) as influenced by soil texture, temperature, and moisture levels. J. Amer. Soc. Sugar-Beet Tech. 11:287.

Cowan, C. T., and D. White. 1958. The mechanisms of exchange reactions occurring between sodium montmorillonite and various N-primary aliphatic amino salts. Trans. Faraday Soc. 34:691.

Crafts, A. S. 1961. Translocation in Plants. Holt, Reinehardt, and Winston, New York. 182pp.

Crafts, A. S. 1962. Movement of herbicides in soils and plants. Proc. Western Weed Control Conf. 18:43-47.

Crank, J. 1965. The Mathematics of Diffusion. Oxford University Press, London. 347pp.

Dawson, J. H. 1969. Longevity of dodder control by soil-applied herbicides in the greenhouse. Weed Sci. 17:295-298.

Dodd, C. G., and R. Satyabrata. 1960. Semiquinone cation adsorption on montmorillonite as a function of surface acidity. Clays and Clay Minerals 8:237.

Doyle, R. C. 1979. The effect of dairy manure and sewage sludge on pesticide degradation in soil. Ph.D. Dissertation. Univ. of Md., College Park, MD.

Doyle, R. C., D. D. Kaufman, G. W. Burt. 1978. Effect of dairy manure and sewage sludge on ^{14}C-pesticide degradation in soil. J. Agric. Food Chem. 26:987-989.

Duffield, P. C. 1952. Combination insecticide-fungicide seed treatments for corn. J. Econ. Entomol. 45:672-674.

Edwards, C. A. 1964. Factors affecting the persistence of insecticides in soils. Soils Fertilizers 27:451-454.

Edwards, C. A. 1966. Insecticides residues in soils. Residue Rev. 13:83-132.

Edwards, C. A. 1970. Persistent pesticides in the environment. CRC Press, The Chem. Rubber Co., Cleveland, OH

Engelhardt, G., P. R. Wallnofer, and G. Plapp. 1973. Purifications and properties of an aryl acylamidase of Bacillus sphaericus, catalyzing the hydrolysis of various phenylamide herbicides and fungicides. Appl. Microbiol. 26:709-718.

Engst, R., and M. Kujawa. 1967. Enzymic decomposition of DDT by a fungus. II. The course of enzymic DDT decomposition. Nahrung 11:751-760.

Farmer, W. J., W. F. Spencer, R. A. Shepherd, and M. M. Claith. 1974. Effect of flooding and organic matter applications on DDT residues in soil. J. Environ. Quality 3:343-346.

Farqui, F. A., S. Okuda, and W. O. Williamson. 1967. Chemisorption of methylene blue by kaolinite. Clay Minerals 7:19.

Ferris, I. G., and E. P. Lichtenstein. 1980. Interactions between agricultural chemicals and soil microflora and their effects on the degradation of [^{14}C] parathion in a cranberry soil. J. Agric. Food Chem. 28:1011-1019.

Fewson, C. A., and D. J. D. Nicholas. 1961. Utilization of nitrate by microorganisms. Nature (London) 190:2.

Finlayson, D. G., and H. R. MacCarthy. 1965. The movement and persistence of insecticides in plant tissue. Residue Rev. 9:114-152.

Firestone, D. 1972. Determination of polychloro-dibenzo-p-dioxins and related compounds in commercial chlorophenols. J. Assoc. Offic. Anal. Chem. 55:85-92.

Fleming, W. E., L. B. Parker, W. W. Maines, E. L. Plasket, and P. J. McCabe. 1962. Bioassay of soil containing residues of chlorinated hydrocarbon insecticides with special reference to control of Japanese beetle grubs. ARS, U.S. Dept. Agr. Tech. Bull. No. 1266. 44pp.

Foote, C. S. 1968. Mechanisms of photosensitized oxidation. Science 174:407-408.

Foster, J. W. 1962. Bacterial oxidation of hydrocarbons. pp. 241-271. In O. Hayaishi (ed.) The Oxygenases. Academic Press, New York.

Foster, A. C., V. R. Boswell, R. D. Chisholm, R. H. Canter, G. L. Gilpin, W. S. Anderson, and M. Geiger. 1956. USDA Tech. Bull., 1149, 36pp.

Foy, C. D., G. E. Coats, and D. W. Jones. 1971. Translocation of growth regulators and herbicides: vascular plants. p. 743-791. In P. L. Altman and D. S. Dittner (eds.), Respiration and Circulation. Biological Handbooks, Fed. Amer. Soc. for Ex. Biol., Bethesda, MD.

Franzke, C., M. Kujawa, and R. Engst. 1970. Enzymatischer Abbau des DDT durch Schimmelpilze. 4. Mitt. Einfluss des DDT auf das Wachstum von Fusarium oxysporum sowie auf die Pilzesterase. Die Nahrung 14:339-346.

Freed, V. H., J. Vernutti, and M. Montgomery. 1962. The soil behavior of herbicides as influenced by their physical properties. Proc. W. Weed Control Conf. 19:21.

Fries, G. F., and G. S. Marrow. 1981. Chlorobiphenyl movement from soil to soybean plants. J. Agric. Food Chem. 29:757-759.

Frissel, M. J. 1961. The adsorption of some organic compounds, especially herbicides on clay minerals. Verslag. Landbouwk. Onderzoek., 76:3.

Fryer, J. D., and K. Kirkland. 1970. Field experiments to investigate long-term effects of repeated applications of MCPA, triallate, simazine and linuron: Report after 6 years. Weed Res. 10:133-158.

Funderburk, H. H., Jr. 1969. Diquat and paraquat. pp. 283-209. In P. C. Kearney and D. D. Kaufman (eds.) Degradation of Herbicides. Dekker, New York.

Funderburk, H. H., Jr., and G. A. Bozarth. 1967. Review of metabolism and decomposition of diquat and paraquat. J. Agric. Food Chem. 15:563-567.

Getzin, L. W., and I. Rosefield. 1968. Organophosphate insecticide degradation by heat labile substances in soil. J. Agric. Food Chem. 16:598-601.

Gieseking, J. E. 1939. The mechanism of cation exchange in the montmorillonite-bedidellite-nontronite type of clay mineral. Soil Sci. 47:1-13.

Gingell, S. M., R. Campbell, and M. H. Martin. 1976. The effect of zinc, lead, and cadmium pollution on the leaf surface microflora. Environ. Pollut. 11:25-37.

Goldman, P. 1972. Enzymology of carbon-halogen bonds. pp. 147-165. In Dedgradation of Synthetic Organic Molecules in the Biosphere. National Academy of Science, Washington, D.C.

Good, N. E. 1961. Inhibitors of the Hill Reaction. Pl. Physiol., Lancaster, 36:788-803.

Goring, C. A. I. 1962. Theory and principles of soil fumigation. Advan. Pestic. Control Res. 5:47-84.

Goring, C. A. I. 1967. Physical aspects of soil in relation to the action of soil fungicides. Ann. Rev. Phytopathol., 5:285-318.

Gottlieb, D., and P. Siminoff. 1952. The production and role of antibiotics in the soil. Chloromycetin. Phytopathology 42:91.

Gottlieb, D., P. Siminoff, and M. M. Martin. 1952. The production and role of antibiotics in soil. IV. Actidione and clavacin. Phytopathology 42:483.

Gray, R. A. 1962. Rate of vapam decomposition in different soils and other media. Phytopathology 52:734.

Greenland, D. J. 1965. Interactions between clays and organic compounds in soils. Part I. Mechanisms of interaction between clays and defined organic compounds. Soils and Fertilizers 28:415-425.

Greenland, D. J., and J. P. Quirk. 1962. The adsorption of alkyl-pyridinium bromides by montmorillonite. Clay and Clay Minerals 9:484.

Grimm, R. E., W. H. Allaway, and F. L. Cuthbert. 1947. Reactions of different clay minerals with organic cations. J. Amer. Ceram. Soc. 30:137.

Guiffrida, L., N. F. Ives, and D. C. Bostwick. 1966. Gas chromatography of pesticides - improvements in the use of special ionization detection systems. J. Assoc. Offic. Anal. Chem. 49:8-21.

Gundersen, K. C., and H. L. Jensen. 1955. A soil bacterium decomposing organic nitrocompounds. Acta Agric. Scand. 5:100-114.

Hacskaylo, J., J. K. Walker, Jr., E. G. Pires. 1964. Response of cotton seedlings to combinations of preemergence herbicides and systemic insecticides. Weeds 12:288-291.

Hadzi, D., C. Klofutar, and S. Oblak. 1968. Hydrogen bonding on some adducts of oxygen bases with acids. Part IV. Basicity in hydrogen bonding and in ionization. J. Chem. Soc. (A) 1968:905.

Hamaker, J. W. 1966. Mathematical prediction of cumulative levels of pesticides in soil. Advan. Chem. Ser. 60:122-131.

Hamaker, J. W., C. R. Youngson, and C. A. I. Goring. 1967. Prediction of the persistance and activity of Tordon herbicide in soils under field conditions. Down to Earth 23:(2).

Hamaker, J. W., and J. M. Thompson. 1972. Adsorption. pp. 49-143. In C. A. I. Goring and J. W. Hamaker (eds.) Organic Chemicals in the Soil Environment. Dekker, New York.

Hamilton, R. H., and D. E. Moreland. 1962. Simazine: degradation by corn seedlings. Science 135:373-374.

Hance, R. J. 1965. The adsorption of urea and some of its derivatives by a variety of soils. Weed Research 5:98.

Hansch, C. 1969. Quantitative approach to biochemical structure-activity relationships. Accounts Chem. Res. 2:232-239.

Hansch, C., and E. W. Deutsch. 1966a. The structure-activity relationships in amides inhibiting photosynthesis. Biochim. Biophys. Acta 112:381-391.

Hansch, C., and E. W. Deutsch. 1966b. The use of substituent constants in the study of structure-activity relationships in cholinesterase inhibitors. Biochim. Biophys. Acta 126:117-128.

Hansch, C., and E. W. Deutsch, and R. N. Smith. 1965. The use of substituent constants in the study of enzymatic reaction mechanisms. J. Am. Chem. Soc. 87:2738-2742.

Hansch, C., and T. Fujita. 1964. P-σ-π Analysis. A method for the correlation of biological activity and chemical structure. J. Am. Chem. Soc. 86:1616-1626.

Hansch, C., K. Kiehs, and G. L. Lawrence. 1965. The role of substituents in the hydrophobic bonding of phenols by serum and mitochondrial proteins. J. Am. Chem. Soc. 87:5770-5773.

Hansch, C., R. M. Muir, T. Fujita, P. P. Maloney, F. Geiger, and M. Streich. 1963. The correlation of biological activity of plant growth regulators and chloromycetin derivatives with Hammett constants and partition coefficients. J. Am. Chem. Soc. 85:2817-2824.

Hanson, W. J., and R. W. Nex. 1953. Diffusion of ethylene dibromide in soils. Soil Sci. 76:209-214.

Harris, C. I. 1964. Movement of dicamba and diphenamid in soils. Weeds 12:112-115.

Harris, C. I. 1967. Fate of 2-chloro-s-triazines in soil. J. Agric. Food Chem. 15:157-162.

Harris, C. I., and G. F. Warren. 1964. Adsorption and desorption of herbicides by soil. Weeds 12:120-126.

Hartley, G. S. 1964. Herbicide behavior in the soil. pp. 111-161. In L. J. Audus (ed.) The Physiology and Biochemistry of Herbicides. Academic Press, New York.

Hartley, G. S. 1976. Physical behavior in the soil. pp. 1-28. In L. J. Audus (ed.) Physiology, Biochemistry, Ecology, Vol. II. Academic Press, London.

Hecht, A., and J. H. Seinfeld. 1972. Development and validation of generalized mechanism for photochemical smog. Environ. Sci. Technol., 6:47-57.

Helling, C. S., D. D. Kaufman, and C. T. Dieter. 1971. Algae bioassay detection of pesticide mobility in soils. Weed Science 19:685-690.

Helling, C. S. 1971a. Pesticide mobility in soils. I. Parameters of soil thin-layer chromatography. Soil Sci. Soc. Amer. Proc. 35:732-737.

Helling, C. S., 1971b. Pesticide mobility in soils. II. Applications of soil thin-layer chromatography. Soil Sci. Soc. Amer. Proc. 35:737-743.

Helling, C. S., D. G. Dennison, and D. D. Kaufman. 1974. Fungicide movement in soils. Phytopathology 64:1091-1100.

Helling, C. S., P. C. Kearney, and M. Alexander. 1971. Behavior of pesticides in soils. Advan. Agron. 23:147-240.

Helling, C. S., and B. C. Turner. 1968. Pesticide mobility: Determination by soil thin-layer chromatography. Science 162:562-563.

Hendricks, S. B. 1941. Base exchange of the clay mineral montmorillonite for organic cations and its dependence upon adsorption due to van der Waals forces. J. Phys. Chem. 45:65.

Hill, G. D., J. W. McGrahen, H. M. Baker, D. W. Finnerty, and C. W. Bingeman. 1955. The fate of substituted urea herbicides in agricultural soils. Agron. J. 47:93-104.

Hilton, H. W., and Q. H. Yuen. 1963. Adsorption of several pre-emergence herbicides by Hawaiian sugarcane soils. J. Agr. Food Chem. 11:230-234.

Hornstein, I. 1955. Determination of lindane in mushrooms. J. Agr. Food Chem. 3:848-849.

Horowitz, M. 1966. Breakdown of endothal in soil. Weed Res. 6:168-171.

Horvath, R. S. 1971. Cometabolism of the herbicide 2,3,6-trichlorobenzoate. J. Agr. Food Chem. 12:291-293.

Horvath, R. S., and M. Alexander. 1970a. Cometabolism of m-chlorobenzoate by an Arthrobacter. Appl. Microbiol. 20:254-258.

Horvath, R. S., and M. Alexander. 1970b. Cometabolism: A technique for the accumulations of biochemical products. Can J. Microbiol. 16:1131-1132.

Hurle, K., and B. Rademacher. 1970. Untersuchungen uber den Einfluss langjahrig wiederholter Anwendung von DNOC und 2,4-D auf ihren Abbau im Boden. Weed Res. 10:159-164.

Hutzinger, O., S. Safe, and V. Zitko. 1972. Photochemical degradation of chlorobiphenyls (PCB's). Environ. Health Perspect. No. 1:15.

Isensee, A. R., and G. E. Jones. 1971. Absorption and translocation of root and foliage applied 2,4-dichlorophenol, 2,7-dichlorodibenzo-p-dioxin, and 2,3,7,8-tetrachlorodibenzo-p-dioxin. J. Agric. Food Chem. 19:1210-1214.

Ivie, G. W., and J. E. Casida. 1971. Sensitized photodecomposition and photosensitizer activity of pesticide chemicals exposed to sunlight on silica gel chromatoplates. J. Agr. Food Chem. 19:405-409.

Iwainsky, H., and I. Sehrt. 1970. Enzyminduktion und Stoffwechsel-regulation bei Mykobakterien: III. Induktion der enzymatischen hydrolyse und der Verwertung von Amiden bei. M. Smegmatis. 2. Bakteriol. Parasitenk Infektion-Krankh Hyg. Abt. I Orig. 2213:222-232.

Jacob, F., and J. Monod. 1961. Genetic regulatory mechanisms in the synthesis of proteins. J. Molec. Biol. 3:318-356.

Jensen, H. L. 1964. Biological decompositon of herbicides in the soil. Tidskr. Pl. Avl. 37:553-571.

Jordan, J. W. 1949. Alteration of the properties of bentonites by reaction with amines. Mineral. Mag. 28:598.

Jordan, L. S., B. E. Day, and W. A. Clerx. 1963. Effect of incorporation and method of irrigation on pre-emergence herbicides. Weeds 11:137.

Jurinak, J. J., and D. H. Volman. 1957. Application of the Brunauer, Emmett, and Teller equation to ethylene dibromide adsorption by soils. Soil Sci. 83:487-496.

Kaufman, D. D. 1964. Microbial degradation of 2,2-dichloropropionic acid in five soils. Can. J. Microbiol. 10:843-852.

Kaufman, D. D. 1966a. Microbial degradation of herbicide combinations: Amitrole and Dalapon. Weeds 14:130-134.

Kaufman, D. D. 1966b. Structure of pesticides and decomposition by soil microorganisms. pp. 85-94. In ASA Special Publication No. 8. Soil Sci. Soc. Amer., Madison, WI.

Kaufman, D. D. 1970. Pesticide metabolism. pp. 73-86. In International Symposium on Pesticides in the Soil. Michigan State Univ., East Lansing, MI.

Kaufman, D. D. 1971. Effect of chemical structure on biodegradation of s-triazine herbicides. Abstr. Weed. Soc. Am. No. 35, p.18.

Kaufman, D. D. 1972. Degradation of pesticide combinations. In A. S. Tahori (ed.) Fate of Pesticides in Environment. Pestic. Chem. 6:175.

Kaufman, D. D. 1974. Degradation of pesticides by soil microorganisms. pp. 133-202. In W. D. Guenzi (ed.) Pesticides in Soil and Water. Soil Sci. Soc. Amer., Madison, WI.

Kaufman, D. D. 1977a. Biodegradation and persistence of several acetamide, acylanilide, azide, carbamate, and organophosphate pesticide combinations. Soil Biol Biochem. 9:49-57.

Kaufman, D. D. 1977b. Soil-fungicide interactions. pp. 1-49. In M. R. Siegel and H. D. Sisler (eds.) Antifungal Compounds, Vol. 2. Marcel Dekker, Inc., New York.

Kaufman, D. D., and J. Blake. 1970. Degradation of atrazine by soil fungi. Soil Biol. Biochem. 2:73-80.

Kaufman, D. D., and J. Blake. 1973. Microbial degradation of several acetamide, acylanilide, carbamate, toluidine, and urea pesticides. Soil Biol. Biochem. 5:297-308.

Kaufman, D. D., J. Blake, and D. E. Miller. 1971. Methylcarbamates affect acylanilide herbicide residues in soil. J. Agr. Food Chem. 19:204-206.

Kaufman, D. D., and P. C. Kearney. 1965. Microbial degradation of isopropyl-N-3-chlorophenylcarbamate and 2-chloroethyl-N-3-chlorophenylcarbamate. Appl. Microbiol. 13:443-446.

Kaufman, D. D., and P. C. Kearney. 1970. Microbial degradation of s-triazine herbicides. Residue Rev. 32:235-265.

Kaufman, D. D., P. C. Kearney, D. W. Von Endt, D. E. Miller. 1970. Methylcarbamate inhibition of phenylcarbamate metabolism in soil. J. Agr. Food Chem. 18:513-519.

Kaufman, D. D. and D. E. Miller. 1969. Microbial degradation of several aniline based herbicides. Weed Sci. Soc. Amer. Abstr. No. 235.

Kaufman, D. D., and D. E. Miller. 1970. Controlled biodegradation and persistance of several amide, carbamate, and urea herbicides. Weed Sci. Soc. Amer. Abstr. No. 158.

Kaufman, D. D., and J. R. Plimmer. 1972. Approaches to the synthesis of soft pesticides. pp. 173-203. In R. Mitchell (ed.) Water Pollution Microbiology. Wiley: Interscience, New York, New York.

Kaufman, D. D., J. R. Plimmer, P. C. Kearney, J. Blake, and F. S. Guardia. 1968. Chemical versus microbial decomposition of amitrole in soil. Weed Sci. 16:266-272.

Kaufman, D. D., and T. J. Sheets. 1965. Microbial decomposition of pesticide combinations. Agron. Abstr. pp. 85.

Kearney, P. C. 1965. Purifications and properties of an enzyme responsible for hydrolyzing phenylcarbamates. J. Agric. Food Chem. 13:561-564.

Kearney, P. C. 1967. Influence of physicochemical properties on biodegradability of phenylcarbamate herbicides. J. Agric. Food Chem. 15:563-571.

Kearney, P. C., and D. D. Kaufman. 1965. Enzyme from soil bacterium hydrolyzes phenylcarbamate herbicides. Science 147:740-741.

Kearney, P. C., and D. D. Kaufman. 1969. Degradation of Herbicides. Dekker, New York.

Kearney, P. C., D. D. Kaufman, and M. L. Beall. 1964. Enzymatic dehalogenation of 2,2-dichloropropionate. Biochem. Biophys. Res. Commun., 14:29-33.

Kearney, P. C., D. D. Kaufman, D. W. VonEndt, and F. S. Guardia. 1969. TCA-metabolism by soil microorganisms. J. Agric. Food Chem. 17:581-584.

Kearney, P. C., R. G. Nash, and A. R. Isensee. 1969. pp. 54-67. In M. W. Miller and G. G. Berg (eds.) Chemical Fallout: Current Research on Persistent Pesticides. Charles C. Thomas, Springfield, IL

Kearney, P. C.,, and J. R. Plimmer. 1970 Relation of structure to pesticide decomposition. pp. 65-172. In Pesticides in the Soil: Ecology, Degradation and Movement (International Symposium). Michigan State Univ., East Lansing, MI.

Kearney, P. C., T. J. Sheets, and J. W. Smith. 1964. Volatility of seven s-triazines. Weeds 21:83-87.

Kesaree, M., T. Demirel, and E. Rosauer. 1960. X-ray diffraction studies of quarternary ammonium treated montmorillonite. Proc. Iowa Acad. Sci. 69:384.

Kinter, E. B., and S. Diamond. 1963. Characterization of montmorillonite saturated with short-chain amine cations. II. Interlayer surface coverage by the amine cations. Clays and Clay Minerals 10:174.

Kirkland, K. 1967. Inactivation of MCPA in soil. Weed Res. 7:364-367.

Kirkland, K., and J. D. Fryer. 1966. Pretreatment of soil with MCPA as a factor affecting persistence of a subsequent application. Proc. 8th Br. Weed Control Conf. pp. 616-621.

Koblitsky, L., and R. D. Chisholm. 1949. Determination of DDT in soils. J. Assoc. Offic. Agric. Chem. 32:781-786.

Kohn, G. K. 1969. Attenuation of pesticidal residues on seeds. Residue Rev. 29:3-12.

Kotter, K., J. Willenbrink, and K. Junkmann. 1961. Der Abbau von ^{35}S-markietem methylsenefol in verschiedenen Boden. Z. Pflanzenkrankh. Pflanzenschutz 68:407-411.

Kreutzer, W. A. 1960. Soil treatment. p. 431. In J. G. Horsfall and A. E. Dimond (eds.) Plant Pathology. Vol. 3. Academic Press, New York.

Kul'chitskii, L. I. 1961. A spectrophotometric study of the adsorption of methylene blue by highly dispersed aluminosilicate. Kolloid Zh. 23:63.

Kuritenko. O. D., and R. V. Mikhalyuk. 1959. Adsorption of aliphatic amines on bentonite from aqueous solutions. Kolloid Zh. 20:181.

Lanzilotta, R. P., and D. Pramer. 1970a. Herbicide transformation. I. Studies with whole cells of Fusarium solani. Appl. Microbiol. 19:301-306.

Lanzilotta, R. P., and D. Pramer. 1970b. Herbicide transformation. II. Studies with an acylamidase of Fusarium solani. Appl. Microbiol. 19:307-313.

Leasure, J. K. 1964. The halogenated aliphatic acids. J. Agric. Food Chem. 12:40-43.

Leermakers, P. A., H. T. Thomas, L. D. Weis, and F. C. James. 1966. Spectra and photochemistry of molecules adsorbed on silica gel IV. J. Amer. Chem. Soc. 88:5075-5083.

Leopold, A. C., K. P. VanSchaik, and N. Neal. 1960. Molecular structure and herbicide adsorption. Weeds 8:48.

Lichtenstein, E. P., and K. R. Schulz. 1960. Expoxidation of aldrin and heptachlor in soils as influenced by autoclaving, moisture, and soil types. J. Econ. Entomol. 53:192-197.

Lichtenstein, E. P., and K. R. Schulz. 1964. The effects of moisture and microorganisms on the persistence and metabolism of some organophosphorus insecticides in soil with special emphasis on parathion. J. Econ. Entomol. 57:618-627.

Little, P. 1973. A study of heavy metal contamination of leaf surfaces. Environ. Pollut. 5:159-172.

Little, P., and M. H. Martin. 1972. A survey of zinc, lead, and cadmium in soil and natural vegetation around a smelting complex. Environ. Pollut. 3:241-254.

Loos, M. A. 1969. Phenoxyalkanoic acids. pp. 1-49. In P. C. Kearney and D. D. Kaufman (eds.) Degradation of herbicides. Marcel Dekker, Inc., New York.

Loos, M. A. 1974. Phenoxyalkanoic acids. pp. 1-128. In P. C. Kearney and D. D. Kaufman (eds.) Herbicides: Chemistry, Metabolism, and Mode of Action. Dekker, New York.

Loos, M. A., J. M. Bollag, and M. Alexander. 1967. Phenoxyacetate herbicide detoxication by bacterial enzymes. J. Agric. Food Chem. 15:858-860.

Low, P. F. 1961. Physical chemistry of clay-water interaction. Adv. Agron 13:269-327.

Marth, F. H. 1965. Residues and some effects of chlorinated hydrocarbon insecticides and biological material. Residue Rev. 9:1-89.

Martin, M., and D. Gottleib. 1952. The production and role of antibiotics in the soil. III. Terramycin and aureomycin. Phytopathology 42:294.

Martin, M., and D. Gottleib. 1955. The production and role of antibiotics in soil. X. Antibacterial activity of five antibiotics in the presence of soil. Phytopatholology 45:407.

McAtee, J. L. 1959. Inorganic-organic cation exchange on montmorillonite. Amer. Mineral. 44:1230.

McAtee, J. L. 1962. Organic cation exchange on montmorillonite as observed by ultraviolet analyses. Clays and Clay Minerals 10:153.

McAtee, J. L., and J. R. Hachman. 1964. Exchange equilibria on montmorillonite involving organic cations. Amer. Mineral. 49:1569.

Menzie, C. M. 1969. Metabolism of pesticides. U.S. Fish Wildl. Serv., Special Scientific Report, Wildlife No. 127. Bureau of Sport Fisheries and Wildlife, Washington, D. C.

Metcalf, R. L. 1967. Absorption and translocation of systemic insecticides. Agrochimica. 11:105-123.

Miller, P. M., and R. J. Lukens. 1966. Deactivation of sodium N-methyldithiocarbamate in soil by nematocides containing halogenated hydrocarbons. Phytopathology 56:967.

Mills, P. A., J. Onley, and R. A. Gaither. 1963. Rapid method for chlorinated pesticide residues in nonfatty goods. J. Assoc. Offic. Agric. Chem. 46:186-191.

Milnes, M. H. 1971. Formation of 2,3,7,8-tetrachlorodioxin by thermal decomposition of sodium 2,4,5-trichlorophenate. Nature. 232:395.

Miyazaki, S., G. M. Boush, and F. Matsumura. 1969. Metabolism of ^{14}C-chlorobenzilate and ^{14}C-chloropropylate by Rhodotorula gracilis. Appl. Microbiol. 18:972-976.

Moje, W. 1959. Structure and nematocidal activity of allylic and acetylenic halides. J. Agric. Food Chem. 7:702-707.

Moje, W. 1960. The chemistry and nematocidal activity of organic halides. Advan. Pestic. Control Res. 3:181-217.

Moje, W., J. P. Martin, and R. C. Baines. 1957. Structural effect of some organic compounds on soil organisms and citrus seedlings grown in an old citrus soil. J. Agric. Food Chem. 5:32-36.

Moore, F. W. 1949. The utilizatin of pyridine by microorganisms. J. Gen. Microbiol. 3:143-147.

Moreland, D. E., and K. L. Hill. 1959. The action of alkyl N-phenylcarbamates on the photolytic activity of isolated chloroplasts. J. Agric. Food Chem. 7:832-837.

Mortland, M. M., and K. V. Raman. 1967. Catalytic hydrolysis of some organic phosphate pesticides by copper (II). J. Agric. Food Chem. 15:163-167.

Munnecke, D. E. 1958. The persistence of non-volatile diffusible fungicides in soil. Phytopathology 48:581-585.

Munnecke, D. E. 1967. Fungicides in the soil environment. pp. 509-559. In D. C. Torgenson (ed.) Academic Press, New York.

Munnecke, D. E. 1961. Movement of non-volatile diffusible fungicides through columns of soil. Phytopathology 51:593-599.

Munnecke, D. E., and J. Ferguson. 1953. Methyl bromide for nursery soil fumigation. Phytopathology 43:375-377.

Munnecke, D. E., and J. P. Martin. 1964. Release of methylisothiocyanate from soils treated with Mylone (3,5-di-methyl-tetrahydro-1,3,5-2H-thiadiazine-2-thione). Phytopathology 54:941-945.

Nash, R. G. 1967. Phytotoxic pesticide interactions in soil. Agron. J. 59:227-230.

Nash, R. G. 1968. Synergistic phytotoxicities of herbicide-insecticide combinations in soil. Weed Sci. 16:74-77.

Nash, R. G. 1974. Plant uptake of insecticides, fungicides, and fumigants from soils. pp. 257-313. In W. D. Guenzi (ed.) Pesticides in Soil and Water. Soil Sci. Soc. Amer., Inc., Madison, WI.

Nash, R. G., and W. G. Harris. 1969. Dexon fungicide antagonism toward herbicidal activity of s-triazines. Weed Soc. Amer. Abstr. No. 240.

Newman, A. S., and J. R. Thomas. 1949. Decomposition of 2,4,-dichlorophenoxyacetic acid in soil and liquid media. Proc. Soil Sci. Soc. Am. 14:160-164.

Newman, A. S., J. R. Thomas, and R. L. Walker. 1952. Disappearance of 2,4-dichlorophenoxyacetic acid and 2,4,5-trichlorophenoxyacetic acid from soil. Proc. Soil Sci. Soc. Am. 16:21-24.

Newsom, L. D. 1967. Consequences of insecticide use on nontarget organisms. Ann. Rev. Entomol. 12:257-286.

Pardee, A. B. 1962. The synthesis of enzymes. pp. 577-630. In "The Bacteria", Vol. 3 (I.C. Gunsalus and R. Y. Stamier, eds), Academic Press, London and New York.

Peachey, J. E., and M. R. Chapman. 1966. Chemical control of plant nematodes. Commonwealth Bur. Helminthology, G. Brit. Tech. Commun. No. 36. 119 pp.

Phillips, F. T. 1964. The aqueous transport of water-soluble nematacides through soils. III. Natural factors modifying the chromatographic leaching of phenol through soil. J. Sci. Food. Agric. 15:458.

Pimentel, D. 1971. Ecological effects of pesticides on nontarget species. Executive Office of the President, Office of Science and Technology. Superintendent of Documents, Washington, D. C.

Pinck, L. A., W. F. Holton, and F. E. Allison. 1961. Antibiotics in soil. I. Physico-chemical studies of antibiotic-clay complexes. Soil Sci. 91:22.

Pinck, L. A., D. A. Soulides, and F. E. Allison. 1961. Antibiotics in soils. II. Extent and mechanism of release. Soil Sci. 91:94.

Plimmer, J. R. 1972. Photochemistry of pesticides: a discussion of some environmental factors. pp. 47-76 In A. S. Tahori (ed.), Fate of Pesticides in Environment, Vol. 6, Gordon and Breach, New York.

Plimmer, J. R. 1974. Photolysis of 2,3,7,8-tetrachloro-dibenzo-p-dioxin. Abstr. 168th Mtg., Amer. Chem. Soc., Pestic. Chem. Div., Atlantic City. No. 29.

Plimmer, J. R., U. I. Klingebiel, and B. E. Hummer. 1970. Photo-oxidation of DDT and DDE. Science 167:67-69.

Plimmer, J. R., and U. I. Klingebiel. 1971. Riboflavin photosensitized oxidation of 2,4-dichlorophenol: assessment of possible chlorinated dioxin formation. Science 174:407-408.

Plimmer, J. R., P. C. Kearney, D. D. Kaufman, and F. S. Guardia. 1967. Amitrole decomposition by free radical-generating systems and by soils. J. Agric. Food Chem. 15:996-999.

Pramer, D. 1959. The status of antibiotics in plant disease control. Advan. Appl. Microbiol. 1:75-85.

Pramer, D. 1961. Eradicant and therapeutic materials for disease control. Recent Advan. Botany. Sec. 5., pp. 452-456.

Ranney, C. D. 1964. A deleterious interaction between a fungicide and systemic insecticides on cotton. Plant Dis. Rept. 48:241-245.

Reay, J. S. S., and R. M. Barrer. 1957. Sorption and intercalation by methylammonium montmorillonites. Trans. Faraday Soc. 53:1253.

Rosen, J. D., and M. Siewierski. 1970. Sensitized photolysis of heptachlor. J. Agr. Food Chem. 18:943.

Roth, W., and E. Knusli. 1961. Beitrag zur Kenntis der Resistenz - phenomene einzelner Pflanzen gegenuber dem phototoxischen Wirkstoff Simazin. Experientia 17:312-313.

Rowland, R. L., and E. J. Weiss. 1963. Bentonite-methyl-amine complexes. Clays and Clay Minerals 10:460.

Sanderman, W., H. Stockman, and R. Caston. 1957. Pyrolysis of pentachlorophenol. Chem. Ber. 90:960.

Sandmann, E. R. I. C. 1974. Aromatic metabolism by phenoxy herbicide-degrading soil bacteria. M. Sc. Thesis, Univ. Natal. South Africa.

Schechter, M. S., and I. Hornstein. 1952. Colorimetric determination of benzene hexachloride. Anal. Chem. 24:544-547.

Schuldt, P. H., H. P. Burchfield, and H. Bluestone. 1957. Stability and movement studies on the new experimental nematocide 3,4-dichlorotetrahydrothiophene-1-1-dioxide in soil. Phytopathology 47:534.

Sethunathan, N. 1973. Microbial degradation of insecticides in flooded soil and in anaerobic cultures. Residue Rev. 47:143-165.

Shaw, W. C., and C. R. Swanson. 1953. The relation of structural configuration to the herbicidal properties and phytotoxicity of several carbamates and other chemicals. Weeds 2:43-65.

Sheets, T. J. 1964. Review of disappearance of substituted urea herbicides from soil. J. Agric. Food Chem. 12:30-33.

Sheets, T. J., and C. I. Harris. 1965. Herbicide residues in soils and their phytotoxicities to crops grown in rotations. Residue Rev. 11:119-140.

Siminoff, P., and D. Gottlieb. 1951. The production and role of antibiotics in the soils. I. The fate of streptomycin. Phytopathology 41:420-430.

Skujins, J. J. 1967. Enzymes in soil. pp. 371-414. In A. D. McLaren and G. H. Peterson (eds.) Soil Biochemistry. Dekker, New York.

Smith, C. R. 1934. Base exchange reactions of bentonite and salts of organic bases. J. Amer. Chem. Soc. 56:1561.

Spanis, W. C., D. E. Munnecke, and R. A. Solberg. 1962. Biological breakdown of two organic mercurial fungicides. Phytopathology 52:435-462.

Spencer, E. Y. 1965. The significance of plant metabolites of insecticide residues. Residue Rev. 9:153-168.

Stanier, R. Y. 1947. Simultaneous adaptation: a new technique for the study of metabolic pathways. J. Bacteriol. 54:339-348.

Steenson, T. I., and N. Walker. 1958. Adaptive patterns in the bacterial oxidation of 2,4-dichloro-, and 4-chloro-2-methyl-phenoxyacetic acid. J. Gen. Microbiol. 18:692-697.

Stipes, R. J., and D. R. Oderwald. 1970. Soil thin-layer chromatography of fungicides. Phytopathology 60:1018. (Abstr).

Sund, G. W. 1956. Residual activity of 3-amino-1,2,4-triazole in soils. J. Agric. Food Chem. 4:57-60.

Sutherland, H. H., and D. M. C. MacEwan. 1961. Organic complexes of vermiculite. Clay Minerals Bull. 4:229.

Swanson, C. R., and H. R. Swanson. 1968. Inhibition of degradation of monuron in cotton leaf tissue by carbamate insecticides. Weed Sci. 16:481-484.

Talbert, R. E., and O. H. Fletchall. 1965. The adsorption of some s-triazines in soils. Weeds 13:46-52.

Talbert, R. E., R. L. Runyan, and H. R. Baker. 1970. Behavior of amiben and dinoben derivatives in Arkansas soils. Weed Sci. 18:10-15.

Templeman, W. G., and W. A. Sexton. 1945. Effect of some arylcarbamic esters and related compounds upon cereals and other plant species. Nature, Lond. 156:630.

Tewfik, M. S., and W. C. Evans. 1966. The metabolism of 3,5-dinitro-o-creosol (DNOC) by soil microorganisms. Biochem. J. 99:31-32.

Thomas, G. W. 1963. Kinetics of chloride desorption from soils. J. Agr. Food Chem. 11:201-203.

Thomas, S., and T. H. Grainger. 1952. Bacteria. Blakiston, New York. 623 pp.

Tiedje, J. M., and M. Alexander. 1969. Enzymatic cleavage of the ether bond of 2,4-dichlorophenoxyacetate. J. Agric. Food Chem. 17:1080-1084.

Tiedje, J. M., J. M. Duxbury, M. Alexander, and J. E. Dawson. 1969. 2,4-D metabolism: Pathway of degradation of chlorocatechols by *Arthrobacter* sp. J. Agric. Food Chem. 17:1021-1026.

Tonomura, K., and F. Kanzaki. 1969a. The decomposition of organic mercurials by cell-free extract of a mercury-resistant Pseudomonas. J. Ferment. Technol. 47:430-432.

Tonomura, K., and F. Kanzaki. 1969b. The reductive decomposition of organic mercurials by cell-free extract of a mercury-resistant pseudomonad. Biochim. Biophys. Acta. 184:227-229.

Torgeson, D. C., D. M. Yoder, and J. B. Johnson. 1957. Biological activity of mylone breakdown products. Phytopathology 47:536.

Turner, N. J., and M. E. Corden. 1963. Decomposition of sodium N-methyldithiocarbamate in soil. Phytopathology 53:1388-1394.

Turner, N. J., M. E. Corden, and R. A. Young. 1962. Decomposition of vapam in soil. Phytopathology 52:756.

Upchurch, R. P., and W. C. Pierce. 1957. The leaching of monuron from lakeland soil. I. The effect of amount, intensity, and frequency of simulated rainfall. Weeds 5:321.

Upchurch, R. P., and W. C. Pierce. 1958. The leaching of monuron from lakeland sand soil. II. The effect of temperature, organic matter, soil moisture, and amount of herbicides. Weeds 6:24.

Von Endt, D. W., P. C. Kearney, and D. D. Kaufman. 1968. Degradation of monosodium methancarsonic acid by soil microorganisms. J. Agric. Food Chem. 16:17-20.

Wade, P. 1954. The sorption of ethylene dibromide by soils. J. Sci. Food Agric. 5:184-192.

Wade, P. 1955. Soil fumigation. III. The sorption of ethylene dibromide by soils at low moisture contents. J. Sci. Food Agric. 6:1-3.

Walker, A. 1970. Effects of quintozene on the persistence and phytotoxicity of chlorpropham and sulfallate in soil. Horticultural Res. 10:45-49.

Ward, T. M., and R. P. Upchurch. 1965. Role of the amino group in adsorption mechanisms. J. Agric. Food Chem. 13:334-340.

Warren, H. V., and R. E. Delavault. 1962. Lead in some food crops and trees. J. Sci. Food Agric. 13:96-98.

Weber, J. B., and H. D. Coble. 1968. Microbial decomposition of diquat adsorbed on montmorillonite and kaolinite clays. J. Agric. Food Chem. 16:475-478.

Wedemeyer, G. 1967. Dechlorination of 1,1,1-trichloro-2,2-bis(p-chlorophenyl)ethane by Aerobacter aerogenes. I. Metabolic products. Appl. Microbiol. 15:569-574.

Weiss, A. 1963. Mica-type layer silicates with alkylammonium ions. Clays and Clay Minerals 10:191.

Wheatley, G. A. 1965. The assessment and persistence of residues of organochlorine insecticides in soils and their uptake by crops. Ann. Appl. Biol. 55:325-329.

Willis, G. H., R. C. Wander, and L. M. Southwick. 1974. Degradation of trifluralin in soil suspensions as related to redox potential. J. Environ. Qual. 3:262-265.

Wolf, D. E., R. S. Johnson, G. D. Hill, and R. W. Varner. 1958. Herbicidal properties of neburon. Proc. N. C. Weed Control Conf. 15:7.

Woodcock. D. 1964. Microbial degradation of synthetic compounds. Ann. Rev. Phytopathology 2:321-340.

Wright, S. J. L., and A. Forey. 1972. Metabolism of the herbicide Barban by a soil penicillium. Soil Biol. Biochem. 4:207-213.

Young, H. Y. 1971. Pesticide and growth regulator residues in pineapple. Residue Rev. 35:81-101.

Youngson, C. R., R. B. Baker, and C. A. I. Goring. 1962. Diffusion and pest control by methyl bromide and chloropicrin applied to covered soil. J. Agric. Food Chem. 10:21-25.

Yuen, Q. H., and H. W. Hilton. 1962. Soil adsorption of herbicides. The adsorption of monuron and diuron by Hawaiian sugarcane soils. J. Agric. Food Chem. 10:386-392.

Potential Effects of Waste Constituents on the Food Chain

R.L. Chaney

Persistent negative impacts on humans, agriculture, or environment from water-borne toxic chemicals would be the basis for prohibiting treatment of a waste, or a management or closure technique for a land treatment site. This chapter considers food-chain pathways for effects of hazardous waste land treatment, based on long experience with natural toxicants and recent experience with land application of sewage sludge and other wastes. We must separate our interpretation of this information for clearly distinct aspects of land treatment: 1) during site operation; 2) during the closure period; and 3) persistent beyond closure. It seems clear that many wastes can be managed at land treatment sites if the wastes are sufficiently characterized, and site personnel appropriately trained.

Macronutrients

Potential Beneficial Animal Responses--

Many sludges formed during treatment of industrial wastewater will contain sufficient amounts of macronutrients (N, P, K, Ca, and Mg) to act as fertilizers. Plant yields reach maximum levels when macronutrients are provided at optimum levels; hence, when wastes are used to supply the required nutrients, the wastes are fertilizers. A recent USDA report (USDA, 1978) discussed N, P, and K supply from wastes.

Besides increase in crop yields, crops grown with adequate supply of a nutrient have higher levels of that nutrient in their shoots than do plants whose yield is limited by inadequate supply of that nutrient (Bates, 1971).

Crop species and cultivars differ in response to nutrient supply and in forage nutrient levels (Hill and Guss, 1976; Reid and Horvath, 1980). Animal response (weight gains and health) to ingesting a crop is greater when that crop supplies needed macronutrients (Reid and Jung, 1974); further, forage protein content is dependent on N supply. Thus, land application of wastes to supply the macronutrient fertilizer requirements of forage crops can benefit livestock.

Potential Toxicity and Mineral Imbalances--

In contrast to the situation described above in which good management practices allow use of wastes at rates to supply crop macronutrient requirements, poor waste, soil, crop or animal management can lead to several problems. The problems fit a general group: excess supply of one nutrient causes imbalance in soil, crop, and/or animals leading to lower rate of weight gain and even specific animal health problems. The recent review by Reid and Horvath (1980) discusses these phenomena fairly comprehensively.

Grass tetany--Cattle and sheep with magnesium deficiency (hypomagnesemia) suffer a potentially lethal disease called grass tetany. This is a complex animal reponse in which soil, fertilizers, environment, and plant and animal characteristics all influence severity. Several valuable reviews discuss these many factors (Grunes, Stout, and Brownell, 1970; Grunes, 1973; Reid and Jung, 1974; Rendig and Grunes, 1979; Wilkinson et al., 1972; Reid and Horvath, 1980). Forages with Mg less than 0.20%, and with the ratio K:(Ca+Mg) (on an equivalent basis) greater than 2.2, with K greater than 2.5%, and high N generally cause grass tetany in susceptible animals.

The disease occurs on selected soil types (Kubota, Oberly, and Naphan, 1980). Many forage grasses can reach a nutrient composition which will cause the disease; further, cultivars of a crop also differ in tetanigenic potential (Thill and George, 1975; Grunes, Stout, and Brownell, 1970; Sleper et al., 1980; Brown and Sleper, 1980; Mayland, Grunes, and Lazar, 1976; Karlen et al., 1978). Fertilization with N to increase forage yield sometimes increases tetanigenic potential by increasing K uptake (Follett et al., 1975).

Environmental conditions play a dominant role in grass tetany. Nearly all cases result from forage grasses grown on seasonally wet or poorly drained, cool soils. Several plant and soil factors are altered under these conditions. First, wet soils have a higher K:(Ca+Mg) ratio in their soil solution due to Donnan equilibrium control of cation exchange reactions; thus wet soils provide more K and less Mg than do dry soils (Karlen et al., 1978, 1980a, 1980b). Second, wet soils have low O_2 supply, which limits cation absorption by roots and also limits root exploration of the soil (further limited by cool temperatures). Wet, low O_2 soils provide more K and less Mg to forage grasses than do dry soils (Elkins et al., 1978). Cultivar differences in Mg level and K:(Ca+Mg) are much greater under low O_2 conditions, even when the soil has normal water content (Elkins et al., 1978; Haaland, Elkins, and Hoveland, 1978).

Both low O_2 and cool, wet soils act to increase K:(Ca+Mg) in forages and thus their tetanigenic potential. This happens even though soil Mg status is at recommended levels for crop production. Use of dolomitic (high Mg) limestone seldom counteracts grass tetany, although MgO application may help if K supply is not excessive. Adequacy of soil Mg is generally judged by a consideration of both exchangeable Mg and exchangeable Mg as a fraction of the total exchangeable cations (Doll and Lucas, 1973). Soil K status is similarly evaluated, with greater emphasis on exchangeable K as a fraction of the total exchangeable cations. Obviously, high soil K status is more likely to cause tetanigenic forage than is normal soil K status.

Repeated annual application of poultry manure (high in K and N) to fertilize tall fescue led to high incidence of grass tetany in cattle (Stuedemann et al., 1975). Soil K levels rose while soil Mg was decreased by crop removal (Jackson, Leonard, and Wilkinson, 1975).

Application of other wastes, lower in K, would not be expected to cause grass tetany easily; soil, environmental, and crop factors should be managed carefully to avoid tetany. If poorly drained soils are selected for a land treatment site, greater management is clearly needed.

Cattle often reject Mg supplements offered _ad lib._ to correct forage K:(Ca+Mg) imbalance. Stuedemann _et al._ (1974) reported a practical method to correct this problem; a slurry of MgO and bentonite is sprayed on the forage, with the bentonite added to improve adherence (see also Stuedemann et al., 1975). Cattle grazing treated pastures maintained normal Mg status while cattle on the control treatment experienced grass tetany.

Fat necrosis--A disease called fat necrosis reduces weight gain of cattle grazing high K status tall fescue (Williams, Tyler, and Papp, 1969). Fat necrosis occurred at much higher rate in cattle grazing poultry manure fertilized, than NH_4NO_3 fertilized, tall fescue (Stuedemann et al., 1973, 1975). The crop species which have caused fat necrosis are very limited (mostly tall fescue); further, it does not need cool, wet soils which are needed for expression of grass tetany. Land treatment sites using tall fescue as cattle feed will need to manage soil K status reasonably.

Others--Domestic livestock often suffer Ca:P imbalance problems if they do not receive concentrated supplements when ingesting Ca:P imbalanced forages (Reid and Jung, 1974; Reid and Horvath, 1980; Baker and Chesnin, 1976). Bone development and arthritis problems result if Ca:P imbalance is not corrected. Research has been conducted to select crop cultivars which prevent imbalance (Hill and Jung, 1975).

Wastes very high in $Fe(OH)_3$ (e.g., pickle liquor neutralization sludge) greatly increase P adsorption capacity in acidic soils. This in turn would reduce forage P concentration, and even reduce crop yield (Nelson, D. W. 1980. Lafayette, Ind., personal communication). Some industrial biological sludges are low in P, and supplemental P fertilizer would be required to achieve optimum crop yield and P concentration. On the other hand, P concentration in sludge fertilized crops is not increased to excessive levels which would cause crop or animal P toxicity even when sewage sludge supplies quite high available P in soils.

Pathways for Transfer of Toxic Chemicals in Wastes to the Food-chain

Liquid sludges can be spray-applied to cropland and tilled into the soil. Alternatively, liquid sludge can be sprayed onto forage or pasture land where it can contact plants and/or remain on the soil surface. Dewatered or dried sludges or composted wastes can be applied and mixed with or remain on the soil surface. These management options allow substantially different quantities of waste-borne toxic chemicals to enter the food chain, by quite different routes. Some options allow animals to directly ingest sludges, while other options use reactions in soils and properties of plants to largely prevent exposure.

Sludge Adherence to Existing Crops--

When liquid sludges (0-10% solids) are sprayed on pastures or forage crops, a thin film of the sludge coats the plant foliage. Research has found that some wastes dry and adhere strongly while others dry and flake off upon weathering. The first records of organic waste adherence came from a study of land application of high copper pig manure slurry (Batey, Berryman, and Line, 1972); forage grasses were enriched in Cu due to adhering manure.

Based on these findings, research was begun on sewage sludge adherence to forage crops and effects on grazing cattle. Chaney and Lloyd (1979) found that once liquid digested sludge dried on tall fescue forage it was not readily washed off by rainfall. Growth of the crop biomass diluted the sludge percentage in harvested forage. Sludge adherence was greater at higher application rates. Jones et al. (1979) found that sludge could be washed off forages before it dried, but not after. They also found that the amount of adhering sludge was approximately a linear function of the %-solids of the applied liquid sludge.

Sludge has adhered to all crops studied (Chaney and Lloyd, 1979; Lloyd and Chaney, unpublished; Jones et al., 1979; Bertrand et al, 1981). Sludge adherence is easily characterized since the levels of some microelements in sludge-contaminated forage are much greater than levels ordinarily possible by uptake-translocation by forage plants. Plant uptake and translocation to shoots of Cu, Pb, Cr, Fe, etc., is so limited that high levels of these elements indicates direct sludge contamination (see Chaney and Lloyd, 1979). Many early reports on uptake of microelements from surface applied sludges presumed uptake when in fact sludge adherence fully explains their observations (Boswell, 1975; Fitzgerald, 1978). Industrial aerobic sludges adhere to forages in a manner similar to that of sewage sludge (Chaney and Hornick, unpublished results).

Another route for entry of microelements into the food chain is through farm equipment. Studies with pig manure indicate that organic wastes on the soil surface can be lifted and mixed into baled hay (Dalgarno and Mills, 1975).

When increased levels of microelements in forage indicate sludge adherence, all constituents present in the sludge contaminate the forage. Not only microelements, but also macroelements, pathogens (Brown, Jones, and Donnelly, 1980), and toxic organics (Fitzgerald, 1978) are increased.

Ingestion of Sludge-Amended Soil or Sludge on the Soil Surface--

Several research programs have established that grazing animals consume soil as a part of the normal grazing process. Teeth of sheep and cattle wear out more rapidly when the forage is contaminated with soil (Healy and Ludgwig, 1965; Nolan and Black, 1970). Study of the teeth wear problem led Healy (1968) to more fully develop Field and Purves' (1964) method of soil ingestion measurement in which the Ti level in forage and feces is compared to that of soil. Titanium present in soil is not appreciably absorbed and translocated by plants. Forage Ti level thus becomes a label for soil in/on forages. Healy, Rankin, and Watts (1974) found that wet weather and excessive stocking rates caused forages to be trampled into the soil, thereby increasing soil adherence to forages. Although soil was normally 1-2% of sheep's diet, it reached 24% in the worst cases. In other research, Mayland et al. (1975) and Mayland, Shewmaker, and Bull (1977) found that cattle grazing on dryland-grown crested wheatgrass consumed considerable quantities of soil. Because the cattle consumed plants complete with soil-laden roots, the ingested diet contained 20% soil. Silage contains soil as well, and the soil can interfere with microelement availability (Lamand, 1979). Fries et al. (1982) have recently reviewed soil ingestion by dairy cattle.

Ingested soil can cause Pb poisoning of livestock when cattle graze soil naturally high in Pb (Egan and O'Cuill, 1970; Harbourne, McCrea, and Watkinson, 1968; Thornton and Kinniburgh, 1978). Even after closure of a smelter, Pb enriched crop residues remain on the soil surface, exposing cattle to possible Pb poisoning. Reclamation of Pb-smelter-polluted rangeland required incorporation of the organic sward thatch into the soil to prevent ingestion by cattle (Edwards and Clay, 1977).

Similarly, sewage sludge or composted sludge are ingested from the soil surface. Decker et al. (1980) found 6.5% (1977) and 2.0% (1978) compost in feces of cattle grazing sludge compost fertilized pastures. Compost did not adhere to the plant surfaces but lay on the soil surface.

Soil ingestion can also expose humans to waste-applied microelements in land treatment sites subsequently developed for housing. Some children and adults deliberately consume soil in a practice called "pica". If the soil is high in Pb (over 500-1000 ppm), individuals may absorb excessive amounts of Pb (Wedeen et al., 1978; Shellshear et al., 1975). Children also ingest soil and dust due to hand-to-mouth play activities and by mouthing of toys, etc. This phenomenon is discussed in detail in the section on lead later in this chapter.

Soil or sludge ingestion can be an important process which allows entry of a sludge-borne microelement into the food chain especially when the element is normally not absorbed by plants (plant level $\leq$soil level). For some elements (Zn, Cd, Mn, Se, etc.), plant levels often exceed soil levels, and plant uptake is a more important process than soil ingestion. However, soil ingestion is a potential route for allowing excessive Pb, Fe, Cu, F, As, Hg, Cu, Co, Mo, Se, and other elements into the food chain. Further, soil ingestion can interfere with availability of microelements in plants to animals. Individual elements will be discussed below.

Soil ingestion is an especially important pathway for persistent lipophilic toxic organic compounds. Harrison, Mol, and Healy (1970) found increased DDT in sheep grazing pastures where DDT was on the soil surface. They also studied lindane (Harrison, Mol, and Rudman, 1969; Collett and Harrison, 1968). Bergh and Peoples (1977) noted PCB movement from surface applied dewatered sludge to milk of a grazing cow, but did not estimate sludge ingestion. Hansen et al. (1981) noted PCB retention by swine grazing a field where the surface soil was largely sewage sludge.

"Soil-Plant Barrier" to Microelements in the Food-Chain--
As discussed in the text regarding plant uptake of microelements, some elements are easily absorbed and translocated to food-chain plant tissues (e.g. Zn, Cd, Mn, Mo, Se, B), while others are not. These other elements are strongly bound to soil or retained in plant roots, and are not translocated to plant foliage in injurious amounts, even when soils are greatly enriched (e.g. Fe, Pb, Hg, Al, Ti, Cr^{3+}, Ag, Au, Sn, Si, Zr). Even though an element may be easily or relatively easily absorbed and translocated to plant foliage, phytotoxicity may limit plant levels of these elements to levels safe for animals (e.g. Zn, Cu, Ni, Mn, As, B).

During the last 40 years, these concepts were developed by many researchers. Important reviews of the research supporting these concepts have been prepared but had not named the general theory (Underwood, 1977; Allaway, 1968, 1977a, 1977b; Bowen, 1966, 1979; Baker and Chesnin, 1976; Chaney, 1980; Lisk, 1972; Kienholz, 1980; Loneragen, 1975; Reid and Horvath, 1980; Cataldo and Wildung, 1978; Leeper, 1978; Ammerman et al., 1977; Shacklette et al., 1978; Beckett and Davis, 1979; Page, 1974; and Walsh, Sumner, and Corey, 1976). Chaney (1980) introduced the term "Soil-Plant Barrier" to describe these concepts when considering waste-soil-plant-animal relationships of toxic microelements. A "Soil-Plant Barrier" protects the food chain from toxicity of a microelement when one or more of these processes limit maximum levels of that element in edible plant tissues to levels safe for animals: 1) insolubility of the element in soil prevents uptake, 2) immobility of an element in fibrous roots prevents translocation to edible plant tissues, or 3) phytotoxicity of the element occurs at concentrations of the element in edible plant tissues below that injurious to animals.

Unfortunately, the "Soil-Plant Barrier" does not protect animals from toxicities of all elements. The exceptions important in assessing risk from land application of municipal sludge are Cd, Se, and Mo; a few more elements may have to be considered for land application of industrial wastes (Be, Co). Ingestion of amended soil or sludge can circumvent the "Soil-Plant Barrier". Many elements are so insoluble or non-toxic that animal health is not influenced even if ingested soil or waste contains the element (e.g., Cr^{3+}, Zr, Ti, Al, Sn, Si). However, direct ingestion of soil or wastes rich in some elements (e.g., Cu, F, Zn, Pb, Fe^{2+}, As, Co, and Hg) allows risk to livestock when risk would have been insignificant if the sludge were mixed with the surface soil (0-15cm). These concepts will be applied in considering risk to the food chain from individual microelements applied in wastes (see below).

Interactions Among Dietary and Sludge Constituents Influence Microelement Impact on Food-Chain--

Evaluation of the potential impact of microelements on animals via their consumption of sludge, sludge-amended soil, or crops grown on sludge-amended soil, is very complex. Animal species differ in tolerance of microelements. Tolerance to microelements is also influenced by age; younger animals are generally more sensitive than older. Crop species absorb unequal amounts of microelements. Total and relative microelement uptake is affected by crop species and cultivar, soil pH, organic matter, soil temperature and other factors. Wastes differ in levels of elements and ratios among elements. Individual potentially toxic elements interact with other elements in the diet, often reciprocally. These interactions are often the basis for physiological toxicity; hence, interactions are of great importance in assessing risk.

Interactions affecting Cu deficiency in ruminant animals were among the first studied, and have been intensely examined because of their practical significance. Animals can experience simple Cu deficiency, Mo-induced, sulfate-induced, or Zn-, Cd-, or Fe-induced Cu deficiency. Among the most complex is the 3-way Cu-Mo-S interaction. Dietary sulfate is reduced to sulfide in the rumen; sulfide reacts with Mo to form a thiomolybdate. Thiomolybdate reacts with Cu to form an insoluble compound which is unavailable and is excreted; this leads to depletion of liver Cu reserves and subsequently to clinical Cu deficiency (Mills et al., 1978; Bremner, 1979; Spence et al., 1980). Copper is of lower bioavailability in young forage plants than mature plants, and in fresh forages than in dried hay (Hartmans and Bosman, 1970). Forage species differ in bioavailability of Cu (Stoszek, et al., 1979). Soil consumed with forages reduces Cu absorption by sheep, perhaps due to soil Mo, Zn, or Fe but probably due to Cu sorption by soil constituents preventing Cu absorption in the intestine (Suttle, Alloway, and Thornton, 1975).

After the Cu-Mo-S interaction in ruminants was identified, it became clear that Zn, Cd, and Fe also interact with Cu bioavailability to both ruminants and monogastric animals (Bunn and Matrone, 1966; Hill et al., 1963; Matrone, 1974; McGhee, Creger, and Couch, 1965; Mills, 1974, 1978; Standish et al., 1971; Standish and Ammerman, 1971; Suttle and Mills, 1966). Reciprocally, high dietary Cu interacts to reduce absorption and toxicity of Zn, Fe, and Cd (Bunn and Matrone, 1966; Grant-Frost and Underwood, 1958; Cox and Harris, 1960; Lee and Matrone, 1969; L'Estrange, 1979; McGhee, Creger, and Couch, 1965). Other elemental interactions have been studied and found to be important in assessing risks (Underwood, 1977; Matrone, 1974; Levander, 1979; NRC, 1980b; Mills and Dalgarno, 1972; Mills et al., 1980; Mahaffey and Vanderveen, 1979; Fox, 1974, 1979; Fox et al., 1979; Bremner, 1979.)

In many cases, food chain toxicity is a result of microelement imbalance as much as it is a result of increased supply of one potentially toxic element. When one element is so increased that the ratio of it to other elements or dietary constituents is great enough to induce a deficiency of another, then animal weight gain declines and a health effect is observed. Chaney (1980) noted that domestic sewage sludge contains a mixture of potentially toxic elements. Consumption of sludge or sludge-amended soil is a very different case for risk assessment than standard toxicological studies where a soluble salt of one element is added at rates to cause health effects (and often to purified rather than practical diets). With sludge ingestion, increased levels of dietary Zn are balanced by increased levels of Cu and Fe. Industrial wastes and sludge are a somewhat different case than domestic sewage sludge, because they are more likely to be enriched in only one or a few elements. The potential for microelement imbalances is greater with sludges rich in one element. Also, a wider range of elemental interactions must be considered to evaluate food chain impact of industrial wastes. A number of elements are considered below, in regard to sludge-soil-plant-animal interactions influencing the food chain.

Evaluation of Potential for Food-Chain Impacts of Individual Microelements Applied in Industrial Wastes--

A number of processes must be considered herein. Phytotoxicity and other aspects of the "Soil-Plant Barrier" must be considered in relation to maximum plant foliage levels of the microelement in question. Interactions should be considered through the complete sludge-soil-plant-animal system. Although not every element will be considered in detail, elements commonly present in sewage sludges and industrial wastes, and those EPA considered hazardous constituents will be discussed.

An important reference for tolerance of microelements by animals has recently been published by the National Research Council (NRC, 1980b). The NRC committee considered increased levels of only the element being evaluated, although they discuss interactions. Their tolerance levels are shown in Table 8. Unfortunately, these levels may not be valid for sludge fertilized crops or for ingestion of sludge or soil because of the noted interactions.

Arsenic-- Arsenic is not essential for plants; however, since rats fed a highly purified diet show a response to low levels of As, the essentiality of As to animals is being intensely studied. Experience with As phytotoxicity comes mostly from pesticide residues in soil; sewage sludge is seldom a rich source of As. Substantial yield reduction due to As toxicity in most plants occurs under conditions which do not allow appreciable increase in As in edible crop tissues (Walsh and Keeney, 1975; Woolson, 1973). Rice and other crops grown on submerged soils are especially sensitive to As phytotoxicity (Deuel and Swoboda, 1972) because arsenate is reduced to the more toxic arsenite. Arsenic is somewhat increased in the peel of root crops at increased soil As, apparently due to soil contamination of the peel.

The food chain is protected from excess soil As applied in sludges because food and feed crops are not As-tolerant. Some rare plant species tolerate and accumulate As from As-rich soils (Rocovich and West, 1975; Porter and Peterson, 1975, 1977a, 1977b; Benson, Porter, and Peterson, 1981). Ingestion of As-rich industrial wastes from the soil surface could cause As toxicity to grazing animals and wildlife. This topic has not been studied although Porter and Peterson (1977b) noted animal deaths have occurred near As mine wastes.

In summary, phytotoxicity of soil As to crops prevents plant-borne As impacts on the food chain even in the garden soil scenario. Wastes rich in As are thus poor candidates for land treatment, due to phytotoxicity, and risks from direct ingestion.

Cadmium--Cadmium is not essential for plants. Although one study indicated Cd was essential for rats (Schwartz and Spallholz, 1978), it is not generally agreed that Cd is essential for animals (NRC, 1980b; Fox et al., 1979).

It now appears that Cd activity in most soils is controlled by adsorption rather than by formation of crystalline inorganic compounds (Street, Lindsay, and Sabey, 1977; Soon, 1981). Street, Lindsay, and Sabey (1977) found that $CdCO_3$

TABLE 8. MAXIMUM TOLERABLE LEVELS OF DIETARY MINERALS FOR DOMESTIC LIVESTOCK IN COMPARISON WITH LEVELS IN FORAGES

Element	"Soil-Plant Barrier"	Level in Plant Foliage[1/]		Maximum Levels chronically tolerated[2/]			
		Normal	Phytotoxic	Cattle	Sheep	Swine	Chicken
		---mg/kg dry foliage---		----------mg/kg dry diet--------			
As, inorganic	yes	0.01-1	3-10	50.	50.	50.	50.
B	yes	7-75	75	150.	(150.)	(150.)	(150.)
Cd[3/]	Fails	0.1-1	5-700	0.5	0.5	0.5	0.5
Cr^{+3}, oxides	yes	0.1-1	20	(3000.)	(3000.)	(3000.)	3000.
Co	Fail?	0.01-0.3	25-100	10.	10.	10.	10.
Cu	yes	3-20	25-40	100.	25.	250.	300.
F	yes?	1-5	-	40.	60.	150.	200.
Fe	yes	30-300	-	1000.	500.	3000.	1000.
Mn	?	15-150	400-2000	1000.	1000.	400.	2000.
Mo	Fails	0.1-3.0	100	10.	10.	20.	100.
Ni	yes	0.1-5	50-100	50.	(50.)	(100.)	(300.)
Pb[3/]	yes	2-5	-	30.	30.	30.	30.
Se	Fails	0.1-2	100	(2.)	(2.)	2.	2.
V	yes?	0.1-1	10	50.	50.	(10.)	10.
Zn	yes	15-150	500-1500	500.	300.	1000.	1000.

1/ Based on literature summarized in text.
2/ Based on NRC (1980b). Continuous long-term feeding of minerals at the maximum tolerable levels may cause adverse effects. Levels in parentheses were derived by interspecific extrapolation by NRC.
3/ Maximum levels tolerated based on human food residue consideration.

can form in low cation exchange capacity, low organic matter, calcareous soils. Under anaerobic conditions, CdS forms in soil; CdS has very low solubility and is unavailable to plants (Takijima and Katsumi, 1973; Bingham et al., 1976b) but is readily oxidized in aerobic soil. Unfortunately, formation of CdS is not a practical management practice to minimize Cd uptake for crops other than rice.

A recent consensus review of Cd relationships in sewage sludge, soil, and plants summarized this complex topic (CAST, 1980). Of all soil properties affecting Cd level in plants, soil pH has the greatest effect. Increasing soil pH causes stronger adsorption of Cd by soil and reduces Cd uptake. Of other soil chemical properties, soil organic matter has been shown to have some effect; since higher organic matter reduces Cd uptake (e.g., White and Chaney, 1980). Other soil factors which affect Cd uptake include: temperature, soluble salts, chelators, and water status (Haghiri, 1974; Giordano, Mays, and Behel, 1979; Wallace et al., 1977; Bingham, 1980; Shaeffer et al., 1979).

The CAST (1980) report also summarized evidence which indicates that soil Cd remains crop available for a prolonged period after application. Availability to crops decreases only in calcareous soils. These conclusions are based on sludge field plots, sludge utilization farms, (CAST, 1980) and natural high Cd soils (Lund et al., 1981). Recent studies by Lloyd et al. (1981) indicated that sludge applied Cd remained nearly 100% labile many years after application.

Crops differ remarkably in their Cd accumulation, Cd tolerance, and translocation of Cd to edible plant parts (CAST, 1980; Bingham, 1979; Bingham et al., 1975, 1976a, 1976b; MacLean, 1976; Furr et al., 1976a, 1976b; Dowdy and Larson, 1975; Chaney and Hornick, 1978). Figure 7 shows Cd concentration in leaves and edible plant tissues of many crops grown on a neutral pH sludge amended soil containing 10 ppm Cd (based on data from Bingham et al., 1975, 1976a, 1976b). Tobacco, lettuce, spinach, chard, endive, cress, and turnip accumulate much higher foliar Cd levels than other leafy crops (e.g., kale, collards, cabbage). Although Cd in edible root of radish, turnip, and beet is only a small fraction of the Cd level in the shoots of the plants, carrot root Cd is about half of carrot leaf Cd. Similarly, the ratio (Cd in grain):(Cd in leaf) ranges from very low for corn to relatively high for wheat, oat, and soybean; Chaney, White and Tienhoven (1976) found that this ratio in soybean was reduced from >1 to <0.2 by increasing soil Zn.

The wide variation in crop tolerance of Cd causes difficulty in assessing the impact of soil Cd on the food

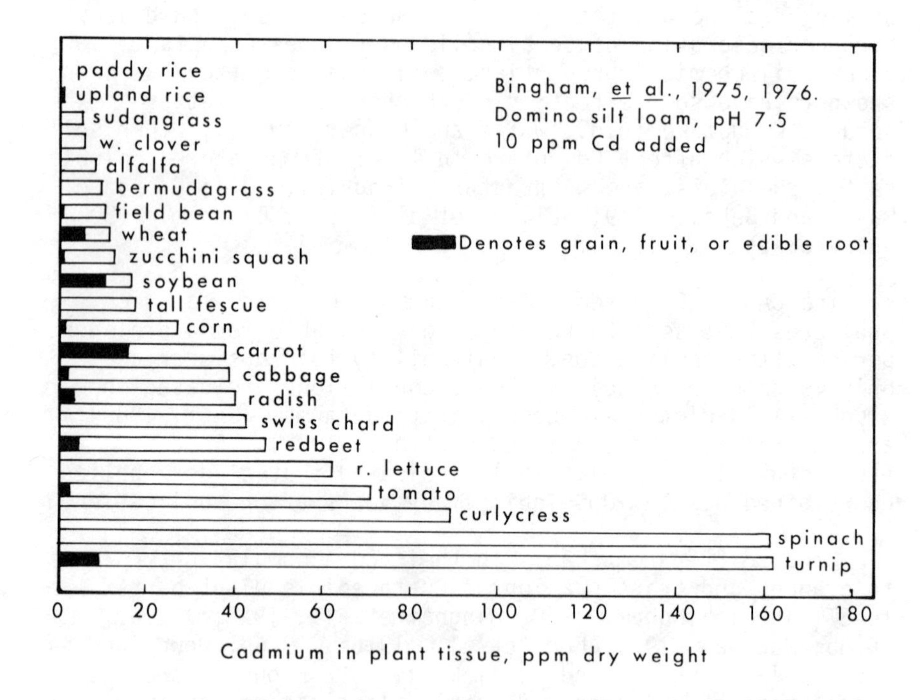

FIGURE 7. Crop differences in Cd accumulation. Crops were grown on calcareous Domino silt loam amended with 1% of a Cd-enriched sewage sludge (1000 ppm Cd) such that the amended soil contained 10 ppm Cd. Where a plant tissue other than leaves is normally eaten, its Cd concentration is shown by the black bar; foliar Cd for each plant is the full open bar (turnip leaves = 163 ppm Cd). (From Chaney and Hornick, 1978, based on Bingham et al., 1975, 1976a, 1976b).

chain. The foliar Cd associated with phytotoxicity (25% yield reduction) varies in different crops from 7 to 160 ppm dry weight (Bingham, 1979). Further, the foliar Cd concentration causing 50% yield reduction in lettuce and chard is greater in acidic soils (470 ppm in lettuce; 714 ppm in chard) than in calcareous soils (160 ppm in lettuce; 203 ppm in chard) (Mahler, Bingham, and Page, 1978). Some plants are unusually tolerant of Cd; Simon (1977) and Wigham, Martin, and Coughtrey (1980) have reported tolerance of Cd by ecotypes of grasses adapted to Cd-enriched Zn and Pb mining wastes. In summary, phytotoxicity of Cd does not limit crop Cd to acceptable levels.

Cadmium is an unusual and difficult case for evaluation of risk to the food chain. In contrast to other elements, Cd has a quite long biological half-life in humans -- generally considered 20 years. Absorbed Cd is bound to a low molecular weight protein to form metallothionein which is accumulated and retained in the kidney for a long period. High metallothionein-Cd in the kidney can lead to adverse health effects in the kidney.

Over one's lifetime, chronic food chain Cd exposure can cause different health problems than those experienced from acute exposure. Long-lived animals (e.g., humans) are at greater risk of this health effect than are short-lived animals (wildlife; domestic animals). Accumulation of Cd in organ meats (liver, kidney) was the basis for suggesting a low dietary Cd tolerance in domestic animals rather than a direct health effect to the animals (NRC, 1980b).

The potential risk of excess soil Cd to humans has been clearly documented. Adverse health effects resulted from prolonged consumption of foods grown locally on Cd enriched soils (Tsuchiya, 1978; Friberg et al., 1974; Fulkerson and Goeller, 1973; Hammons et al., 1978; Yamagata and Shigematsu, 1970; Kobayashi, 1978; Nogawa, 1978). A large number of Japanese farmers suffered Cd health effects after long-term ingestion of Cd-enriched rice grown in paddies polluted by Zn- and Pb-mining wastes or Zn-, Pb-, and Cu-smelter emissions in at least 7 different areas of Japan (Kobayashi, 1978; Takijima and Katsumi, 1973; Shigematsu et al., 1979; Kjellstrom, Shiroishi, and Evrin, 1977; Kojima et al., 1979; Saito et al., 1977; Nogawa, 1978; Nogawa, Ishizaki, and Kawano, 1978; Nogawa and Ishizaki, 1979; Nogawa et al., 1975; 1980). Rice Cd concentration and number of years exposure were both strongly related to the incidence rate of Cd health effects. A smelter enriched area in Belgium may have caused Cd-induced renal disease (Roel et al., 1981a) although route for exposure and increased kidney Cd have not yet been demonstrated.

The name "itai-itai" disease (translated as ouch-ouch disease) came from expressions of pain by elderly women suffering repeated bone fractures due to Cd-induced osteomalacia. Although the osteomalacia brought attention to this environmental Cd disease, severe osteomalacia does not frequently result in humans ingesting excessive Cd. Renal proximal tubular dysfunction (Franconi syndrome) is the first health effect of excessive chronic Cd exposure. The renal disease had high incidence in areas where Cd exposure was increased, and showed a dose-response relationship with Cd exposure (expressed as "Cd level in rice-times-years ingested"). All individuals with advanced itai-itai disease had severe proteinuria characteristic of the kidney disease. Renal disease subsequently proceeded to osteomalacia in some workers who ceased exposure when the kidney disease was identified (Kazantzis, 1979). However, this aspect of Cd disease is poorly understood. Sub-clinical osteomalacia is found in many of the Japanese farmers who experience renal disease (Mulawa, Nogawa, and Hagino, 1980).

Renal tubular dysfunction (Franconi syndrome) resulting from Cd ingestion is quite different from classic kidney failure. Franconi syndrome seldom proceeds to kidney failure requiring dialysis. Kjellstrom (1978) indicated that Franconi syndrome (low molecular weight proteinuria, glucosuria, aminoaciduria, phosphaturia, etc.) is the first Cd health effect; if Cd-exposure (rate-times-duration) is increased, kidney stones and osteomalacia/osteoporosis may result. Kjellstrom, Friberg, and Rahnster (1979) found greater mortality (shorter life span) in Cd exposed workers, but this may not be relevant to ingested Cd. Neither hypertension nor prostate cancer incidence are increased even when proteinuria is severe (Friberg et al., 1974; Doyle, 1977; Hammons et al., 1978; Tsuchiya, 1978; Ryan et al., 1979; Commission of the European Communities (CEC), 1978; Kjellstrom and Nordberg, 1978; Kjellstrom, Friberg, and Rahnster, 1979; Pahren et al., 1979; Lauwreys et al., 1980; Nogawa, 1978; Shigematsu et al., 1979). Although laboratory studies with rats and other animals have shown that anemia, enteropathy, and teratogenesis (due to Cd-induced Zn or Cu deficiency in the fetus) can result from ingested Cd, these are very unlikely with practical diets.

A number of researchers and groups have attempted to clarify the dose-effect and dose-response relationships for Cd (CEC, 1978; Friberg et al., 1974; Kjellstrom and Nordberg, 1978; Ryan et al., 1979; Tsuchiya, 1978; Hammons et al., 1976). The first sign of renal tubular dysfunction (increased excretion of B_2-microglobulin, a specific proteinuria characteristic of Cd injury) is generally agreed to occur at about 200 mg Cd/kg wet kidney cortex. Some research indicates that the critical kidney cortex Cd level may be as high as 300 mg/kg (Roels et al., 1981b), but 200 mg/kg is the level generally accepted for use in risk analysis.

Kjellstrom and Nordberg (1978) developed a sophisticated multicompartmental dose-effect model for Cd metabolism in humans: "This present model predicted that a daily intake corresponding to 440 μg at age 50 would give 200 μg Cd/g of (wet) kidney cortex at age 45-50." These results were obtained by assuming a high, constant Cd concentration per unit calories, and that calorie (hence Cd) varied with age in the manner of the average diet of the Swedish population. The "best fit" calculated 4.8% lifetime average absorption of dietary Cd, 440 μg Cd/d at age 50, and a 12 year biological half-life for Cd to achieve the 200 μg Cd/g wet kidney cortex at age 45-50.

Other researchers have used different ways to express Cd-exposure information, thus complicating interpretation of results from these many sources. In the U.S., the Food and Drug Administration (FDA) has measured food Cd concentrations and average Cd ingestion (FDA, 1977). Food consumption was based on USDA's dietary intake survey; FDA, USDA, and EPA agreed to use a food consumption model based on teenage males (highest food consuming group) in a pesticide residue survey program. Thus, for the same food supply, a mean food Cd ingestion of 39 μg/day from FDA corresponds to about 23 μg/day intake at age 50 in Kjellstrom and Nordberg's (1978) model. Their model reflected 3430 cal/d for Swedish teenage males vs. 2045 cal/d for 50-year-old Swedish individuals (Fig. 4.32 and 4.34 in Friberg et al., 1974). Thus, the critical 440 μg Cd/day ingestion rate for 50-year-old individuals in Kjellstrom and Nordberg's (1978) model corresponds to approximately 738 μg Cd/d ingestion in U.S. teenage male diets. The present exposure is only 5.2% of the critical exposure (23 ÷ 440 or 39 ÷ 738).

Chaney (1980) and Ryan et al. (1979) discuss difficulties in interpreting dose-response relationships for dietary Cd. Individuals vary widely in self-selected diet and dietary Cd (Yost, Miles, and Parsons, 1980), in Cd absorption rate (Flanagan et al., 1978), and in sensitivity to absorbed Cd. These phenomena are generally assumed to vary in a log-normal fashion in a population. Kjellstrom (1978) extended the 440 μg/d model "critical" level to a population by arbitrarily using a geometric standard deviation of 2.35 based on studies of Cd in autopsy tissues (see Ryan et al., 1979 for details). However, Kjellstrom's (1978) model would require greater than 100% absorption of dietary Cd by the most sensitive individuals (see Figure 1 in Chaney, 1980).

The highest Cd absorption rate observed for humans is 25% reported by Flanagan et al. (1978) for a woman with mild anemia; Fe stress strongly increases Cd absorption. Several researchers (Chaney, 1980; Ryan et al., 1979) argued that it was unreasonable to extrapolate the 440 g Cd/d "Average Human" model result to an assumed maximum sensitivity group with greater absorption of Cd than ever observed in humans. Further, individuals are unlikely to be in this greatest risk group for their whole lifetime.

Ryan et al. (1979) concluded that a 200 g/d threshold model (based on average lifetime daily Cd intake) was more appropriate for dose-response considerations, as did the CEC (1978) workgroup. This value corresponds to about 10.6% lifetime Cd absorption rate for the most sensitive individuals [4.8(440/200)].

Cadmium absorption by animals is strongly influenced by other dietary factors (Fox, 1976, 1979; Fox et al., 1978, 1979; Jacobs et al., 1978a, 1978b, 1978c; Flanagan et al., 1978; Welch, House, and Van Campen, 1978; Welch and House, 1980; Neathery and Miller, 1975; Kostial et al., 1979; Cousins, 1979; Kobayashi, 1978; Washko and Cousins, 1977). Iron status of the animal appears to be the most important control of %-absorption of Cd. Zinc status of the animal and dietary Zn level is the next most important factor, followed by dietary Ca. Protein and fiber in the diet and age of animal also influence Cd retention. These factors should allow a greater %-absorption rate for women than men. Women as a group showed greater Cd absorption (Flanagan et al., 1978), and women's kidney Cd exceeds men's in autopsy kidney studies.

Dietary interactions can thus influence bioavailability of Cd. Leafy and root vegetables which are enriched in Cd may also be a good dietary supply of Zn, Fe, and Ca. Leafy vegetables have been shown to provide bioavailable Fe and Zn (Welch, House, and Van Campen, 1977, 1978; Van Campen and Welch, 1980; Wien, Van Campen, and Rivers, 1975). Chaney (1980) suggested that for leafy and root vegetables grown on soils enriched in Cd from being fertilized by low Cd, low Cd:Zn sewage sludges comprise a separate scenario. In this case, consuming sufficient food Cd to pose a risk to susceptible individuals would result in increased dietary Fe, Zn, and Ca, thereby shifting the individual to a less susceptible population group.

Feeding studies have been conducted with sludge and with crops grown on sludge-fertilized soil. Ingestion of sludge Cd has been evaluated in ruminant and monogastric animals with most work done with cattle. When sludges with high Cd and high Cd:Zn were fed, kidney Cd was significantly increased (Kienholz, 1980; Baxter, Johnson, and Kienholz, 1980; Hansen and Hinesly, 1979; Hinesly et al., 1979; Edds et al., 1980; Fitzgerald, 1980). However, when sludges with lower Cd and low Cd:Zn were fed, kidney Cd was not significantly increased (Decker et al., 1980; Kienholz, 1980; Baxter, Johnson, and Kienholz, 1980; Smith et al., 1977; Smith, Kiesling, and Sivinski, 1978; Edds et al., 1980; Smith, Kiesling, and Ray, 1979; Smith et al., 1980). Sludge Cd was less bioavailable to swine than equal Cd added as $CdCl_2$ (Osuna et al., 1979; Edds et al., 1980). Food products of animals are unchanged in Cd except for liver and kidney (e.g., Sharma et al., 1979). Kienholz (1980) noted that dietary interactions could avoid even this impact of sludge Cd. Thus, risk analysis for ingested sludge Cd requires evaluation of several factors other than dietary Cd concentration.

Similarly, risk analyses for ingestion of Cd in foods grown on Cd-enriched soils requires careful evaluation of factors other than Cd. Far too little research has been conducted to characterize bioavailability of food Cd. Further, very little of the completed research conforms with the experimental designs which Fox et al. (1978, 1979) and Fox (1976) indicated were needed to allow interpretation. Dietary Cd level should correspond to the range of nutritional relevance to humans. Intrinsically Cd labelled foods should be fed in the state ordinarily ingested by humans (e.g., fresh leafy vegetables). Nutritional status of the experimental diet should be adequate for all known essential factors or varied as part of the experiment. The feeding period should be of sufficient length to allow nutritional status of animals to be under control of experimental diet for the bulk of the experimental period. Several animal species should be studied. Bioavailability of Cd in a food or a sludge grown food can only be determined experimentally.

Tobacco is an especially high risk crop in terms of potential for Cd effects on humans. Among all crops studied to date, tobacco accumulates more Cd per unit soil Cd than any other (Chaney et al., 1978a; MacLean, 1976). Tobacco is normally grown on strongly acid soils to prevent crop loss from root diseases. This soil pH management leads to maximum Cd uptake under normal crop production conditions. In contrast, most other crops are best grown at pH 6.5 to 7. Tobacco is normally high in Cd compared to leaves of other crop plants, and high leaf Cd levels in some production areas are being studied (Frank et al., 1977; Westcott and Spincer, 1974). When tobacco is grown on sewage sludge-amended soils, crop Cd level can be increased from 1 to as high as 44 ppm Cd in dry leaves (Chaney et al., 1978a) with only 1 ppm soil Cd.

Cadmium in tobacco is an important source of Cd for humans. Individuals who smoke one pack of cigarettes per day have about 50% higher Cd in kidney cortex than non-smokers (Lewis et al., 1972; Elinder et al., 1976). About 15% (5-25%) of cigarette Cd enters the mainstream smoke (Szadkowski et al., 1969; Menden et al., 1972; Westcott and Spincer, 1974). Filters can remove much of this Cd and reduce Cd exposure of smokers (Westcott and Spincer, 1974; Franzke, Ruick, and Schmidt, 1977). Based on the potential of sludge-applied Cd to increase risk of chronic kidney disease in smokers if sludge were applied to tobacco cropland, EPA (1979) regulated and discouraged this practice.

Several food crops are of especial importance to evaluating Cd-risk for humans. While grains supply much Cd to individuals in the general population (Braude, Jelinek, and Corneliussen, 1975; Jelinek and Braude, 1978; Ryan et al., 1979), gardeners are unlikely to grow a significant portion of their food grains. Rather, individuals are likely to grow leafy and root vegetables, legume vegetables, garden fruits, and potatoes. If the Cd:Zn ratio of an acidic Cd-enriched garden soil is high, edible crop tissues of leafy, root, and legume vegetables, garden fruits, and potatoes can be greatly increased in Cd concentration with no injury to the crop and provide excessive bioavailable Cd. If the Cd:Zn ratio of an acidic Cd-enriched garden soil is low ($\leq$ 0.010), these crops are not greatly increased in Cd when Zn phytotoxicity limits crop yield, and bioavailable Cd would be only slightly increased. The difference in risk from low Cd:Zn and high Cd:Zn gardens is due to: 1) Zn-phytotoxicity at low pH in the low Cd:Zn garden causing the gardener to add limestone which reduces crop Cd or have little yield (hence, reduced exposure), 2) interactions between Cd and Zn in plant uptake and translocation to edible plant tissues (Chaney, White, and Tienhoven, 1976; Chaney and Hornick, 1978); and 3) interactions in the diet which influence Cd bioavailability (Chaney, 1980).

It is much more difficult to evaluate Cd bioavailability from foods grown on waste-amended soils than from Cd-amended purified diets. Freeze-dried lettuce and chard grown on acidic soils amended with domestic sludge were fed at a high % of diet to mice or guinea pigs (Chaney et al., 1978b, 1978c). Although dietary Cd was increased by up to 5-fold by lettuce or chard grown on acidic, domestic sludge-amended soil, kidney Cd was not increased. In other studies with high Cd and/or higher Cd:Zn sludges, feeding sludge-grown crops has caused increased kidney Cd (Chaney et al., 1978b; Miller and Boswell, 1979; Bertrand et al., 1980; Williams, Shenk, and Baker 1978; Hinesley, Ziegler, and Tyler, 1976). Clearly, many more sludge-soil-plant-animal studies are needed to characterize the bioavailability of Cd in crops grown on waste-amended soils. It seems very likely that factors besides background soil pH, and annual and cumulative Cd application will eventually have to be considered in setting allowed Cd loadings on land treatment sites (EPA, 1979, 1980b; Chaney, Hornick and Parr, 1980).

Much of the potential risk from Cd in waste-amended soils has now come under regulation, although these regulations do not have to be enforced for several more years. The highest risk case, application of sludges to gardens as fertilizers or soil conditioners, has not yet been regulated (Comptroller General, 1978; Chaney, Hornick and Parr, 1980). Further, pretreatment of Cd-bearing industrial wastes, segregation of waste streams, and avoidance of Cd use for non-critical applications offer great opportunity to avoid all Cd health effects (Dage et al., 1979; Gurnham et al., 1979; Chaney and Hundemann, 1979).

In the process of developing Federal Regulations for land application of sewage sludge (EPA, 1979), EPA prepared a "worst case" scenario relating sludge-applied soil Cd to potential for kidney dysfunction (EPA, 1979b). The worst case which may occur appears to be the acid garden case. Individuals in the U.S. do not grow their own food grain on acidic, Cd-enriched soils. Similarly, consumption of liver and kidney enriched in Cd from sludge utilization, is a minor source of dietary Cd.

Thus, the acidic garden scenario was used. It presumed that 1) the garden contains the full allowed Cd application, 5 kg/ha; 2) the garden is continuously acidic, about pH 5.5; 3) the gardener obtains 50% of his annual supply of garden vegetables from the acidic, sludge-amended garden, including potatoes, leafy, root, and legume vegetables, and garden fruits; 4) the individual eats these amounts of garden vegetables for 50 years from the acidic sludge-amended garden; and 5) the individual is part of the sensitive-to-cadmium portion of the population. Further, EPA relied on the FDA teenage male diet model, which supplies 39 μg Cd/day. They subtracted this 39 μg Cd/d from the 71 μg/d WHO-FAO (1972) provisional maximum daily Cd ingestion to obtain a maximum allowed increase due to sludge use. Others have noted that U.S. adult dietary Cd is about 20 μg/d (Ryan et al., 1982).

It appears now that several linked assumptions of EPA's acidic garden scenario may well be mutually exclusive, and provide excessive protection. First, individuals who grow 50% of their garden vegetables have such a large time and work investment in their gardens that they learn about the effects of acid soils on yield of vegetable crops, and carefully manage soil pH at 6.5 to 7. Second, presuming that a low Cd, low Cd:Zn ratio sludge applied the soil Cd, and that soil pH declines slowly due to fertilizer use, phytotoxicity in sensitive crops will cause a "50% gardener" to learn about soil pH management and interrupt the necessary 50 year acid garden exposure. Third, vegetables supply microelements which counteract Zn, Fe, and Ca deficiencies; these deficiencies are the identified basis for sensitive individuals. Thus, consumption of the vegetables which comprise the minimal Cd risk to sensitive individuals may push them out of the sensitive population. Recall that increased Cd in "domestic" sludge grown chard and lettuce did not cause increase in kidney Cd (Chaney et al., 1978b, 1978c). In their discussion of Cd dose-response models, Ryan et al., (1982) concluded that U.S. sensitive individuals are protected at the 150 μg Cd/day level of exposure (150-20 = 130 μg/d _vs_ 71-39 = 32 μg/d). Based on the above discussion, it seems _clear_ that the EPA (1979, 1980b) limits are _very_ protective of the worst recognized case when _recommended_ low Cd sludges are managed by land treatment. As a result of these newer understandings discussed above, the regulatory and advisory Federal agencies developed a policy statment on utilization of sewage sludge on cropland for production of fruits and vegetables (EPA-FDA-USDA, 1981).

In summary, the "Soil-Plant Barrier" does not protect the food chain from excessive Cd. Unregulated application of Cd-bearing wastes can cause health effects in humans. Cadmium is not easily kept out of food crops; conversion of treated land to gardens is a worst case scenario upon which regulations to limit Cd applications were based (EPA, 1979a, 1980b). Recent research on gardens polluted with Cd by mining wastes or smelter emissions support the view that gardens can provide much Cd in locally grown foods to the family maintaining the garden for many years (Davies and Ginnever, 1979; Chaney et al., 1980. Unpublished.). Many aspects of the waste-soil-plant-animal food chain are not well established, and research is needed to avoid unnecessarily restrictive limits in the regulations.

Chromium--Chromium is not essential for plants (Huffman and Allaway, 1973). Chromium is essential for animals (Underwood, 1977; Mertz, 1969; NRC, 1980b). An organic chelated Cr^{+3} compound (Toepfer et al., 1977) is a cofactor in insulin hormone response controlling carbohydrate metabolism. Human diets are often deficient in this Cr compound, and some older humans presently experience Cr deficiency (NRC, 1980b; Underwood, 1977).

Chromium exists in two redox forms in nature: chromic (Cr^{3+}) and chromate (Cr^{6+}). Bartlett and Kimble (1976a) show an Eh-pH diagram for Cr in water. Chromium in soil solution is further decreased since chromic is strongly adsorbed and chelated by soils at all practical pH levels (Bartlett and Kimble, 1976a; Cary, Allaway, and Olson, 1977b; Grove and Ellis, 1980). Chromate is rapidly reduced to chromic in soils; reduction is more rapid in acidic soils (Bartlett and Kimble, 1976b; Cary, Allaway, and Olson, 1977b; Grove and Ellis, 1980; Bloomfield and Pruden, 1980).

Bartlett and James (1979) discovered that soluble chromic could be oxidized to chromate in soils and that air drying soils prevented this reaction. Although oxidation may be important in plant uptake of Cr, it is not yet clear that chromic oxidation is significant in the natural environment since chromic is so insoluble and strongly sorbed. Recent research has shown that freshly precipitated (less crystalline) MnO_2 oxidizes the chromic, producing chromate and manganous (Bartlett and James, 1979; Amacher and Baker, 1980; Slater and Reisenauer, 1979; Bartlett, 1979). Bloomfield and Pruden (1980) recently found that soils have substantial chromate adsorption capacity, using a method improved from that of Bartlett and James (1979).

Plant uptake of Cr to plant shoots is generally very limited. Even when soluble chromate is supplied, it is reduced to chromic in plant roots and kept there as chelates, precipitates, or adsorbed Cr. Chromate can be quite phytotoxic (Turner and Rust, 1971; Chapman, 1966; Gemmell, 1973; Breeze, 1973; Mortvedt and Giordano, 1975; Slater and Reisenauer, 1979). However, even under conditions of chromate phytotoxicity, plants contain <10 ppm Cr (normal plant shoots are <2 ppm Cr).

Because humans are often deficient in Cr, research has been conducted on methods to increase Cr in plants. Although chromate existence is more favorable at alkaline pH, adjusting soil pH had little effect. Repeated sub-phytotoxic additions of chromate to soils, and adding 1% Cr as freshly precipitated $Cr(OH)_3$, did increase Cr in plant leaves, but not in grain or fruits (Cary, Allaway, and Olson, 1977a,b). Heating soils to 300°C to simulate effects of forest fires caused soil chromate to increase; corn grown in heated soil had greater Cr in roots than plants grown on unheated soils, but not in shoots (Hafez, Reisenauer, and Stout, 1979). Most plants growing on naturally high Cr serpentine soils have only normal levels of Cr (Cary, Allaway, and Olson, 1977b; Proctor and Woodell, 1975; Anderson, Meyer, and Mayer, 1973). However, some plants occurring naturally on these soils are tolerant of the low Ca status and high Ni and Cr.

Both endemic and ecotypic serpentine tolerant plant species include some Cr accumulators (Lyon et al., 1968, 1970, 1971; Wild, 1974; Shewry and Peterson, 1976).

Lyon, Peterson, and Brooks (1969) found that *Leptospermum scoparium* (a Cr accumulator) translocated Cr in xylem as chromate. However, Cr in roots and shoots was chromic (Lyon, Brooks, and Peterson, 1969); the major identifiable soluble Cr anionic chelate was trisoxalatochromium(III) in all tissues except xylem exudate. Shewry and Peterson (1974), Skeffington, Shewry, and Peterson (1976), and Lahouti and Peterson (1979) examined Cr uptake and translocation by barley. In contrast to the results with *L. scoparium*, chromate remained chromate in the roots; further, some chromic was oxidized to chromate by the roots, and the trisoxalatochromium(III) was not found in roots or shoots. Lahouti and Peterson (1979) found new evidence for trisoxalatochromium-(III) in shoot and root extracts of several species supplied Cr^{3+} or Cr^{6+}. Chromate may limit plant growth on high Cr serpentine soils. Based on the study of Anderson, Meyer, and Mayer (1973), chromate in the displaced soil solution was thought to be too low to be toxic to any plants. However, this work used air-dried soils, and air-drying temporarily inhibits a soil's ability to oxidize chromic to chromate (Bartlett and James, 1979). Thus, the whole subject must be re-evaluated with soils not air-dried before studying.

The food chain is well protected from excess Cr by the "Soil-Plant Barrier." Animals tolerate high levels of insoluble Cr components in their diet (NCR, 1980b). Chromic oxide has actually been fed as a dietary marker for animals (Raleigh, Kartchner, and Rittenhouse, 1980; Dansky and Hill, 1952; NRC, 1980b; Underwood, 1977) and even humans (Kreula, 1947; Irwin and Crampton, 1951; Schurch, Lloyd and Crampton, 1950; Whitby and Lang, 1960). Recently, McLellan et al. (1978) showed that $^{51}Cr^{3+}$ mixed in a diet is an effective label to indicate when all (99.9% of ingested ^{51}Cr by 16 days) of a meal has finally transited the digestive system and been excreted. NRC (1980b) indicated that 3000 ppm Cr as insoluble Cr_2O_3 or 1000 ppm Cr as other chromic salts were tolerated by domestic animals. Because crop plants have such low Cr levels even when grown on soils very high in Cr, the food chain is protected against excess plant absorbed Cr. Animal ingestion of organic wastes rich in Cr appears to offer a similarly low hazard based on sewage sludge (Kienholz, 1980) and tannery sludge (Dilworth and Day, 1970; Knowlton et al., 1976; Waldrup et al., 1970) feeding studies. Hydrolyzed leather meal ($\leq$ 2.75% Cr) is an approved feed ingredient (Anonymous, 1967).

Chromate has been found to be mutagenic (Petrilli and de

Flora, 1977) and carcinogenic, particularly if inhaled; humans and experimental animals exposed to chromate dusts and mists are quite susceptible to nasal cancer (Towill et al., 1978; NRC, 1980b). Chromate in feed has not caused similar cancer in experimental animals, and chromic in feed was not carcinogenic or mutagenic. Chromate in land-applied wastes is rapidly converted to chromic, greatly reducing the chance for a carcinogenic risk. In contrast, when chromate-rich inorganic wastes are placed on the land, chromate may remain for a long period (Breeze, 1973; Gemmell, 1973).

Land application of refuse compost, sewage sludge, tannery sludge, cooling tower blowdown, and sewage effluent can cause Cr in the surface soil to be greatly enriched. However, these practices have not yet been shown to affect Cr in soil below the tilled zone, and do not increase plant Cr after the first crop year (Mortvedt and Giordano, 1975; El-Bassam, Poelstra, and Frissel, 1975; Taylor, 1980; Grove and Ellis, 1980; Kick and Braun, 1977; Silva and Beghi, 1978; Cunningham, Keeney, and Ryan, 1975; Dowdy et al., 1979, 1980; Dowdy and Ham, 1977). Gemmell (1974) and Breeze (1973) found that sewage sludge and refuse compost were very effective in reducing chromate toxicity in smelter wastes.

In summary, Cr in organic wastes is very unlikely to cause food chain problems because animals tolerate high levels of insoluble Cr compounds in diets, and plants do not absorb and translocate high levels of Cr to edible plant tissues even when soils are greatly enriched in Cr.

Cobalt--Cobalt is essential for plants which are dependent on N fixed by nodule bacteria (See Loneragen, 1975). Cobalt is essential for animals; ruminants can use plant Co to make vitamin B_{12}, while monogastric animals require B_{12} (Underwood, 1977; NRC, 1980b).

Cobalt in soils is strongly sorbed and even co-precipitated with the manganese oxides (McKenzie, 1972). Availability of soil Co to plants is controlled by two major factors, MnO_2 in soil and soil pH. Because MnO_2 adsorbs Co so strongly, soils high in MnO_2 often produce forages with deficient Co for ruminants. Cobalt fertilizers can not be successfully used to increase plant Co on these soils (Adams et al., 1969; Loneragen, 1975). Poorly drained soils generally produce plants having higher Co concentration than nearby well drained soils, apparently because MnO_2 surfaces can not effectively adsorb Co under these conditions (Kubota, Lemon, and Allaway, 1963; Allaway, 1968; Loneragen, 1975). Cobalt deficiency of livestock is found on a limited number of soil types in the U.S. (Kubota and Lazar, 1958; Kubota and Allaway, 1972).

Soil pH is very important in Co uptake and phytotoxicity. More acidic soils sorb Co less strongly, allowing plants to absorb it (Adriano, Delaney, and Paine, 1977; Fujimoto and Sherman, 1950; Wallace and Mueller, 1973). Cobalt reactions are similar to those of Ni in soils and plants except for the strong effect of soil MnO_2 on Co uptake.

Phytotoxicity from soil Co results in plants containing 50-100 ppm when foliar symptoms are apparent (Vergnano and Hunter, 1952; Hunter and Vergnano, 1953; Wallace and Mueller, 1973; Austenfeld, 1979; Wallace, Alexander, and Chaudry, 1977). Only a few plant species accumulate Co above the 100 ppm which causes severe phytotoxicity. Hyperaccumulators of Co have been found which contain over 1% Co in dry leaves (Yamagata and Murakami, 1958; Kubota, Lazar, and Beeson, 1960; Brooks, 1977; Brooks, McCleave, and Malaise, 1977). Co-tolerant non-accumulators have also been found (Hogan and Rauser, 1979). The physiological mechanisms of Co, Ni, and Zn phytotoxicity have been studied in some detail by Rauser and co-workers. They recently showed that phytotoxicity inhibited pumping sugars from mesophyll cells into the phloem and veins, severely altering energy transport to rapidly growing tissues (Rauser and Samarakoon, 1980).

Cobalt phytotoxicity has not been of practical importance. Cobalt is seldom used by itself in industrial processes. Pinkerton et al. (1981) studied phytotoxicity and excessive plant Co on a Co-contaminated petroleum landfarm where the soil was Co-enriched by catalyst wastes. Uses of some industrial enzymes require Co as a co-factor. Soils naturally rich in Co are richer in Ni, and Ni phytotoxicity usually dominates (Anderson, Meyer, and Mayer, 1973).

Cobalt toxicity in livestock has not been reported under field conditions. Experimentally, adding over 10 ppm Co to the diet of cattle and sheep has caused Co toxicity (NRC, 1980b; Becker and Smith, 1951; Keener, Baldwin, and Percival, 1951; Keener et al., 1949; Dunn, Ely, and Huffman, 1952; MacLaren, Johnston, and Voss, 1964).

Humans experienced Co toxicity in a few cities when local breweries added Co to beer to stabilize the foam. Heavy beer drinkers died from sudden heart attacks (Anonymous, 1968). Research has not fully clarified this disease. Monogastric animals, not drinking alcoholic beverages, appear to tolerate about 50 ppm Co in their diet (NCR, 1980b; Burch, Williams, and Sullivan, 1973; Huck and Clawson, 1976).

Based on these findings with soil, plants, and animals, Co should be able to cause toxicity to ruminants grazing healthy Co-rich forages. Monogastric animals should not be injured because grain and fruit tissues are only somewhat Co-enriched when Co-phytotoxicity causes visual crop damage. Individuals consuming large amounts of leafy vegetables grown on Co-toxic soils may be at risk. Because none of these risks have been observed in nature, they should be studied to learn whether even the worst case can cause Co-toxicity in any animals.

In summary, the food-chain appears not to be protected from Co toxicity by the "Soil-Plant Barrier." Research is needed on all aspects of Co in waste-soil-plant-animal systems to determine whether this is an actual risk even under worst case conditions.

Copper--Copper is an essential element for plants and animals. Phytotoxicity occurs in most plants at about 25-40 mg Cu/kg dry foliage (Chaney and Giordano, 1977; Page, 1974; Chapman, 1966; Walsh, Erhardt, and Seibel, 1972). Copper toxicity to sensitive sheep and cattle occurs at 25-100 mg Cu/kg dry diet (NRC, 1980b). Cu toxicity in animals can be counteracted by increased dietary Zn, Cd, Fe, or Mo (Suttle and Mills, 1966; Mills, 1978; McGhee, Creger, and Couch, 1965; Standish et al., 1971; Mills and Dalgarno, 1972; Campbell and Mills, 1979; Bremner, 1979). Accidental Cu toxicity in sheep has resulted from feeding pig rations high in Cu; high levels of Cu are added to promote gains in the absence of antibiotics (Baker, 1974; Davis, 1974). Mills (1978), Schwarz, and Kirchgessner (1978), L'Estrange (1979), and Bremner, Young, and Mills (1976) suggested that increasing Zn in these pig rations would reduce the consequences of feeding them to sheep.

Crops with 25-40 ppm Cu could be grown on strongly acid soils rich in Cu from sewage sludge; these crops would have high Zn and normal Fe. Thus, Cu toxicity to even sensitive ruminants from crops fertilized with domestic sludges appears extremely unlikely. Wastes unusually high in only Cu might cause toxicity through ingestion of the waste or soil. For the case of direct ingestion of "Domestic" sludges, the interactions with other dietary and sludge constituents (Fe, Zn, fiber, sulfide) was so pronounced that liver Cu of cattle was depleted rather than increased to toxic levels (Decker et al., 1980; Kienholz, 1980); for high Cu sludges, liver Cu may rise (Bertrand et al., 1980; Johnson et al., 1981). It may be possible to have Cu toxicity expressed as induced-Mo-deficiency if forages are marginal or deficient in Mo, and the sludge is rich in Cu and S, but low in Mo.

As noted in the "pathways" and "interactions" sections above, the potential for effects from land application of high Cu swine manure has caused concern. As in the sewage sludge case (Chaney, 1980), earlier research indicated that sheep would be easily poisoned by applied Cu but field studies found no injury to sheep grazing high Cu swine manure fertilized pastures. These manures contain 500 to 1500 mg Cu/kg dry weight (Aumaitre, 1981). Studies showed extensive plant contamination from adhering manure, and enrichment of the surface organic layer with manure Cu (e.g. McGrath, 1981). Even under the conditions of multi-year practical manure application, sheep were not injured (Bremner, 1981; Poole, 1981; Unwin, 1981). Although manure Cu is "bioavailable" in controlled research studies, the presence of other minerals (Zn, Fe), and of ingested surface soil reduce the practical bioavailable Cu to non-toxic exposures. Authorities do recommend a waiting period after spray application to reduce Cu and pathogen exposure.

A recent study by Gupta and Haeni (1981) illustrates a principle underlying all heavy metal reactions with soils, especially more strongly adsorbed elements such as Cu. When Cu was mixed with sewage sludge to comprise varied proportions of the Cu-saturation capacity, and then applied to sand at varied Cu levels, excessive plant uptake and phytotoxicity was easily caused by sludge of higher Cu-saturation levels and $CuSO_4$, but not by lower Cu sludge. Repeated application of high Cu manure causes Cu accumulation in the soil; although great concern has been voiced regarding Cu phytotoxicity (Lexmond, 1981; Lexmond and DeHaan, 1980), corn Cu has been affected little (Mullens, 1982). These results are corroborated by research on Cu toxicity from sewage sludge (Webber et al., 1981; Davis, 1981a; Beckett, 1981). Soil and manure in the diet of sheep greatly increases the Cu-sorption capacity of the diet, and can induce Cu-deficiency in diets of normal Cu content (Suttle, Allaway, and Thornton, 1975). Thus risk to sensitive livestock from Cu in wastes is much lower than expected, and eventual Cu-phytotoxicity from excessive soil Cu accumulation is likely the dominant risk.

It is likely that land treatment of industrial wastes rich in Cu but very low in Zn and Mo could lead to Cu toxicity in grazing ruminants; Underwood (1977, p. 96) noted this result on Cu rich, low Mo Australian soils. Monogastric animals are much more tolerant of dietary Cu than ruminants. Further, Cu is generally much lower in fruits and grain than leaves when soils are high in Cu. Thus, humans are at no risk from sludge applied Cu even in the worst case acid garden scenario.

Fluorine--Fluorine is not an essential element for plants; it has been shown to be essential for animals through feedings of highly purified diets low in F. Practical diets provide adequate F.

Plant uptake of F is strongly controlled by precipitation of inorganic F compounds in soil; CaF_2 formation had been presumed, but the exact chemical species controlling F^- activity has not been demonstrated (could be fluoroapatite). Plant F is increased by F additions to soil only where soil Ca activity is very low (strongly acid soils) (Chapman, 1966; Prince et al., 1949). Sewage sludges are not rich in F, and sludge use has not been reported to increase crop F unless extremely F contaminated sludge was used (Davis, 1980). Phosphate fertilizers have added F to cropland (Gilpin and Johnson, 1980). The "Soil-Plant Barrier" is very effective in limiting impact of soil F via plant uptake; monogastric animals are fully protected from soil F.

Direct ingestion of F rich sludges or soil could affect animals depending on the chemical forms of F present (Davis, 1981). Kienholz et al. (1979) observed significant increases in bone F in cattle fed sewage sludge. Sludge F should be CaF_2 or fluoroapatite, forms of low bioavailability (NRC, 1980b; Underwood, 1977) compared to NaF. Most fluorosis in domestic animals results from high F water supplies. Another important cause of fluorosis is industrial air-borne F pollution which contaminates forages allowing direct ingestion of F. Industrial wastes rich in F are poor candidates for surface application on land treatment sites which will be used as pastures for cattle or sheep.

Iron--Iron is an essential element for plants and animals. Phytotoxicity of Fe is not an agronomic problem except when rice is grown on some reduced soils (Foy, Chaney, and White, 1978). Leaves of most crops contain 30-300 mg Fe/kg dry weight. Fertilizing with sludges does not cause foliar Fe to rise above these normal levels; sludge can increase foliar Fe concentration from deficient levels (30-40 ppm in chlorotic leaves) to normal, and may be a valuable Fe fertilizer (McClaslin and Rodriguez, 1978; McClaslin and Sivinski, 1979).

Plant Fe can reach unusually high levels due to interactions with other plant micronutrients (Olsen, 1972). Deficiencies of Cu (DeKock, Cheshire and Hall, 1971), Mn, or Zn (Ambler and Brown, 1969; Warnock, 1970) can cause foliar Fe to exceed 1000 ppm. Excessive Fe in leaves is stored as phytoferritin (Seckbach, 1968, 1969).

The levels noted above are for leaves free of soil. Soil dust ($>1\%$ dry weight) on leaves can supply more total Fe than internal plant Fe. Trampled forages become "soiled"

(Healy, Rankin, and Watts, 1974). Spray-applied sludges may greatly increase Fe in/on forages (Chaney and Lloyd, 1979), allowing forage Fe to exceed even 1% dry weight (Decker et al., 1980). This external Fe (from anaerobic sludge) can injure domestic animals, while internal plant Fe has shown no evidence of toxicity to animals.

Animals tolerate higher levels of Fe than normally occur in feedstuffs (NRC, 1980b). However, chronic Fe toxicity is complex, and toxicity of sludge-borne Fe depends on many interactions. Dietary Fe interacts strongly with Cu and Zn; 1000 ppm Fe added to a forage diet reduced Cu in sheep and cattle liver (Standish et al., 1969, 1971; Standish and Ammerman, 1971; Grun et al., 1978). On low Cu diets, Fe toxicity is expressed as an Fe-induced Cu-deficiency. However, when sewage sludge supplies the excessive Fe, it also supplies Cu (which would tend to correct Cu deficiency) and Zn, Cd, S, Mo, and fiber which could further interfere with Cu utilization (Anke, 1973; Bremner, 1979).

The chemical form of Fe influences Fe bioavailability. Iron in $FeSO_4$ is 3 to 5 times more available to cattle than Fe in ferric oxide (Ammerman et al, 1967). Much of the Fe in anaerobically digested sludge is in the ferrous form; this quickly oxidizes as the sludge becomes aerobic. Iron in soils (and sludges on the soil surface) should have different bioavailability depending on the degree of crystallinity of the ferric oxides present.

Iron toxicity was suspected in cattle grazing tall fescue pastures sprayed with a digested sewage sludge containing 11% Fe (Decker et al., 1980). When lower Fe anaerobic sludges, or when sludge composts are applied, iron toxicity did not result (Decker et al., 1980; Bertrand et al., 1981). That sludge was low in Cu. Liver Cu was at marginal or deficient levels in the cattle on sludge treatments. Liver Cu has been low in most studies of sludge feeding (Kienholz, 1980). In two studies, with sludge Cu at 1000 ppm, the cattle showed Cu balance or a slight increase in liver Cu (Kienholz, 1980; Baxter, Johnson, and Kienholz, 1980; Johnson et al., 1981; Bertrand et al., 1980).

Thus, high Fe wastes which are low in Cu can promote Fe-induced Cu-deficiency and high Fe accumulation in liver, spleen, and intestine. Even the erosion of cartilage in joints of cattle (Decker et al., 1980) may be due to a severe Cu deficiency. Zinc-induced Cu deficiency in horses (Willoughby et al., 1972; Willoughby and Oyaert, 1973; Gunson et al., 1982) caused erosion of cartilage in joints very similar to that observed in cattle grazing pastures fertilized with high Fe sludge.

Industrial sludges and other wastes high in Fe may cause Fe-induced Cu-deficiency if they are surface applied to pasture land. Zinc and Mo in the waste can also contribute to the toxicity, and Cu in the waste can counteract it. Supplemental Cu may be needed for ruminants grazing pastures where high Fe wastes remain on the soil surface.

Lead-- Lead is not essential for either plants or animals. Phytotoxicity due to Pb has been observed only under unusual conditions of phosphate deficiency. Because sewage sludges contain high amounts of P, Pb in domestic sludge has not caused increased plant Pb. Plants do not ordinarily translocate high amounts of Pb to their shoots (<5 ppm) because an insoluble Pb-phosphate is formed in the roots. However, high foliage Pb has been reported for grasses growing on infertile, strongly acidic, Pb-mine spoils (Johnson, McNeilly, and Putwain, 1977; Johnson and Proctor, 1977).

Lead toxicity to animals occurs at about 30 mg Pb/kg diet (NRC, 1980b); Pb interacts strongly with dietary Fe, Ca, Zn, P, fiber, etc. (Mahaffey, 1978; Mahaffey and Vanderveen, 1979; Levander, 1979). Thus, animals appear to be protected from excess Pb in crops grown on land treatment sites. A seasonal increase in foliar Pb occurs in forages growing during mild winters (Mitchell and Reith, 1966). This seasonal increase was not exaggerated in crops grown on soils enriched in Pb by sewage sludge (Haye et al., 1976).

As noted above, consumption of soil can comprise a risk for elements not readily accumulated by plants. Consumption of soils with naturally high Pb concentrations and mine wastes has caused Pb toxicity in domestic livestock (Egan and O'Cuill, 1970; Harbourne, McCrea, and Watkinson, 1968; Edwards and Clay, 1977; Thornton and Kinniburgh, 1978). Ingestion of domestic sewage sludges has not caused Pb toxicity, probably because sludge Fe, P, Ca, etc. counteract Pb absorption (Kienholz, 1980). Industrial wastes high in Pb comprise a very different case than domestic sewage sludge. Conversion of land treatment sites to uses allowing children to ingest soil should require consideration of whether this exposure could allow Pb poisoning if the soils are too high in Pb.

As noted earlier, children ingest soil and dust by 1) a deliberate practice of consuming non-food items, called "pica", and 2) normal hand-to-mouth play activities and mouthing of toys etc. (Lepow et al., 1975; Sayre et al., 1974; Roels et al., 1980). High Pb soils can be carried into homes on dusty clothing and shoes (e.g. Archer and Barratt, 1976), thereby causing housedust to become rich in Pb (Jordan and Hogan, 1975). Several research programs have shown that soil Pb and housedust Pb contribute to Pb exposure of children (Angle and McIntire, 1979; Charney, Sayre, and Coulter, 1979; Galke et al., 1977; Roels et al., 1980, Barltrop et al., 1974; Hammond et al., 1980). Paint Pb, air Pb, soil Pb, and housedust Pb comprise a multiple source model for Pb exposure of children.

Perhaps the clearest example of the possibility of human health effects from inadvertent ingestion of high Pb housedust is found in recent cases in which children of Pb-industry workers experienced excessive blood Pb. The only identifiable source of Pb was the industrial dusts carried into the house on the workers' clothing and shoes (Baker et al., 1977; Giguere et al., 1977; Rice et al., 1978; Fergusson, Hibbard, and Ting, 1981).

Soil Pb can only cause excessive blood Pb if it is at least somewhat bioavailable to humans. Studies have shown that soil Pb is largely bioavailable to rats (Dacre and Ter Haar, 1977; Stara et al., 1973), but that soil properties can reduce Pb bioavailability (Chaney et al, 1980). At the pH of the stomach of monogastric animals, soil Pb is quite soluble (Day et al., 1979). Table 9 shows results from a study of the bioavailability of soil Pb to rats. Although soil properties do reduce apparent risk from soil Pb, these Pb rich garden soils would add to Pb absorption risk to children. Garden soils are enriched in Pb (in descending order of impact) from exterior and interior paint; application of ashes, sludges, and pesticides; and automotive emissions (Davies, 1978; Preer et al., 1980; Spittler and Feder, 1979; Ter Harr and Aronow, 1974). Many research questions must be addressed to allow setting of maximum acceptable levels for soil Pb.

Recent research indicates a different health risk from Pb than those traditionally relied on in developing regulations, the effects on heme synthesis and frank lead encephalopathy. This new clinical result of low level lead exposure is neurobehavioral impairment (Needleman et al., 1979; Needleman, 1980; Silbergeld and Hruska, 1980; David et al., 1978). Lower IQ and school achievment, and problem classroom behavior appear in populations of Pb-exposed children whose exposure was not great enough to require chelation theraphy (60 μg Pb/dL blood), or source identification (30 to 40 μg Pb/dL). A recent National Research Council review considered the potential importance of soil and dust ingestion in Pb poisoning and low level lead exposure of children (NRC, 1980a).

Molybdenum--Molybdenum is essential for both plants and animals. Molybdenum is present in soils and plants as

TABLE 9. EFFECT OF SOIL ON AVAILABILITY OF Pb TO RATS, AND BIOAVAILABILITY OF Pb IN URBAN GARDEN SOILS 1/

Diet 2/	Tibia Ash mg/tibia	Pb in Tibia mg/kg tibia ash ± S.E.
Basal	102	0.3 ± 0.3 e 3/
Basal + 5% Soil	101	0.0 e
Basal + PbOAc	92	247. ± 10.1 a
Basal + PbOAc + 5% Soil	105	130. ± 29.5 bc
706 ppm Pb Soil	98	40.0 ± 6.1 de
995 ppm Pb Soil	96	108. ± 26.3 c
1078 ppm Pb Soil	100	37.1 ± 7.3 de
1265 ppm Pb Soil	107	53.6 ± 7.4 d
10240 ppm Pb Soil	92	173. ± 21.8 b

1/ Levander, O. A., R. L. Chaney, S. O. Welsh, C. H. Gifford, and H. W. Mielke (unpublished results, USDA, Beltsville, MD).

2/ Fed to Fisher rats for 30 days. A purified casein-based complete diet was fed; Pb acetate and garden soils were added to supply 50 mg Pb/kg dry diet.

3/ Means followed by the same letter are not significantly different ($P<.05$) according to Duncan's Multiple Range Test.

anionic molybdate. This causes Mo to differ markedly in its reactions with soils and plants from those of the potentially toxic cations (e.g. Zn^{2+}) for which liming to alkaline pH reduces plant uptake (see also review by Jarrell, Page, and Elseewi, 1980). Soils adsorb Mo more strongly under more acidic conditions; inorganic compounds of Mo have not been identified in soils (Vlek and Lindsay, 1977a). The strong Mo sorption at low soil pH holds Mo against leaching and reduces plant Mo uptake. Alkaline soil pH weakens soil sorption of Mo, promoting plant uptake of Mo, and allowing Mo to leach through the root zone with leaching water. Sorption of Mo is further decreased by reducing conditions (low E_h). Molybdenum uptake is greatest in alkaline, poorly drained soils with Mo enrichment (Kubota, 1977). The chemical or biological basis for greater Mo concentration in plants growing on poorly drained soils remains unclear. It has been suggested that a higher percentage of plant roots in the surface soil, greater Mo solubility (lower sorption) in reduced soils, or reduced plant biomass diluting absorbed Mo may explain these higher concentrations (Allaway, 1977c; Kubota et al., 1961; Kubota, Lemon, and Allaway, 1963; Vlek and Lindsay, 1977b).

Plants tolerate very high levels of Mo, and translocate Mo to edible plant tissues. Legume crops accumulate substantially more Mo than do grasses (Chapman, 1966; Allaway, 1977c; Jensen and Lesperance, 1971; Kubota, 1977). Vegetable crops are also Mo accumulators (Hornick, Baker and Guss, 1977; Gupta, Chipman, and Mackay, 1978).

Ruminant animals are especially susceptible to Mo toxicity (molybdenosis) because Mo interacts to induce a Cu deficiency (Mills, et al, 1978; Underwood, 1977) Dietary Mo interacts also with sulfate (increased sulfate competitively reduces Mo absorption), Mn, Zn, Fe, tungstate, sulfur-containing amino acids, etc. Dietary sulfate and Cu are the primary factors interacting with excessive dietary Mo. Normal forages contain 0.1 to 3.0 ppm Mo (dry weight basis). Levels over 10 ppm Mo are generally high enough to cause molybdenosis; lower levels can cause molybdenosis depending on dietary Cu availability. Plants grown on poorly drained, alkaline, Mo enriched soils can exceed 100 ppm Mo, far surpassing the levels necessary for molybdenosis (>10 ppm). Industrial air pollution from Mo-smelters, oil refineries and steel mills have caused molybdenosis (Hornick, Baker and Guss, 1977; Alary et al., 1981; Gardner and Hall-Patch, 1962, 1968; Buxton and Allcroft, 1955; Parker and Rose, 1955). It is clear that the "Soil-Plant Barrier" does not protect the ruminant animal food-chain from excessive soil Mo.

A few soils in the U.S. have naturally high Mo levels, and cause molybdenosis in cattle and sheep (Kubota, 1977). Soil and crop management can reduce the problem; draining the area and growing grasses rather than legumes reduces potential molybdenosis considerably. Land treatment sites for Mo-rich wastes can be managed to remove the Mo from the surface soil before ruminants are allowed to graze the area; drainage and crop selection are important. Adequate or high copper in crops and surface soil reduces Mo risk. Animals can be supplemented with Cu by injection, or salt licks or feed supplements containing Cu.

Sewage sludges are very seldom high in Mo (median = ca 10 ppm). Historically, sludge use has not been reported to cause molybdenosis in ruminants through either plant uptake or direct ingestion of sludge. High Mo sludges do exist, and would be a cause for concern (Table 5) (Lahann, 1976; Sterritt and Lester, 1981). Sludge Cu would counteract any effects of considerable amounts of Mo in most ingested sludges. Fly ash is richer in Mo than sludges and has been evaluated as a Mo-fertilizer; fly ash Mo is about as available to plants as Na molybdate (Doran and Martens, 1972). Molybdenosis could result where high rates of fly ash are applied on alkaline pastures (Gutenmann et al., 1979).

Monogastric animals are protected from excess soil Mo by the "Soil-Plant Barrier"; Mo concentration in grain is lower than in leaves, monogastric animals are more tolerant of Mo than ruminants, and monogastric animals are seldom fed grain crops grown only on a restricted area of land. It is conceivable (but very unlikely) that humans could be at some risk of molybdenosis by consuming large amounts of garden crops grown on high Mo alkaline soils which had previously been used for land treatment of Mo rich wastes. Sites managed at lower soil pH to avoid excessive crop Mo can subsequently cause molybdenosis after limestone is applied (Hornick, Baker and Guss, 1977).

Nickel--Nickel is an essential element for animals and probably is for plants (Polacco, 1977; Welch, 1981). Phytotoxicity causes visible symptoms (interveinal chlorosis of young leaves) at about 50-100 mg/kg dry leaves. These levels are phytotoxic for grasses, legumes, and leafy vegetables (Chapman, 1966; Chaney et al., 1978a; Hunter and Vergnano, 1952; Roth, Wallihan and Sharpless, 1971; Halstead, Finn, and MacLean, 1969; Anderson, Meyer, and Mayer, 1973). Some plants are very tolerant of soil and plant Ni and are Ni-accumulators; plants tolerating over 1000 mg Ni/kg dry leaves have been labelled hyperaccumulators (Brooks et al., 1979; Jaffre et al., 1976; Wild, 1970; Morrison, Brooks and Reeves, 1980). These Ni-accumulator species are rare and are not reported among crop plants.

Nickel toxicity in domestic animals occurs at 50-100 mg Ni/kg feed where Ni is added as soluble salts (NRC, 1980b). This range of Ni tolerance in livestock overlaps the range of Ni present in plants at phytotoxicity. However, when $NiCO_3$ was added to cattle diets, no toxicity was observed at 250 ppm Ni (O'Dell et al., 1970). Greater tolerance in animals could also occur to Ni in green forage or hay grown on land treatment sites. Little study has been done with Ni bioavailability in foods or feeds grown to contain high levels of Ni. Alexander et al. (1979) found no health effect or Ni bioaccumulation in voles fed sludge-fertilized soybeans containing 30 ppm Ni; this is the Ni level in soybean grain when yield is significantly reduced.

Based on these results, it seems very unlikely that Ni toxicity would occur in ruminants grazing Ni phytotoxic forages on worst case acidic, high Ni land treatment sites.

The availability of ingested soil Ni has not been reported. Grain and vegetable crops do not accumulate Ni to levels which would injure monogastric animals. Although wildlife might be affected by Ni-rich sludges on the soil surface, they would not be by Ni in crops.

Selenium--Selenium is not essential for plants, but is for animals. The forms of Se in soil depend on soil pH and redox (Geering et al., 1968). At equilibrium, most soil Se should be elemental Se. Selenate additions are weakly adsorbed by soils; selenite is strongly sorbed on hydrous iron oxide surfaces; elemental Se is quite inert compared to elemental S, but some oxidation occurs; selenide is readily oxidized in soils (Allaway, 1977a; Carter, Brown and Robbins, 1969; Cary and Allaway, 1969). Plant residues can supply seleno-amino acids to soil (Olson and Moxon, 1939). Some Se in soil and plants is converted to dimethylselenide and volatilized (Lewis, Johnson and Broyer, 1974).

Selenate in soils is readily absorbed by plants, while selenite is not as plant available since it is sorbed strongly by soil. After plant uptake, selenite is more easily converted to seleno-amino acids than selenate. Elemental Se is so inert that it will not supply enough soluble Se to plants to correct Se deficiency in animals. In the year that elemental Se (or flyash) is applied to soil, plant Se is somewhat increased. Selenite is a useful slow-release Se fertilizer which corrects Se deficiency of animals for a number of years while not allowing plant Se to reach acutely toxic levels (>3 ppm) during the year of application (Cary and Allaway, 1973; Carter, Brown and Robbins, 1969). Neutral soil pH favors both plant uptake and leaching of Se through soils.

Once Se is absorbed by plants, it is freely translocated to edible plant tissues, including grain (Anderson et al., 1961; Allaway, 1968, 1977a). Selenium follows the sulfate pathway in uptake, translocation, and metabolism (Asher, Butler, and Peterson, 1977). Plant species differ widely in relative Se uptake (Hamilton and Beath, 1963; Allaway, 1968). Although most plant species convert only a small fraction of absorbed Se to seleno-amino acids, a few convert Se extensively. These species are endemic to high Se soils (Shrift and Virupaksha, 1965; Trelease and Trelease, 1939; Anderson et al., 1961; Trelease, 1945). When these plants die, seleno-amino acids are returned to the soil, greatly increasing the amount of plant available Se in the soil (Olson and Moxon, 1939).

Small areas of the Great Plains have naturally Se toxic soils (Kubota et al., 1967; Kubota and Allaway, 1972;

Anderson et al., 1961). Domestic livestock were poisoned by the high Se crops. Acute poisoning can result from ingestion of plants which convert Se to seleno-amino acids (e.g., *Astragalus* sp.) (Allaway, 1968; Anderson et al., 1961). Chronic poisoning can also result from ingestion of Se-accumulating normal forages and grains grown on the Se-rich soils. Humans were not shown to suffer serious health effects in these areas, although mild Se effects were observed (Anderson et al., 1961); medical science was not well developed at that time and renewed research might detect human health effects. These natural Se poisoning incidents clearly show that the "Soil-Plant Barrier" fails to protect animals from excessive soil Se (Allaway, 1968; Anderson et al., 1961; Trelease, 1945).

Sewage sludges are usually low in Se and this has prevented toxicity from sludge-borne Se. Sludge ingestion research has also shown no Se impacts (Kienholz, 1980). However, fly ash from coal burning power plants can be quite rich in Se when Western coals are burned (Furr et al., 1977). Crops grown on fly ash amended soils are high in Se, and animals can obtain excessive Se by this route (Furr et al., 1975, 1977; Gutenmann et al., 1976). On the other hand, fly ash can be used at lower application rates as a Se-fertilizer to amend the large areas of land in the U.S. which yield crops with Se too low to satisfy the needs of animals (> 0.1 ppm) (Stoewsand, Gutenmann, and Lisk, 1978; Combs, Barrows, and Swader, 1980, Gutenmann et al., 1979). Similarly, fly ash can be a Se-feed supplement (Furr et al., 1978b; Hogue et al., 1980).

In summary, the food chain is not protected from excessive soil Se. Land treatment of wastes high in Se will require management designed to avoid food-chain impacts of soil Se (e.g., use at low rates as a Se fertilizer). Selenium toxic land in South Dakota was purchased by the Federal Government to prevent or minimize Se-poisoning risks (Anderson et al., 1961). Use of selected management practices has allowed some productive use of these lands.

Tin--Tin is not essential for plants; one study suggested Sn was essential for rats fed a highly purified diet, but it is not yet generally agreed that Sn is essential for animals (NRC, 1980b; Underwood, 1977).

Tin forms very insoluble minerals in soils (SnO_2), and this appears to prevent plants from accumulating much Sn regardless of soil pH. Studies by Romney, Wallace and Alexander (1975) and Kick and Warnusz (1972) found that, for addition of soluble Sn salts to soils, soil of very low pH may allow some phytotoxicity; for soil pH near neutral, 500 ppm Sn had no effect on several crop species, and did not increase foliar Sn. Several biogeochemical studies of Sn, and studies of Sn in plants growing on Sn mine wastes, showed very little uptake of Sn by plants even when soil Sn was quite high (Millman, 1957; Peterson et al., 1976). The latter authors found a few plant species which had increased foliar Sn when found growing on Sn mine wastes. The levels were only 20 ppm, far below levels tolerated by animals.

Tin has very low toxicity when mixed with practical diets. Rats tolerate several hundred ppm with no effects; at higher dietary Sn, supplemental Fe and Cu counteract effects of Sn (deGroot, 1973; deGroot, Feron, and Til, 1973). Tin-contaminated fruit juices had little effect on animals (Benoy, Hooper, and Schneider, 1971). Studies of the bioavailability of soil Sn or effects of ingesting Sn-rich soils have not been reported; it appears that this would be very unlikely to affect grazing animals.

Sewage sludges contain low levels of Sn. No effects of sludge Sn on animals, or residues of Sn in animals fed sludge have been reported.

Thus, the food chain is protected from excess Sn by the "Soil-Plant Barrier." Tin in industrial wastes should have little influence on plants or animals when applied to a land treatment site.

Vanadium--Vanadium is essential for some algae, but apparently not for higher plants (Welch and Huffman, 1973). Vanadium is essential for animals, although the requirement is low. Practical diets supply adequate V for domestic livestock (NRC, 1980b; Underwood, 1977).

Vanadium is a constituent of many petroleum materials, coals, and some rock phosphates. Some industrial wastes are enriched in V from these sources.

Little is known about forms of V in soils. Many redox species of V are possible, but forms in soil have not been reported. Soils naturally contain 30-100 ppm V (and higher near ore deposits), but only a small fraction of this is extractable. Crops growing on most soils contain low V, less than 1 ppm dry weight (Welch and Cary, 1975). Supplying soluble forms of V in nutrient solution did increase V in shoots of most plants; however, V was poorly translocated from roots to shoots (Welch and Huffman, 1973; Wallace, Alexander, and Chaudhry, 1977). Tyler (1976) suggested that V could inhibit soil enzymes, at least when soluble V is added. Soluble V can be phytotoxic, but this may be environmentally irrelevant depending on soil reactions of V.

So little is known about V reactions in soil and plant uptake of V from soil that it is difficult to evaluate risks from land treatment/application of industrial wastes rich in V. The NRC (1980b) indicated that ruminants would chronically tolerate 50 ppm V, and monogastric animals 10 ppm V; forage V would be lower than 10 ppm V when V is phytotoxic. One study on V-accumulator plants growing on V-mineralized soils indicated plant foliage could reach 150 ppm V under unusual circumstances (Cannon, 1963, 1964). The bioavailability of V in forages and grains remains ill-defined.

As noted above, little was known about soil-plant relationships for V. Furr et al. (1977) applied fly ash from 15 sources at 7% in a neutral pH soil. Vanadium in the ash varied from 68 to 442 ppm, and V in the cabbage was only 0.02 to 0.4 ppm. Further study with other crops and one fly ash showed no change in crop V due to fly ash application. Sewage sludge also caused no change in plant V (Chaney et al., 1978c).

In summary, the "Soil-Plant Barrier" appears to protect the food chain from excess soil V. So little research has been done on V that no conclusion is possible.

Zinc--Zinc is an essential element for plants and animals. Phytotoxicity (ca. 25% yield reduction) occurs in most plants at about 500 mg Zn/kg dry foliage (Chapman, 1966; Boawn, 1971; Boawn and Rasmussen, 1971; Walsh et al., 1972b). Leafy vegetables such as chard and spinach may be more tolerant of foliar Zn; chard does not show phytotoxicity in acid soils until foliar Zn is about 1500 mg/dry kg (Baxter, Chaney, and Kinlaw, 1974). Zn toxicity to domestic livestock occurs at 300 to 1000 mg/kg dry diet (NRC, 1980b). This Zn toxicity appears to result from Zn-induced Cu deficiency (Grant-Frost and Underwood, 1958; Lee and Matrone, 1969; Campbell and Mills, 1979; Bremner, 1979; Gunson et al., 1982; Bremner and Campbell, 1980). Sheep are more sensitive to excessive dietary Zn if Cu in the feed is marginal to deficient (Campbell and Mills, 1979).

Several factors influence risk of soil Zn. Crop uptake of Zn is strongly reduced as soil pH increases. Forage crops are poor accumulators of Zn compared to vegetables and crop plants. Forage crops grown on sewage sludge amended soil would have normal to somewhat enriched Cu concentration to counteract the Zn; however, industrial wastes high in Zn but low in Cu would comprise a greater Zn risk to the food chain than sewage sludge.

Ruminant animals could consume Zn rich sludges from the

soil surface. For alkaline soils, sludge and soil ingestion could supply more Zn than the forage crops grown on the amended soil; for acidic soils, crop uptake would be the predominant Zn source. For animals consuming domestic sewage sludge, dietary Cu and Fe would be substantially increased along with Zn and would counteract the risk of Zn toxicity. For high Zn sludges or industrial wastes, these protective factors might not be increased.

Forage crops comprise the worst case for evaluation of excess soil Zn. Under conditions of high Zn supply, grain and fruit contain lower Zn concentrations than leaves. Research by Ott et al. (1966), Campbell and Mills (1979), L'Estrange (1979), and Bremner, Young, and Mills (1976) indicate that crop uptake of Zn is very unlikely to cause Zn poisoning of cattle or sheep. Although humans could consume appreciable amounts of high Zn leafy vegetables (acid soils; phytoxicity), it is very unlikely that Zn toxicity would occur in humans in even this worst case. Leafy vegetables supply appreciable Cu and Fe, especially if grown on sludge-amended soils, and these counteract the Zn risk. Individuals are unable to consume high Zn leafy vegetables as the bulk of their diet the whole year. Many humans presently consume low or deficient amounts of Zn (Hambridge et al., 1972); increased food Zn would be beneficial in many cases.

Wildlife comprise a greater risk case than do domestic animals because wildlife could chronically consume Zn-phytotoxic foliage as most of their diet. It seems likely that high Zn sludges and industrial wastes could cause adverse health effects (due to Zn-induced Cu-deficiency) in wildlife such as that resulting from Zn smelters (Gunson et al., 1982; Bremner and Campbell, 1980).

In summary, land application/treatment of sewage sludges and industrial wastes is very unlikely to cause Zn toxicity in domestic livestock or humans even under worst case conditions. Wastes unusually high in Zn but low in Cu may cause Zn toxicity to animals if mismanaged.

Potential Food-chain Impacts of Toxic Organic Compounds Applied in Industrial Wastes

Introduction--

Animals can be exposed to toxic organic compounds (TO's) present in wastes by the pathways described above: 1) direct ingestion of wastes, wastes adhering to forages, wastes lying on the soil surface, or soil treated with wastes; 2) ingestion of plant tissues which are increased in TO content after plant uptake or volatilization from the soil to the plant; or 3) consumption of animal products enriched in TO by other routes. The chemical and physical properties of a TO control its adsorption by soil, volatilization, plant uptake and translocation, biodegradation (in soil, plant or animal), and accumulation in animal tissues. Because each TO is chemically and pharmacologically unique, each compound will have its unique behavior in waste-soil-plant-animal systems (Fries, 1982; Majeta and Clark, 1981; Dacre, 1980). Another chapter in this book describes the fate of organic compounds in land-applied wastes, including sorption by soils, plant uptake and translocation to shoots, volatilization from soils and contamination of plant shoots or edible roots, and chemical-, photo-, and bio-degradation.

Although much research has been conducted on insecticides, fungicides, and herbicides, insufficient information is available to assess food chain risk of waste-borne TO's. Environmentally relevant research on waste-borne TO's is quite limited even among pesticides. Very little is known about fate and potential for food-chain effects of industrial TO wastes and byproducts that may be considered for application to land treatment sites.

Thus, this subsection will describe the processes which influence movement of TO's in waste-soil-plant-animal food chains. This should illustrate the research needs for assessing potential impacts of a TO or a TO-enriched waste. PCB's in sewage sludge will provide a particularly relevant example, as regulations were developed based on the available research (EPA, 1979a).

Bioavailability of Ingested Waste-Borne Toxic Organics--

Lipophilic toxic organics in ingested sludges and soil are bioavailable. DDT and lindane in ingested soil were absorbed by sheep and stored in their fat (Harrison, Mol, and Healy, 1970; Healy, Mol, and Rudman, 1969; Collett and Harrison, 1968). PCB's and other compounds in ingested sludge were absorbed and stored in fat of cattle (Kienholz, 1980; Baxter, Johnson, and Kienholz, 1980; Fitzgerald, 1978, 1980), cow's milk (Bergh and Peoples, 1977), and fat of swine (Hansen et al., 1981). In general, PCB residues in fat reached levels 5-fold higher than in dry feed.

Based on these studies and basic research on bioaccumulation of PCB's, Fries (1982) concluded that PCB's should not exceed 2.0 mg/kg dry sludge if milk cows are to be allowed to graze pastures under worst-case conditions which allow 14% sludge in their diet. This was based on a biomagnification from diet to milk fat of 5-fold, and FDA tolerances of 1.5 mg PCB/kg milk fat (FDA, 1979). Forages grown on soils containing PCB's have PCB residues about 0.1 that of the soil, or lower. Good management practices (delay grazing for 30 days after surface application of sludge, and supply feed concentrates during periods of low forage availability) reduce sludge ingestion so that 10 ppm PCB's could be allowed in sludge surface applied at 10 metric tons/ha/yr. Injection of sludge below the soil surface would further reduce exposure.

A seldom considered concentration step involves soil fauna. Earthworms accumulate Cd (Helmke et al., 1979; Beyer, Chaney, and Mulhern, 1982), and lipophilic toxic organics. Beyer and Gish (1980) noted substantial residues of DDT, dieldrin, and heptachlor in earthworms many years post application. Birds and shrews consume appreciable earthworm biomass and are thereby exposed to Cd and pesticides. More study is needed to assess the importance of this unusual foodchain pathway in relation to land treatment of industrial wastes and potential effects on wildlife.

Plant "Uptake" of Toxic Organics in the Soil--
Toxic organics can enter edible parts of plants by two processes: 1) uptake from the soil solution, with translocation from roots to shoots, or 2) adsorption by roots or shoots of TO's volatilized from the soil. "Systemic" acting pesticides are applied to the soil, absorbed and translocated by the plant, and act to protect the plant leaves. These compounds are quite water soluble and would probably not appear in industrial wastewater treatment sludges at appreciable levels. Some systemic TO's are prohibited from use on food crops (other than seed protectants) since residues of the compound or its metabolites on or in food may be unacceptable. The EPA-approved label for each compound lists acceptable uses.

The lipophilic halogenated pesticides represent the case for water insoluble compounds which are largely sorbed by plants from the soil air or the pesticide-enriched air near the soil surface. Beall and Nash (1971) developed a method to discriminate between movement of a TO through the plant vascular system (uptake-translocation) vs. vapor phase movement. They found soybean shoots were contaminated by soil-applied dieldrin, endrin, and heptachlor largely by uptake-translocation. Vapor transport predominated for DDT, and was equal to uptake-translocation for endrin. Using this method, Fries and Marrow (1981) found PCB's reached shoots via vapor transport, while the less volatile PBB's did not contaminate plant shoots by either process (Chou et al., 1978; Jacobs, Chou and Tiedje, 1976). Suzuki et al. (1977) found that PCB's with a low number of chlorines could be absorbed and translocated at low rates by soybean seedlings from sand treated with high levels of PCB's.

Root crops are especially susceptible to contamination by the vapor-transport route. Carrots have a lipid-rich epidermal layer (the "peel") which serves as a sink for volatile lipophilic TO's. Depending on the water solubility and vapor pressure of the individual compound, it may reside nearly exclusively in the peel layer of carrots, or penetrate the storage root several mm (Lichtenstein, Myrdal, and Schulz, 1964, 1965; Jacobs, Chou, and Tiedje, 1976; Lichtenstein and Schulz, 1965; Iwata and Gunther, 1976; Iwata, Gunther, and Westlake, 1974; Fox, Chisholm, and Stewart, 1964; Landrigan et al., 1978).

Carrot cultivars differ in uptake, and in peel vs. pulp distribution of the chlorinated hydrocarbon pesticides endrin and heptachlor (Lichtenstein, Myrdal, and Schulz, 1965; Hermanson, Anderson, and Gunther, 1970). Other root crops (sugar beet, onion, turnip, rutabaga) are much less effective in accumulating lipophilic TO's in their edible roots, possibly because the surface of the peel is lower in lipids (Moza, et al., 1979; Moza, Wiesgerber, and Klein, 1976; Fox, Chisholm, and Stewart, 1964; Chou, et al., 1978; Lichtenstein and Schulz, 1965).

The level of chlorinated hydrocarbon in carrots is sharply reduced by increased organic matter in soil. The increased organic matter adsorbs the TO's and keeps them from being released to the soil solution or soil air (Filonow, Jacobs, and Mortland, 1976; Weber and Mrozek, 1979; Chou, et al., 1978; Strek, et al., 1981). Added sewage sludge increased the ability of soils to adsorb PCB's (Fairbanks and O'Connor, 1980). At some low level of PCB's in sludge, the increased sorption capacity may fully counteract the increased PCB's.

Assessing risk from environmental exposure to PCB's, or other TO's is difficult. PCB's and other persistent chlorinated hydrocarbons seldom occur at excessive levels in present municipal sludges (Sprague et al., 1981). The residue of PCB's in waste products is depleted of the relatively more volatile lower chlorinated compounds, but most research is conducted with the commercial mixture. It is clear that plant contamination by the higher chlorinated compounds is much less than for the lower chlorinated ones at equal soil levels (Iwata and Gunther, 1976; Suzuki et al., 1977; Moza, Weisgerber, and Klein, 1976; Moza et al., 1979; Fries and Marrow, 1981). Recently, research has begun with the individual ^{14}C-labelled PCB's; risk evaluation should focus on the 5, 6, and more highly chlorinated compounds which remain in wastes and soils. Assuming peeling of carrots, the only significant exposure to these higher chlorinated PCB's is to grazing ruminants through soil ingestion. One field research study with a "domestic" sludge (contained 0.93 ppm PCB's) evaluated PCB uptake by carrots; Lee et al. (1980) were unable to detect PCB's in the carrots even though they applied sludge at 224 mt/ha and immediately grew the crops.

Another research effort centered on assessing risk from polycyclic aromatic hydrocarbons (PAH's). Some PAH's are carcinogenic (e.g., benzo(a)pyrene). Researchers found PAH's in composted municipal refuse, and that carrots roots (but not mushrooms) accumulated many PAH's from compost-amended soils (Muller, 1976; Linne and Martens, 1978; Wagner and Siddiqi, 1971; Siegfried, 1975; Siegfried and Muller, 1978; Neudecker, 1978; Ellwardt, 1977; Borneff et al., 1973). The level of 3,4-benzypyrene in carrot roots declined with successive cropping of compost amended soil. Multi-generation feeding studies of control and compost grown carrots found no risk to rats (Neudecker, 1978). Most of the PAH's in human diets result from deposition on plant foliage; root vegetables are a minor source.

Many other carcinogenic or toxic compounds may be present in wastes, and contaminate the food chain through plant uptake, volatile contamination of crop root or shoots, or soil ingestion. Very little information is available on these. Nitrosamines have been found in sewage wastes (Yoneyama, 1981; Green et al., 1981) and are accumulated from nutrient solution and soil by plants (Brewer, Draper, and Wey, 1980; Dean-Raymond and Alexander, 1976). However, nitrosamines appear to be rapidly degraded in soils and plants. Research on N-nitrosodimethylamine and N-nitrosodiethylamine found rapid degradation in soil; plant uptake did occur but these compounds were rapidly degraded there (Dressel, 1976a, 1976b). Traces of nitrosamines are found in nitroanaline based herbicides. These compounds are rapidly degraded and no detectable nitrosamine was found in soybean shoots (Kearney et al., 1980a). An IUPAC committee assessed the environmental consequences of these trace nitrosamines, and found no risk to the food chain (Kearney, et al., 1980b).

Aflatoxin comprises another useful example on fate of toxic organic compounds. Aflatoxin contaminated agricultural wastes are usually tilled into cropland. Aflatoxin is readily decomposed or transformed to nonextractable forms in soil, although detectable aflatoxin remained for about 50 days when 2 ppm was applied (Angle and Wagner, 1980). If present in nutrient solution or freshly amended soil, aflatoxin can be absorbed by corn or lettuce (Mertz et al., 1980; 1981). Thus, although it is possible for plants to absorb aflatoxin from aflatoxin amended land treatment sites, none would remain after closure and little would remain at the time of crop growth after preparing the soil for seeding.

Bioassay techniques for plant residues of soil-borne TO's include 1) field studies, 2) controlled pot studies, 3) vapor barrier pot studies, and 4) small pot intensive uptake (many plants to observe maximum possible uptake) techniques. Results from growing many plants in a small volume of pesticide treated soil (Fuhremann and Lichtenstein, 1980) provide valuable information on possible pathways of TO movement. However, little information obtained is relevant to actual environmental exposures.

Land treatment appears to be an effective method for destruction of mutagens present in sludges. Sewage sludge, feces, and some crop residues contain mutagenic activity (Hopke et al., 1982). Donnelly and Brown (1981), and Brown, Donnelly, and Scott (1982), have characterized reduction in concentration of mutagens during land treatment of petroleum refinery and other industrial sludges. Angle and Wagner (1980) reported biodegradation of aflatoxin, a potent mutagen, when it was added to soil. These studies indicate that mutagens in land-applied sludge should be rapidly degraded. Recently Babich et al. (1981) have voiced concern about TO's in land-applied sewage sludge. These concerns seem to rely on mismanagement of land treatment sites, and presume very high (unlawful under EPA, 1979) application rates followed by immediate cropping with food crops (usually considered prohibited under a 18 month waiting period to prevent pathogen contamination of foods). Boyd (1981) recently grew 4 vegetables (snapbeans, beets, cabbage, and squash) on a soil amended with 112 mt/ha sludge from Syracuse, N.Y.; this industrially contaminated sludge was applied in the Fall, and crops grown the next season. The edible portion of the sludge-grown and control crops were fed to rats at 25% of their diet. Mutagen assays were conducted on the crops and the rat urine; liver enzyme changes were followed; and alpha-fetoprotein (indicates pre-neoplastic transformations) was assayed. Weight gain was comparable from control and sludge-grown crops. No evidence of change in alpha-fetoprotein was observed in rats consuming the 4 sludge-grown vegetables. The liver mixed-function-oxidase enzymes (aminopyrine-N-demethylase and p-nitroanisole-O-demethylase) were affected by type of crop, but no additional changes were observed due to growing the vegetables on sludge-amended soils. No ultrastructural abnormalities were observed in rat liver cells as a consequence of sludge-grown vegetables. Rat urine may have shown increased amounts of mutagens when the urine extract from rats fed sludge-grown crops was activated with mammalian microsomes. The extracts of control and sludge-grown vegetables did not show significant "sludge effects" in normal or activated assays. Thus, although this topic has received little study to date, land treatment appears to provide sufficient biodegradation and adsorption to protect the food-chain from mutagenic compounds present in sludges applied to land under well managed programs. Food crops would not be grown during active land treatment periods.

Land treatment sites can be managed to avoid all unacceptable effects on the food-chain from waste-borne TO's. Wastes can be injected below the soil surface. Mechanically harvesting fresh forages or feed grain crops avoids soil contamination of food chain. Crops can be returned to the treatment site soil until plant absorbable TO's no longer reach unacceptable concentrations. Knowledge of the biodegradation and transport route for compounds in industrial wastes is required to prepare hazardous waste land treatment site management and closure plans (EPA, 1980b). And, if necessary, waste streams can be segregated and/or pretreatment used to remove TO's which are not acceptable in wastes subjected to land treatment.

Summary and Research Needs

Pathways to and potential effects on the food chain of macro- and microelements, and toxic organic compounds in wastes applied on land treatment sites are considered in detail. Wastes usually provide enough N, P, or K to be used as fertilizers. Wastes may cause nutrient imbalance if applied at excessive rates, or if all required nutrients are not supplied by the waste. Excess K (with or without adequate Mg) can cause magnesium deficiency (grass tetany) in cattle and sheep when wastes are used to produce forage grasses on poorly-drained soils. Management to avoid grass tetany is always needed, but wastes with high K/Mg require careful soil-crop-animal management.

Microelements can be placed into several groups based on their potential to cause health effects in livestock and humans. Waste-borne toxic chemicals can enter the food chain in 1) ingested wastes, wastes adhering to forages, or waste-amended soil, or 2) in plants which have increased levels of the toxic chemical when grown on waste-amended soil. Some waste-borne elements are unable to injure animals even when the waste is ingested (e.g., Ti, Cr^{+3}, Zr, Si, Sn, Au, Ag), either because animals tolerate high exposures or the element is bound very strongly by the ingested soil or waste. Other chemicals can injure animals which ingest the waste or soil, but not when animals ingest plants grown on the amended soil (Fe, Pb, Ni, Zn, Cu, Be, As and bioaccumulated toxic organic compounds), because phytotoxicity kills the plant before the metal concentration in the plant reaches toxic levels for the animal. On the other hand, risk may result from ingestion of forage crops containing phytotoxic levels of Co or excessive non-phytotoxic levels of Se, Mo, or Cd. Cumulative applications of only a few elements (Cd, Pb) have to be restricted to protect humans from the worst-case situations. These include 1) the long-term vegetable garden use scenario (chronic Cd-injury to kidney), or 2) the inadvertent soil ingestion scenario (Pb poisoning of children). If land treatment sites were not allowed to be converted to normal unregulated land use patterns until toxic organic compounds were degraded, management practices which avoided adverse food-chain effects could be developed for nearly any organic waste. Under those conditions it may be wise to avoid land treating on one site wastes which contain cationic toxic elements (Zn, Cd) best managed in alkaline soils and anionic toxic elements (Mo, Se) best managed in acidic soils.

Research Needs--

Some microelements have been insufficiently studied to provide adequate waste-soil-plant-animal management information. Too little is known about plant tolerance, plant uptake in relation to tolerance, and/or bioavailability of intrinsic plant Co to animals. Bioavailability information is needed for high levels of Zn, Cd, and Mn in vegetable crops. Bioavailability of microelements in ingested waste-amended soils has not been extensively studied. For Pb, this pathway is of overriding importance. It is also of interest, for Fe, As, Se, Zn, and Cu. Further research in this area is also needed for Zn, Cu, Ni, and Mn since few phytotoxicity studies have been conducted in the field. These are often present at substantial levels in wastes. Also, interactions among microelements, wastes, soils, plants, and animals need further study to develop management practices which prevent adverse effects on the food chain regardless of land use after closure.

REFERENCES

Adams, S. N., J. H. Honeysett, K. G. Tiller, and K. Norrish. 1969. Factors controlling the increase of cobalt in plants following the addition of a cobalt fertilizer. Aust. J. Soil Res. 7:29-42.

Adriano, D. C., M. Delaney, and D. Paine. 1977. Availability of cobalt-60 to corn and bean seedlings as influenced by soil type, lime, and DTPA. Commun. Soil Sci. Plant Anal. 8:615-628.

Alary, J., P. Bourbon, J. Esclassan, J. C. Lepert, J. Vandaele, J. M. Lecuire, and F. Klein. 1981. Environmental molybdenum levels in industrial molybdenosis of grazing cattle. Sci. Total Environ. 19:111-119.

Alexander, J., R. Koshut, R. Keefer, R. Singh, D. J. Horvath, and R. L. Chaney. 1979. Movement of nickel from sewage sludge into soil, soybeans, and voles. pp. 377-388. In D. D. Hemphill (ed.) Trace Substances in Environ. Health-12. Univ. Missouri, Columbia, MO.

Allaway, W. H. 1968. Agronomic controls over the environmental cycling of trace elements. Adv. Agron. 20:235-274.

Allaway, W. H. 1977a. Soil and plant aspects of the cycling of chromium, molybdenum, and selenium. Proc. Intern. Conf. Heavy Metals in the Environment 1:35-47.

Allaway, W. H. 1977b. Food chain aspects of the use of organic residues. pp. 282-298. In L. F. Elliott and F. J. Stevenson (eds.) Soils for Management of Organic Wastes and Wastewaters. Amer. Soc. Agron., Madison, WI.

Allaway, W. H. 1977c. Perspectives on molybdenum in soils and plants. pp. 317-339. In W. R. Chappell and K. K. Petersen (eds.) Molybdenum in the Environment. Vol. 2. Marcel Dekker, Inc., New York.

Amacher, M. C., and D. E. Baker. 1980. Kinetics of Cr oxidation by hydrous manganese oxides. Agron. Abstr. 1980:136.

Ambler, J. E., and J. C. Brown. 1969. Cause of differential susceptibility to zinc deficiency in two varieties of navy beans (Phaseolus vulgaris L.). Agron. J. 61:41-43.

Ammerman, C. B., S. M. Miller, K. R. Fick, and S. L. Hansard, III. 1977. Contaminating elements in mineral supplements and their potential toxicity: A review. J. Anim. Sci. 44:485-508.

Ammerman, C. B., J. M. Wing, B. G. Dunavant, W. K. Robertson, J. P. Feaster, and L. R. Arrington. 1967. Utilization of inorganic iron by ruminants as influenced by form of iron and iron status of the animal. J. Anim. Sci. 26:404-410.

Anderson, A. J., D. R. Meyer, and F. K. Mayer. 1973. Heavy metal toxicities: Levels of nickel, cobalt, and chromium in the soil and plants associated with visual symptoms and variation in growth of an oat crop. Aust. J. Agric. Res. 24:557-571.

Anderson, M. S., H. W. Lakin, K. C. Beeson, F. E. Smith, and E. Thacker. 1961. Selenium in Agriculture. USDA Agricultural Handbook 200. 65pp.

Angle, C. R., and M. S. McIntire. 1979. Environmental lead and children: The Omaha, Nebraska, USA Study. J. Toxicol. Environ. Health 5:855-870.

Angle, J. S. and G. H. Wagner. 1980. Decomposition of aflatoxin in soil. Soil Sci. Soc. Amer. J. 44:1237-1240.

Anke, M. 1973. Disorders due to copper deficiency in sheep and cattle. Monat. Vet. Med. 28:294-298.

Anonymous. 1967. Food additives permitted in the feed and drinking water of animals or for the treatment of food-producing animals. Fed. Reg. 32(157):11734.

Anonymous. 1968. Epidemic cardiac failure in beer drinkers. Nutr. Rev. 26:173-175.

Archer, A., and R. S. Barratt. 1976. Lead levels in Birmingham dust. Sci. Total Environ. 6:275-286.

Asher, C. J., G. W. Butler, and P. J. Peterson. 1977. Selenium transport in root systems of tomato. J. Exp. Bot. 28:279-291.

Aumaitre, A. 1981. Past and present situation in relation to the use of feed additives in diets for piglets; consequences of utilization of copper. pp. 16-41. In P. L'Hermite and J. Dehandtschutter (eds.) Copper in Animal Wastes and Sewage Sludge. Reidel Publ., Boston.

Austenfeld, F.-A. 1979. Effects of nickel, cobalt, and chromium on net photosynthesis of primary and secondary leaves of Phaseolus vulgaris cultivar 'Saxa' (in German). Photosynthetica 13:434-438.

Babich, J. G., D. J. Lisk, G. S. Stoewsand, and C. Wilkinson. 1981. Organic toxicants and pathogens in sewage sludge and their environmental effects. Cornell University Special Report No. 42. 5pp.

Baker, D. E. 1974. Copper: Soil, water, plant relationships. Fed. Proc. 33:1188-1193.

Baker, D. E., and L. Chesnin. 1976. Chemical monitoring of soils for environmental quality and animal and human health. Adv. Agron. 27:305-374.

Baker, E. L., Jr., D. S. Folland, T. A. Taylor, M. Frank, W. Peterson, G. Lovejoy, D. Cox, J. Housworth, and P. J. Landrigan. 1977. Lead poisoning in children of lead workers. Home contamination with industrial dust. New England J. Med. 296:260-261.

Barltrop, D., C. D. Strehlow, I. Thornton, and J. S. Webb. 1974. Significance of high soil lead concentrations for childhood lead burdens. Environ. Health Perspect. 7:75-82.

Bartlett, R. J., 1979. Oxidation-reduction status of aerated soils. Agron. Abstr. 1979:144.

Bartlett, R. J., and B. James. 1979. Behavior of chromium in soils: III. Oxidation. J. Environ. Qual. 8:31-35.

Bartlett, R. J., and J. M. Kimble. 1976a. Behavior of chromium in soils: I. Trivalent forms. J. Environ. Qual. 5:379-383.

Bartlett, R. J., and J. M. Kimble. 1976b. Behavior of chromium in soils: II. Hexavalent forms. J. Environ. Qual. 5:383-386.

Bates, T. E. 1971. Factors affecting critical nutrient concentrations in plants and their evaluation: A review. Soil Sci. 112:116-130.

Batey, T., C. Berryman, and C. Line. 1972. The disposal of copper-enriched pig-manure slurry on grassland. J. Br. Grassland Soc. 27:139-143.

Baxter, J. C., R. L. Chaney, and C. S. Kinlaw. 1974. Reversion of Zn and Cd in Sassafras sandy loam as measured by several extractants and by Swiss chard. Agron. Abstr. 1974:23.

Baxter, J. C., D. E. Johnson, and E. W. Kienholz. 1980. Uptake of trace metals and persistent organics into bovine tissues from sewage sludge - Denver Project. pp. 285-309. In G. Bitton et al. (eds.) Sludge - Health Risks of Land Application. Ann Arbor Sci. Publ., Inc., Ann Arbor, MI.

Beall, M. L., Jr., and R. G. Nash. 1971. Organochlorine insecticide residues in soybean plant tops: Root vs. vapor sorption. Agron. J. 63:460-464.

Becker, D. C., and S. E. Smith. 1951. The level of cobalt tolerance in yearling sheep. J. Anim. Sci. 10:266-271.

Beckett, P. 1981. Copper in sludge -- Are the toxic effects of copper and other heavy metals additive? pp. 204-222. In P. L'Hermite and J. Dehandtschutter (eds.) Copper in Animal Wastes and Sewage Sludge. Reidel Publ., Boston.

Beckett, P. H. T., and R. D. Davis. 1979. The disposal of sewage sludge onto farmland: the scope of the problem of toxic elements. Water Pollut. Contr. 78:419-445.

Benoy, C. J., P. A. Hooper, and R. Schneider. 1971. The toxicity of tin in canned fruit juices and solid foods. Food Cosmet. Toxicol. 9:645-656.

Benson, L. M., E. K. Porter, and P. J. Peterson. 1981. Arsenic accumulation, tolerance, and genotypic variation in plants on arsenical mine wastes in S.W. England. J. Plant Nutr. 3:655-666.

Bergh, A. K., and R. S. Peoples. 1977. Distribution of polychlorinated biphenyls in a municipal wastewater treatment plant and environs. Sci. Total Environ. 8:197-204.

Bertrand, J. E., M. C. Lutrick, H. L. Breland, and R. L. West. 1980. Effects of dried digested sludge and corn grown on soil treated with liquid sludge on performance, carcass quality, and tissue residues in beef cattle. J. Anim. Sci. 50:35-40.

Bertrand, J. E., M. C. Lutrick, G. T. edds, and R. L. West. 1981. Metal residues in tissues, animal performance, and carcass quality with beef steers grazing Pensacola bahiagrass pastures treated with liquid digested sludge. J. Anim. Sci. 53:146-153.

Beyer, W. N., R. L. Chaney, and B. W. Mulhern. 1982. Heavy metal concentrations in earthworms (Annelida: Oligochaeta) from soil amended with sewage sludge. J. Environ. Qual. 11: In press.

Beyer, W. N., and C. D. Gish. 1980. Persistence in earthworms and potential hazards to birds of soil applied DDT, dieldrin, and heptachlor. J. Appl. Ecol. 17:295-307.

Bingham, F. T. 1979. Bioavailability of Cd to food crops in relation to heavy metal content of sludge-amended soil. Environ. Health Perspect. 28:39-43.

Bingham, F. T. 1980. Nutritional imbalances and constraints to plant growth on salt-affected soils. Agron. Abstr. 1980:49.

Bingham, F. T., A. L. Page, R. J. Mahler, and T. J. Ganje. 1975. Growth and cadmium accumulation of plants grown on a soil treated with cadmium-enriched sewage sludge. J. Environ. Qual. 4:207-211.

Bingham, F. T., A. L. Page, R. J. Mahler, and T. J. Ganje. 1976a. Yield and cadmium accumulation of forage species in relation to cadmium content of sludge-amended soil. J. Environ. Qual. 5:57-60.

Bingham, F. T., A. L. Page, R. J. Mahler, and T. J. Ganje. 1976b. Cadmium availability to rice in sludge-amended soil under "flood" and "nonflood" culture. Soil Sci. Soc. Am. J. 40:715-719.

Bloomfield, C., and G. Pruden. 1980. The behavior of Cr (VI) in soil under aerobic and anaerobic conditions. Environ. Pollut. 23:103-114.

Boawn, L. C. 1971. Zinc accumulation characteristics of some leafy vegetables. Commun. Soil Sci. Plant Anal. 2:31-36.

Boawn, L. C. and P. E. Rasmussen. 1971. Crop response to excessive zinc fertilization of alkaline soil. Agron. J. 63:874-876.

Borneff, J., G. Farkasdi, H. Glathe, and H. Kunte. 1973. The fate of polycyclic aromatic hydrocarbons in experiments using sewage sludge-garbage composts as fertilizers. Zbl. Bakt. Hyg. I B, 157:151-164.

Boswell, F. C. 1975. Municipal sewage sludge and selected element applications to soil: Effect on soil and fescue. J. Environ. Qual. 4:267-273.

Bowen, H. J. M. 1966. Trace Elements in Biochemistry. Academic Press, New York. 241pp.

Bowen, H. J. M. 1979. Environmental Chemistry of the Elements. Academic Press, New York. 334pp.

Boyd, J. N. 1981. Changes in blood alpha-fetoprotein concentration in rats fed carcinogens and dietary modifiers of carcinogenesis. Ph.D. Thesis, Cornell Univ. 168pp.

Braude, G. L., C. F. Jelinek, and P. Corneliussen. 1975. FDA's overview of the potential health hazards associated with the land application of municipal wastewater sludges. pp. 214-217. <u>In</u> Proc. 1975 Natl. Conf. Municipal Sludge Management and Disposal. Information Transfer, Inc., Rockville, MD.

Breeze, V. G. 1973. Land reclamation and river pollution problems in the Croal Valley caused by waste from chromate manufacture. J. Appl. Ecol. 10:513-525.

Bremner, I. 1979. The toxicity of cadmium, zinc, and molybdenum and their effects on copper metabolism. Proc. Nutr. Soc. 38:235-242.

Bremner, I. 1981. Effects of the disposal of copper-rich slurry on the health of grazing animals. pp. 245-260. In P. L'Hermite and J. Dehandtschutter (eds.) Copper in Animal Wastes and Sewage Sludge. Reidel Publ., Boston.

Bremner, I., and J. K. Campbell. 1980. The influence of dietary copper intake on the toxicity of cadmium. Ann. N.Y. Acad. Sci. 355:319-332.

Bremner, I., B. W. Young, and C. F. Mills. 1976. Protective effect of zinc supplementation against copper toxicosis in sheep. Brit. J. Nutr. 36:551-561.

Brewer, W. S., A. C. Draper, III, and S. S. Wey. 1980. The detection of dimethylnitrosamine and diethylnitrosamine in municipal sewage sludge applied to agricultural soils. Environ. Pollut. 81:37-43.

Brooks, R. R. 1977. Copper and cobalt uptake by Haumaniastrum species. Plant Soil. 48:541-544.

Brooks, R. R., J. A. McCleave, and F. Malaise. 1977. Copper and cobalt in African species of Crotalaria L. Proc. R. Soc. Lond. B 197:231-236.

Brooks, R. R., R. S. Morrison, R. D. Reeves, T. R. Dudley, and Y. Akman. 1979. Hyperaccumulation of nickel by Alyssum Linnaeus (Cruciferae). Proc. Roy. Soc. Lond. B203:387-403.

Brown, J. R., and D. A. Sleper. 1980. Mineral concentration in two tall fescue genotypes grown under variable soil nutrient levels. Agron. J. 72:742-745.

Brown, K. W., K. C. Donnelly, and B. Scott. 1982. The fate of mutagenic compounds when hazardous wastes are land treated. pp. 383-397. In D. W. Schultz (ed.) Land Disposal of Hazardous Waste. EPA-600/9-82-002.

Brown, K. W., S. G. Jones, and K. C. Donnelly. 1980. The influence of simulated rainfall on residual bacteria and virus on grass treated with sewage sludge. J. Environ. Qual. 9:261-265.

Bunn, C. R., and G. Matrone. 1966. In vivo interactions of cadmium, copper, zinc and iron in the mouse and rat. J. Nutr. 90:395-399.

Burch, R. E., R. V. Williams, and J. F. Sullivan. 1973. Effect of cobalt, beer, and thiamin-deficient diets in pigs. Am. J. Clin. Nutr. 26:403-408.

Buxton, J. C., and R. Allcroft. 1955. Industrial molybdenosis of grazing cattle. Vet. Rec. 67:273-276.

Campbell, J. K., and C. F. Mills. 1979. The toxicity of zinc to pregnant sheep. Environ. Res. 20:1-13.

Cannon, H. L. 1963. The biogeochemistry of vanadium. Soil Sci. 96:196-204.

Cannon, H. L. 1964. Geochemistry of rocks and related soils and vegetation in the Yellow Cat area, Grand County, Utah. U. S. Geol. Surv. Bull. 1176. 127pp.

Carter, D. L., M. J. Brown, and C. W. Robbins. 1969. Selenium concentrations in alfalfa from several sources applied to a low selenium, alkaline soil. Soil Sci. Soc. Am. Proc. 33:715-718.

Cary, E. E., and W. H. Allaway. 1969. The stability of different forms of selenium applied to low-selenium soils. Soil Sci. Soc. Am. Proc. 33:571-574.

Cary, E. E., and W. H. Allaway. 1973. Selenium content of field crops grown on selenite-treated soils. Agron. J. 65:922-925.

Cary, E. E., W. H. Allaway, and O. E. Olson. 1977a. Control of chromium concentration in food plants. 1. Absorption and translocation of chromium by plants. J. Agric. Food Chem. 25:300-304.

Cary, E. E., W. H. Allaway, and O. E. Olson. 1977b. Control of chromium concentration in food plants. 2. Chemistry of chromium in soils and its availability to plants. J. Agr. Food Chem. 25:305-309.

CAST, 1980. Effects of sewage sludge on the cadmium and zinc content of crops. Council for Agricultural Science and Technology Report No. 83. Ames, IA. 77pp.

Cataldo, D. A., and R. E. Wildung. 1978. Soil and plant factors influencing the accumulation of heavy metals by plants. Environ. Health Perspect. 27:145-149.

Chaney, R. L. 1980. Health risks associated with toxic metals in municipal sludge. pp. 59-83. In G. Bitton et al. (eds.) Sludge -- Health Risks of Land Application. Ann Arbor Science Publishers, Inc., Ann Arbor, MI.

Chaney, R. L., and P. M. Giordano. 1977. Microelements as related to plant deficiencies and toxicities. pp. 234-279. In L. F. Elliot and F. J. Stevenson (eds.) Soils for Management of Organic Wastes and Waste Waters. American Society of Agronomy, Madison, WI.

Chaney, R. L., and S. B. Hornick. 1978. Accumulation and effects of cadmium on crops. pp. 125-140. In Proc. First International Cadmium Conference. Metals Bulletin Ltd., London.

Chaney, R. L., S. B. Hornick, and J. F. Parr. 1980. Impact of EPA regulations on utilization of sludge in agriculture. pp. 16-20. In Proc. Natl. Conf. Municipal and Industrial Sludge Utilization and Disposal. Information Transfer, Inc., Silver Spring, MD.

Chaney, R. L., and P. T. Hundemann. 1979. Use of peat moss columns to remove cadmium from wastewaters. J. Water Pollut. Contr. Fed. 51:17-21.

Chaney, R. L., P. T. Hundemann, W. T. Palmer, R. J. Small, M. C. White, and A. M. Decker. 1978a. Plant accumulation of heavy metals and phytotoxicity resulting from utilization of sewage sludge and sludge composts on cropland. pp. 86-97. In Proc. Natl. Conf. Composting Municipal Residues and Sludges. Information Transfer, Inc., Rockville, MD.

Chaney, R. L., O. A. Levander, S. O. Welsh, and H. W. Mielke. 1980. Effect of soil on availability of Pb to rats, and bioavailability of Pb in urban garden soils (unpublished results, USDA, Beltsville, MD.)

Chaney, R. L., and C. A. Lloyd. 1979. Adherence of spray-applied liquid digested sewage sludge to tall fescue. J. Environ. Qual. 8:407-411.

Chaney, R. L., G. S. Stoewsand, C. A. Bache, and D. J. Lisk. 1978b. Cadmium deposition and hepatic microsomal induction in mice fed lettuce grown on municipal sludge-amended soil. J. Agric. Food Chem. 26:992-994.

Chaney, R. L., G. S. Stoewsand, A. K. Furr, C. A. Bache, and D. J. Lisk. 1978c. Elemental content of tissues of guinea pigs fed Swiss chard grown on municipal sewage sludge-amended soil. J. Agr. Food Chem. 26:994-997.

Chaney, R. L., M. C. White, and M. v. Tienhoven. 1976. Interaction of Cd and Zn in phytotoxicity to and uptake by soybean. Agron. Abstr. 1976:21.

Chapman, H. D. 1966. Diagnostic criteria for plants and soils. Univ. Calif., Division Agr. Sci., Riverside. 793pp

Charney, E., J. W. Sayre, and M. Coulter. 1979. Increased lead adsorption in inner city children: Where does the lead come from? Proc. Second Intern. Symp. Environmental Lead Research. In press.

Chou, S. F., L. W. Jacobs, D. Penner, and J. M. Tiedje. 1978. Absence of plant uptake and translocation of polybrominated biphenyls (PBB's). Environ. Health Perspect 23:9-12.

Collett, J. N., and D. L. Harrison. 1968. Lindane residue on pasture and in the fat of sheep grazing pasture treated with lindane prills. N. Z. J. Agr. Res. 11:589-600.

Combs, G. F., Jr., S. A. Barrows, and F. N. Swader. 1980. Biologic availability of selenium in corn grain produced on soil amended with fly ash. J. Agr. Food Chem. 28:406-409.

Commission of the European Communities. 1978. Criteria (dose/effect relationships) for cadmium. Pergamon Press, N York. 202pp.

Comptroller General. 1978. Sewage sludge: How do we cope with it? 38pp. USGAO Report CED-78-152.

Cousins, R. J. 1979. Metallothionein synthesis and degradation: Relationship to cadmium metabolism. Environ. Health Perspect. 28:131-136.

Cox, D. H., and D. L. Harris. 1960. Effect of excess diet Zn on Fe and Cu in the rat. J. Nutr. 70:514-520.

Cunningham, J. D., D. R. Keeney, and J. A. Ryan. 1975. Yi and metal composition of corn and rye grown on sewage sludge-amended soil. J. Environ. Qual. 4:448-454.

Dacre, J. C. 1980. Potential health hazards of toxic orga residues in sludge. pp. 85-102. In G. Bitton et al. (eds. Sludge -- Health Risks of Land Application. Ann Arbor Scie Publ. Inc., Ann Arbor, MI.

Dacre, J. C., and G. L. Ter Harr. 1977. Lead levels in tissues from rats fed soils containing lead. Arch. Environ Contam. Toxicol. 6:111-119.

Dage, E., E. Dyckman, W. Isler, and F. Ogburn. 1979. Proc. Workshop on Alternatives for Cadmium Electroplating in Metal Finishing. EPA 560/2-79-003. 640pp.

Dalgarno, A. C., and C. F. Mills. 1975. Retention by sheep of copper from aerobic digests of pig faecal slurry. J. Agric. Sci. 85:11-18.

Dansky, L. M., and F. W. Hill. 1952. Application of the chromic oxide indicator method to balance studies with growing chickens. J. Nutr. 47:449-459.

David, O. J., S. Hoffman, and B. Kagey. 1978. The sub-clinical effects of lead on children. pp. 29-40. In Lead Pollution -- Health Effects. The Conservation Society Pollution Working Party, London.

Davies, B. E. 1978. Plant-available lead and other metals in British garden soils. Sci. Total Environ. 9:243-262.

Davies, B. E., and R. C. Ginnever. 1979. Trace metal contamination of soils and vegetables in Shipham, Somerset. J. Agric. Sci. 93:753-756.

Davis, G. K. 1974. High-level copper feeding of swine and poultry and the ecology. Fed. Proc. 33:1194-1196.

Davis, R. D. 1980. Uptake of fluoride by ryegrass grown in soil treated with sewage sludge. Environ. Pollut. B1:277-284.

Davis, R. D. 1981. Copper uptake from soil treated with sewage sludge and its implications for plant and animal health. pp. 223-241. In P. L'Hermite and J. Dehandtschutter (eds.) Copper in Animal Wastes and Sewage Sludge. Reidel Publ., Boston.

Day, J. P., J. E. Fergusson, and T. M. Chee. 1979. Solubility and potential toxicity of lead in urban street dust. Bull. Environ. Contam. Toxicol. 23:497-502.

Dean-Raymond, D., and M. Alexander. 1976. Plant uptake and leaching of dimethylnitrosamines. Nature 262:394-396.

Decker, A. M., R. L. Chaney, J. P. Davidson, T. S. Rumsey, S. B. Mohanty, and R. C. Hammond. 1980. Animal performance on pastures topdressed with liquid sewage sludge and sludge compost. pp. 37-41. In Proc. Natl. Conf. Municipal and Industrial Sludge Utilization and Disposal. Information Transfer, Inc., Silver Spring, MD.

deGroot, A. P. 1973. Subacute toxicity of inorganic tin as influenced by dietary levels of iron and copper. Food Cosmet. Toxicol. 11:955-962.

deGroot, A. P., V. J. Feron, and H. P. Til. 1973. Short term toxicity studies of some salts and oxides of tin in rats. Food Cosmet. Toxicol. 11:19-30.

DeKock, P. C., M. V. Cheshire, and A. Hall. 1971. Comparison of the effect of phosphorus and nitrogen on copper-deficient and -sufficient oats. J. Sci. Food Agric. 22:437-440.

Deuel, L. E., and A. R. Swoboda. 1972. Arsenic solubility in a reduced environment. Soil Sci. Soc. Amer. Proc. 36:276-278.

Dilworth, B. C., and E. J. Day. 1970. Hydrolyzed leather meal in chick diets. Poult. Sci. 49:1090-1093.

Doll, E. C., and R. E. Lucas. 1973. Testing soils for potassium, calcium, and magnesium. pp. 133-151. In L. M. Walsh and J. D. Keaton (eds.) Soil Testing and Plant Analysis, Rev. Ed., Soil Sci. Soc. Am. Madison, WI.

Donnelly, K. C., and K. W. Brown. 1981. The development of laboratory and field studies to determine the fate of mutagenic compounds from land applied hazardous wastes. pp. 224-239. In Proc. Seventh Ann. Res. Symp. Land Disposal: Hazardous Waste. EPA-600/9-81-0026.

Doran, J. W., and D. C. Martens. 1972. Molybdenum availability as influenced by applications of fly ash to soil. J. Environ. Qual. 1:186-189.

Dowdy, R. H., C. E. Clapp, D. R. Duncomb, and W. E. Larson. 1980. Water quality of snowmelt runoff from sloping land receiving annual sewage sludge applications. pp. 11-15. In Proc. Nat. Conf. Municipal and Industrial Sludge Utilization and Disposal. Information Transfer, Inc., Silver Spring, MD.

Dowdy, R. H., C. E. Clapp, W. E. Larson, and D. R. Duncomb. 1979. Runoff and soil water quality as influenced by five years of sludge applications on a terraced watershed. Agron. Abstr. 1979:28.

Dowdy, R. H., and G. E. Ham. 1977. Soybean growth and elemental content as influenced by soil amendments of sewage sludge and heavy metals: Seedling studies. Agron. J. 69:300-303.

Dowdy, R. H., and W. E. Larson. 1975. The availability of sludge-borne metals to various vegetable crops. J. Environ. Qual. 4:278-282.

Doyle, J. J. 1977. Effects of low levels of dietary cadmium in animals - a review. J. Environ. Qual. 6:111-116.

Dressel, J. 1976a. Dependence of nitrogen containing constituents which influence plant quality by the intensity of fertilization. Landwirtsch. Forsch. Sondern. 33:326-334.

Dressel, J. 1976b. Relationship between nitrate, nitrite and nitrosamimes in plants and soil at intensive nitrogen fertilization. Qual. Plant. Plant Foods Hum. Nutr. 25:381-390.

Dunn, K. M., R. E. Ely, and C. F. Huffman. 1952. Alleviation of cobalt toxicity in calves by methionine administration. J. Anim. Sci. 11:327-331.

Edds, G. T., O. Osuna, C. F. Simpson, J. E. Bertrand, D. L. Hammell, C. E. White, B. L. Damron, R. L. Shirley, and K. C. Kelley. 1980. Health effects of sewage sludge for plant production or direct feeding to cattle, swine, poultry, or animal tissue to mice. pp. 311-325. In G. Bitton et al. (eds.) Sludge -- Health Risks of Land Application. Ann Arbor Sci. Publ. Inc., Ann Arbor, MI.

Edwards, W. C., and B. R. Clay. 1977. Reclamation of range-land following a lead poisoning incident in livestock from industrial airborne contamination of forage. Vet. Human Toxicol. 19:247-249.

Egan, D. A., and T. O'Cuill. 1970. Cumulative lead poisoning in horses in a mining area contaminated with Galena. Vet. Rec. 86:736-738.

El-Bassam, N., P. Poelstra, and M. J. Frissel. 1975. Chromium and mercury in a soil after 80 years of treatment with urban sewage water. (In German). Z. Pflanzenern. Bodenk. 1975:309-316.

Elinder, C. G., T. Kjellstrom, L. Friberg, B. Lind, and L. Linmann. 1976. Cadmium in kidney cortex, liver, and pancreas from Swedish autopsies. Arch. Environ. Health 28:292-302.

Elkins, C. B., R. L. Haaland, C. S. Hoveland, and W. A. Griffey. 1978. Grass tetany potential of tall fescue as affected by soil O_2. Agron. J. 70:309-311.

Ellwardt, P.-C. 1977. Variation in content of polycyclic aromatic hydrocarbons in soil and plants by using municipal waste composts in agriculture. pp. 291-298. In Soil Organic Matter Studies. Vol II. Int. Atomic Energy Agency, Vienna.

E.P.A. 1979. Criteria for classification of solid waste disposal facilities and practices. Federal Register 44(179):53438-53464.

E.P.A. 1979b. Background Document: Cumulative cadmium application rates. 52pp. Docket 4004, EPA, Washington, D.C.

E.P.A. 1980a. Hazardous Waste Management System: Identification and listing of hazardous waste. Federal Register 45(98):33084-33133.

E.P.A. 1980b. Hazardous waste management system: Standards for owners and operators of hazardous waste treatment, storage, and disposal facilities. Federal Register 45(98):33154-33258.

EPA-FDA-USDA. 1981. Land application of municipal sewage sludge for the production of fruits and vegetables. A statement of federal policy and guidance. U.S. Environmental Agency Joint Policy Statement SW-905. 21pp.

Fairbanks, B. C., and G. A. O'Connor. 1980. Adsorption of polychlorinated biphenyls (PCB's) by sewage sludge amended soil. (Abstract) p. 346. In G. Bitton et al. (eds.) Sludge - Health Risks of Land Application. Ann Arbor Sci. Publ. Inc., Ann Arbor, MI.

Fergusson, J. E., K. A. Hibbard, and R. L. H. Ting. 1981. Lead in human hair: General survey -- Battery factory employees and their families. Environ. Pollut. B2:235-248.

Field, A. C., and D. Purves. 1964. The intake of soil by the grazing sheep. Proc. Nutr. Soc. (London) 23:24-25.

Filonow, A. B., L. W. Jacobs, and M. M. Mortland. 1976. Fate of polybrominated biphenyls (PBB's) in soils. Retention of hexabromobiphenyl in four Michigan soils. J. Agr. Food Chem. 24:1201-1204.

Fitzgerald, P. R. 1978. Toxicology of heavy metals in sludges applied to the land. pp. 106-116. In Proc. Fifth Natl. Conf. Acceptable Sludge Disposal Techniques. Information Transfer, Inc., Rockville, Md.

Fitzgerald, P. R. 1980. Observations on the health of some animals exposed to anaerobically digested sludge originating in the Metropolitan Sanitary District of Greater Chicago system. pp. 267-284. In G. Bitton et al. (eds.) Sludge - Health Risks of Land Application. Ann Arbor Sci. Publ., Inc., Ann Arbor, MI.

Flanagan, P. R., J. S. McLellan, J. Haist, M. G. Cherian, M. J. Chamberlain, and L. S. Valberg. 1978. Increased dietary cadmium absorption in mice and human subjects with iron deficiency. Gastroenterol. 74:841-846.

Follett, R. F., J. F. Power, D. L. Grunes, D. A. Hewes, and H. F. Mayland. 1975. Potential tetany hazard of N-fertilized bromegrass as indicated by chemical composition. Agron. J. 67:819-824.

Food and Drug Administration. 1977. FY 1974 Total Diet Studies - 7320.08. Bureau of Foods, FDA, Washington, D.C.

Food and Drug Administration. 1979. Polychlorinated biphenyls (PCB's); Reduction of tolerances. Fed. Reg. 44:38330-38340.

Fox, C. J. S., D. Chisholm, and D. K. R. Stewart. 1964. Effect of consecutive treatments of aldrin and heptachlor on residues in rutabagas and carrots and on certain soil arthropods and yield. Can. J. Plant Sci. 44:149-156.

Fox, M. R. S. 1974. Effect of esential minerals on cadmium toxicity. A review. J. Food Sci. 39:321-324.

Fox, M. R. S. 1976. Cadmium metabolism - A review of aspects pertinent to evaluating dietary cadmium intake by man. p. 401. In A. S. Prasad and D. Oberleas (eds.) Trace Elements in Human Health and Disease. Vol. 2. Essential and Toxic Elements. Academic Press, New York.

Fox, M. R. S. 1979. Nutritional influences on metal toxicity: Cadmium as a model toxic element. Environ. Health Perspect. 29:95-104.

Fox, M. R. S., R. M. Jacobs, A. O. L. Jones, and B. E. Fry, Jr. 1979. Effects of nutritional factors on metabolism of dietary cadmium at levels similar to those of man. Environ. Health Perspect. 28:107-114.

Fox, M. R. S., R. M. Jacobs, A. O. L. Jones, B. E. Fry, Jr., and R. P. Hamilton. 1978. Indices for assessing cadmium bioavailability from human foods. In M. Kirchgessner (ed.) Trace Element Metabolism in Man and Animals 3:327-331.

Foy, C. D., R. L. Chaney, and M. C. White. 1978. The physiology of metal toxicity in plants. Ann. Rev. Plant Physiol. 29:511-566.

Frank, R., H. E. Braun, M. Holdrinet, K. I. Stonefield, J. M. Elliot, B. Zilkey, L. Vickery, and H. H. Cheng. 1977. Metal contents and insecticide residues in tobacco soils and cured tobacco leaves collected in Southern Ontario. Tob. Sci. 21:74-80.

Franzke, C., G. Ruick, and M. Schmidt. 1977. Investigations on the heavy metal content of tobacco products and tobacco smoke (in German.) Nahrung 21:417-428.

Friberg, L., M. Piscator, G. Nordberg, and T. Kjellstrom. 1974. Cadmium in the Environment. 2nd Ed. CRC Press, Cleveland, OH. 248pp.

Fries, G. F. 1982. Potential polychlorinated biphenyl residues in animal products from application of contaminated sewage sludge to land. J. Environ. Qual. 11:14-20.

Fries, G. F., and G. S. Marrow. 1981. Chlorobiphenyl movement from soil to soybean plants. J. Agr. Food Chem. 29:757-759.

Fries, C. F., G. S. Marrow, and P. A. Snow. 1982. Soil ingestion by dairy cattle. J. Dairy Sci. 65: (In Press).

Fuhremann, T. W., and E. P. Lichtenstein. 1980. A comparative study of the persistence, movement and metabolism of six carbon-14 insecticides in soils and plants. J. Agr. Food Chem. 28:446-452.

Fujimoto, G., and G. D. Sherman. 1950. Cobalt content of typical soils and plants of the Hawaiian Islands. Agron. J. 42:577-581.

Fulkerson, W., and H. E. Goeller. 1973. Cadmium - The Dissipated Element. NTIS No. ORNL-NSF-EP-21. 473pp.

Furr, A. K., W. C. Kelly, C. A. Bache, W. H. Gutemann, and D. J. Lisk. 1976a. Multi-element absorption by crops grown in pots on municipal sludge-amended soil. J. Agric. Food Chem. 24:889-892.

Furr, A. K., W. C. Kelly, C. A. Bache, W. H. Gutenmann, and D. J. Lisk. 1976b. Multi-element absorption by crops grown on Ithaca sludge-amended soil. Bull. Environ. Contam. Toxicol. 16:756-763.

Furr, A. K., T. F. Parkinson, C. L. Heffron, J. T. Reid, W. M. Haschek, W. H. Gutenmann, C. A. Bache, L. E. St. John, Jr., and D. J. Lisk. 1978a. Elemental content of tissues and excreta of lambs, goats, and kids fed white sweet clover growing on fly ash. J. Agr. Food Chem. 26:847-851.

Furr, A. K., T. F. Parkinson, C. L. Heffron, J. T. Reid, W. M. Haschek, W. H. Gutenmann, I. S. Pakkala, and D. J. Lisk. 1978b. Elemental content of tissues of sheep fed rations containing coal fly ash. J. Agr. Food Chem. 26:1271-1274.

Furr, A. K., T. F. Parkinson, R. A. Hinrichs, D. R. VanCampen, C. A. Bache, W. H. Gutenmann, L. E. St. John, Jr., I. S. Pakkala, and D. J. Lisk. 1977. National survey of elements and radioactivity in fly ashes. Absorption of elements by cabbage grown in fly ash-soil mixtures. Environ. Sci. Technol. 11:1194-1201.

Furr, A. K., G. S. Stoewsand, C. A. Bache, W. A. Gutenmann, and D. J. Lisk. 1975. Multielement residues in tissues of guinea pigs fed sweet clover grown on fly ash. Arch. Environ. Health. 30:244-248.

Galke, W. A., D. I. Hammer, J. E. Keil, and S. W. Lawrence. 1977. Environmental determinants of lead burdens in children. Proc. Int. Conf. Heavy Metals in the Environment. III:53-74.

Gardner, A. W., and P. K. Hall-Patch. 1962. An outbreak of industrial molybdenosis. Vet. Rec. 74:113-116.

Gardner, A. W., and P. K. Hall-Patch. 1968. Molybdenosis in cattle grazing downwind from an oil refinery unit. Vet Rec. 82:86-87.

Geering, H. R., E. E. Cary, L. H. P. Jones, and W. H. Allaway. 1968. Solubility and redox criteria for the possible forms of selenium in soils. Soil Sci. Soc. Am. Proc. 32:35-40.

Gemmell, R. P. 1973. Revegetation of derelict land polluted by a chromate smelter. 1: Chemical factors causing substrate toxicity in chromate smelter waste. Environ. Pollut. 5:181-197.

Gemmell, R. P. 1974. Revegetation of derelict land polluted by a chromate smelter. 2: Techniques of revegetation of chromate smelter waste. Environ. Pollut. 6:31-37.

Giguere, C. G., A. B. Howes, M. McBean, W. N. Watson, and L. E. Witherell. 1977. Increased lead absorption in children of lead workers--Vermont. Morbidity Mortality Weekly Report 26(8):61-62.

Gilpin, L., and A. H. Johnson. 1980. Fluorine in agricultural soils of Southeastern Pennsylvania. Soil Sci. Soc. Amer. J. 44:255-258.

Giordano, P. M., D. A. Mays, and A. D. Behel, Jr. 1979. Soil temperature effects on uptake of cadmium and zinc by vegetables grown on sludge-amended soil. J. Environ. Qual. 8:233-236.

Grant-Frost, D. B., and E. J. Underwood. 1958. Zn toxicity in the rat and its interrelation with Cu. Aust. J. Exp. Biol. Med. Sci. 36:339-346.

Green, S., M. Alexander, and D. Leggett. 1981. Formation of N-nitrosodimethylamine during treatment of municipal waste water by simulated land application. J. Environ. Qual. 10:416-421.

Grove, J. H., and B. G. Ellis. 1980. Extractable chromium as related to soil pH and applied chromium. Soil Sci. Soc. Am. J. 44:238-242.

Grun, M., M. Anke, A. Hennig, W. Seffner, M. Partshefeld. G. Flachowsky, and B. Groppel. 1978. Excessive oral iron application to sheep. 2. The influence on the level of iron, copper, zinc, and manganese in different organs (in German). Arch. Tierernahr. 28:341-347.

Grunes, D. L. 1973. Grass tetany of cattle and sheep. pp. 113-140. In A. G. Matches (ed.) Anti-quality Components of Forages. Spec. Publ. 4, Crop Sci. Soc. Am., Madison, WI.

Grunes, D. L., P. R. Stout, and J. R. Brownell. 1970. Grass tetany of ruminants. Adv. Agron. 22:331-374.

Gunson, D. E., D. F. Kowalczyk, C. R. Shoop, and C. F. Ramberg, Jr. 1982. Environmental zinc and cadmium pollution associated with generalized osteochrondrosis, osteoporosis, and nephrocalcinosis in horses. J. Am. Vet. Med. Assoc. 180:295-299.

Gupta, S., and H. Haeni. 1981. Effect of copper supplied in the form of different Cu-saturated sludge samples and copper salts on the Cu-concentration and dry matter yeld of corn grown in sand. pp. 287-304. In P. L'Hermite and J. Dehandtschutter (eds.) Copper in Animal Wastes and Sewage Sludge. Reidel Publ., Boston.

Gupta, U. C., E. W. Chipman, and D. C. Mackay. 1978. Effects of molybdenum and lime on the yield and molybdenum concentration of vegetable crops grown on acid sphagnum peat soil. Can. J. Plant. Sci. 58:983-992.

Gurnham, C. F., B. A. Rose, H. R. Ritchie, W. T. Fetherston, and A. W. Smith. 1979. Control of heavy metal content of municipal wastewater sludge. Report to the National Science Foundation. (NTIS)-PB-295917. 144 pp.

Gutenmann, W. H., C. A. Bache, W. D. Youngs, and D. J. Lisk. 1976. Selenium in fly ash. Science 191:966-967.

Gutenmann, W. H., I. S. Pakkala, D. J. Churey, W. C. Kelly, and D. J. Lisk. 1979. Arsenic, boron, molybdenum, and selenium in successive cuttings of forage crops field grown on fly ash amended soil. J. Agr. Food Chem. 27:1393-1395.

Haaland, R. L., C. B. Elkins, and C. S. Hoveland. 1978. A method for detecting genetic variability for grass tetany potential in tall fescue. Crop. Sci. 18:339-340.

Hafez, A. A. R., H. M. Reisenauer, and P. R. Stout. 1979. The solubility and plant uptake of chromium from heated soils. Commun. Soil Sci. Plant Anal. 10:1261-1270.

Haghiri, F. 1974. Plant uptake of cadmium as influenced by cation exchange capacity, organic matter, zinc, and soil temperature. J. Environ. Qual. 3:180-183.

Halstead, R. L., B. J. Finn, and A. J. MacLean. 1969. Extractability of nickel added to soils and its concentration in plants. Can. J. Soil. Sci. 49:335-342.

Hambridge, K. M., C. Hambridge, M. Jacobs, and J. D. Baum. 1972. Low levels of zinc in hair, anorexia, poor growth, and hypogeusia in children. Pediat. Res. 6:868:874.

Hamilton, J. W., and O. A. Beath. 1963. Selenium uptake and conversion by certain crop plants. Agron. J. 55:528-531.

Hammond, P. B., C. S. Clark, P. S. Gartside, O. Berger, A. Walker, and L. W. Michael. 1980. Fecal lead excretion in young children as related to sources of lead in their environments. Intern. Arch. Occup. Environ. Health 46:191-202.

Hammons, A. S., J. E. Huff, H. M. Braunstein, J. S. Drury, C. R. Shriner, E. B. Lewis, B. L. Whitfield, and L. E. Towill. 1978. Reviews of the environmental effects of pollutants: IV. Cadmium. EPA-600/1-78-026.

Hansen, L. G., and T. D. Hinesly. 1979. Cadmium from soil amended with sewage sludge: Effects and residues in swine. Environ. Health Perspect. 28:51-58.

Hansen, L. G., P. K. Washko, L. G. M. T. Tuinstra, S. B. Dorn, and T. D. Hinesly. 1981. Polychlorinated biphenyl, pesticide, and heavy metal residues in swine foraging on sewage sludge amended soils. J. Agr. Food Chem. 29:1012-1017.

Harbourne, J. F., C. T. McCrea, and J. Watkinson. 1968. An unusual outbreak of lead poisoning in calves. Vet. Rec. 83:515-517.

Harrison, D. L., J. C. M. Mol, and W. B. Healy. 1970. DDT residues in sheep from the ingestion of soil. N. Z. J. Agr. Res. 13:664-672.

Harrison, D. L., J. C. M. Mol, and J. E. Rudman. 1969. DDT and lindane: New aspects of stock residues derived from a farm environment. N. Z. J. Agr. Res. 12:553-574.

Hartmans, J., and M. S. M. Bosman. 1970. Differences in the copper status of grazing and housed cattle and their biochemical backgrounds. pp. 362-366. In C. F. Mills (ed.) Trace Element Metabolism in Animals - 1. Livingstone, Edinburgh.

Haye, S. N., D. J. Horvath, O. L. Bennett, and R. Singh. 1976. A model of seasonal increase of lead in a food chain. pp. 387-393. In D. D. Hemphill (ed.) Trace Substances in Environmental Health - 9. Univ. Missouri, Columbia, MO.

Healy, W. B. 1968. Ingestion of soil by dairy cows. N.Z. J. Agr. Res. 11:487-499.

Healy, W. B., and T. G. Ludwig. 1965. Wear of sheeps' teeth. 1. The role of ingested soil. N. Z. J. Agr. Res. 8:737-752.

Healy, W. B., P. C. Rankin, and H. M. Watts. 1974. Effect of soil contamination on the element composition of herbage. N. Z. J. Agr. Res. 17:59-61.

Helmke, P. A., W. P. Robarge, R. L. Korotev, and P. J. Schomberg. 1979. Effects of soil-applied sewage sludge on concentrations of elements in earthworms. J. Environ. Qual. 8:322-327.

Hermanson. H. P., L. D. Anderson, and F. A. Gunther. 1970. Effects of variety and maturity of carrots upon uptake of endrin residues from soil. J. Econ. Entomol. 63:1651-1654.

Hill, R. R., Jr., and S. B. Guss. 1976. Genetic variation for mineral concentration in plants related to mineral requirements of cattle. Crop Sci. 16:680-685.

Hill, R. R., Jr., and G. A. Jung. 1975. Genetic variability for chemical composition of alfalfa. I. Mineral elements. Crop Sci. 15:652-657.

Hill, C. H., G. Matrone, W. L. Payne, and C. W. Barber. 1963. In vivo interactions of cadmium with copper, zinc, and iron. J. Nutr. 80:227-235.

Hinesly, T. D., L. G. Hansen, E. L. Ziegler, and G. L. Barrett. 1979. Effects of feeding corn grain produced on sludge-amended soil to pheasants and swine. pp. 483-495. In W. E. Sopper and S. N. Kerr (eds.) Utilization of Municipal Sewage Effluent and Sludge on Forest and Disturbed Land. Pennsylvania State University Press, University Park, PA.

Hinesly, T. D., E. L. Ziegler, and J. J. Tyler. 1976. Selected chemical elements in tissues of pheasants fed corn grain from sewage sludge-amended soil. Agro-Ecosystems 3:11-26.

Hogan, G. D., and W. E. Rauser. 1979. Tolerance and toxicity of cobalt, copper, nickel, and zinc in clones of Agrostis gigantea from a mine waste site. New Phytol. 83:665-670.

Hogue, D. E., J. T. Reid, C. L. Heffron, W. H. Gutenmann, and D. J. Lisk. 1980. Soft coal fly ash as a source of selenium in unpelleted sheep rations. Cornell Vet. 70:67-71.

Hopke, P. K., M. J. Plewa, J. B. Johnson, D. Weaver, S. G. Wood, R. A. Larson, and T. Hinesly. 1982. Multitechnique screening of Chicago municipal sewage sludge for mutagenic activity. Environ. Sci. Technol. 16:140-147.

Hornick, S. B., D. E. Baker, and S. B. Guss. 1977. Crop production and animal health problems associated with high soil molybdenum. pp. 665-684. In W. R. Chappell and K. K. Peterson (eds.) Molybdenum in the Environment. Vol. 2. Marcell Dekker, New York.

Huck, D. W., and A. J. Clawson. 1976. Excess dietary cobalt in pigs. J. Anim. Sci. 43:1231-1246.

Huffman, E. W. D., Jr. and W. H. Allaway. 1973. Growth of plants in solution culture containing low levels of chromium. Plant Physiol. 52:72-75.

Hunter, J. G., and O. Vergnano. 1952. Nickel toxicity in plants. Ann Appl. Biol. 39:279-284.

Hunter, J. G., and O. Vergnano. 1953. Trace-element toxicities in oat plants. Ann. Appl. Biol. 40:761-777.

Irwin, M. I., and E. W. Crampton. 1951. The use of chromic oxide as an index material in digestion trials with human subjects. J. Nutr. 43:77-85.

Iwata, Y., and F. A. Gunther. 1976. Translocation of the polychlorinated biphenyl Arochlor 1254 from soil into carrots under field conditions. Arch. Environ. Contam. Toxicol. 4:44-59.

Iwata, Y., F. A. Gunther, and W. E. Westlake. 1974. Uptake of a PCB (Arochlor 1254) from soil by carrots under field conditions. Bull. Environ. Contam. Toxicol. 11:523-52 .

Jackson, W. A., R. A. Leonard, and S. R. Wilkinson. 1975. Land disposal of broiler litter - changes in soil potassium, calcium, and magnesium. J. Environ. Qual. 4:202-206.

Jacobs, L. W., S. F. Chou, and J. M. Tiedje. 1976. Fate of polybrominated biphenyls (PBB's) in soils. Persistence and plant uptake. J. Agr. Food Chem. 24:1198-1201.

Jacobs, R. M., A. O. L. Jones, M. R. S. Fox, and B. E. Fry, Jr. 1978a. Retention of dietary cadmium and the ameliorative effect of zinc, copper, and manganese in Japanese quail. J.Nutr. 108:22-32.

Jacobs, R. M., A. O. L. Jones, B. E. Fry, Jr., and M. R. S. Fox. 1978b. Decreased long-term retention of ^{115m}Cd in Japanese quail produced by a combined supplement of zinc, copper, and manganese. J. Nutr. 108:901-910.

Jacobs, R. M., A. O. L. Jones, M. N. Morra, and M. R. S. Fox. 1978c. Zinc antagonism of cadmium. (Abstract). Fed. Proc. 37:405.

Jaffre, T., R. R. Brooks, J. Lee, and R. D. Reeves. 1976. Sebertia acuminata: a hyperaccumulator of nickel from New Caledonia. Science 193:579-580.

Jarrell, W. M., A. L. Page, and A. A. Elseewi. 1980. Molybdenum in the environment. Residue Rev. 74:1-43.

Jelinek, C. F., and G. L. Braude. 1978. Management of sludge use on land. J. Food Prot. 41:476-480.

Jensen, E. H. and A. L. Lesperance. 1971. Molybdenum accumulation by forage plants. Agron. J. 63:201-204.

Johnson, D. E., E. W. Kienholz, J. C. Baxter, E. Spanger, and G. M. Ward. 1981. Heavy metal retention in tissues of cattle fed high cadmium sewage sludge. J. Anim. Sci. 52:108-114.

Johnson, M. S., T. McNeilly, and P. D. Putwain. 1977. Revegetation of metalliferous mine spoil contaminated by lead and zinc. Environ. Pollut. 12:261-277.

Johnson, W. R., and J. Proctor. 1977. A comparative study of metal levels in plants from two contrasting lead-mine sites. Plant Soil 46:251-257.

Jones, S. G., K. W. Brown, L. E. Deuel, and K. C. Donnelly. 1979. Influence of rainfall on the retention of sludge heavy metals by the leaves of forage crops. J. Environ. Qual. 8:69-72.

Jordan, L. D., and D. J. Hogan. 1975. Survey of lead in Christchurch soils. N. Z. J. Sci. 18:253-260.

Karlen, D. L., R. Ellis, Jr., D. A. Whitney, and D. L. Grunes. 1978. Influence of soil moisture and plant cultivar on cation uptake by wheat with respect to grass tetany. Agron. J. 70:918-921.

Karlen, D. L., R. Ellis, Jr., D. A. Whitney, and D. L. Grunes. 1980a. Soil and plant parameters associated with grass tetany of cattle in Kansas. Agron. J. 72:61-65.

Karlen, D. L., R. Ellis, Jr., D. A. Whitney, and D. L. Grunes. 1980b. Influence of soil moisture on soil solution cation concentrations and the tetany potential of wheat forage. Agron. J. 72:73-78.

Kazantzis, G. 1979. Renal tubular dysfunction and abnormalities of calcium metabolism in cadmium workers. Environ. Health Perspect. 28:115-159.

Kearney, P. C., J. E. Oliver, A. Kontson, W. Fiddler, and J. W. Pensabene. 1980a. Plant uptake of dinitroaniline herbicide-related nitrosamines. J. Agric. Food Chem. 28:633-635.

Kearney, P. C., M. E. Amundson, K. I. Beynon, N. Drescher, G. J. Marco, J. Miyamoto, J. R. Murphy, and J. E. Oliver. 1980b. Nitrosamines and pesticides. A special report on the occurrence of nitrosamines as terminal residues resulting from agricultural use of certain pesticides. Pure and Appl. Chem. 52:499-526.

Keener, H. A., R. R. Baldwin, and G. P. Percival. 1951. Cobalt metabolism studies with sheep. J. Anim. Sci. 10:428-433.

Keener, H. A., G. P. Percival, K. S. Morrow, and G. H. Ellis. 1949. Cobalt tolerance in young dairy cattle. J. Dairy Sci. 32:527-533.

Kick, H., and B. Braun. 1977. The effect of chromium containing tannery sludges on the growth and the uptake of chromium by different crops (in German). Landwirtsch. Forsch. 30:160-173.

Kick, H., and J. Warnusz. 1972. The uptake of tin by crop plants after the use of sewage sludge (in German). Landwirtsch. Forsch. 27:207-211.

Kienholz, E. W. 1980. Effect of toxic chemicals present in sewage sludge on animal health. pp. 153-171. In G. Bitton et al. (eds.) Sludge--Health Risks of Land Application. Ann Arbor Science Publishers, Inc., Ann Arbor, MI.

Kienholz, E. W., G. M. Ward, D. E. Johnson, J. Baxter, G. Braude, and G. Stern. 1979. Metropolitan Denver sewage sludge fed to feedlot steers. J. Anim. Sci. 48:734-741.

Kjellstrom, T. 1978. Comparative study of Itai-Itai disease. pp. 224-231. In Proc. First International Cadmium Conference. Metals Bulletin Ltd., London.

Kjellstrom, T., L. Friberg, and B. Rahnster. 1979. Mortality and cancer morbidity among cadmium-exposed workers. Environ. Health Perspect. 28:199-204.

Kjellstrom, T., and G. F. Nordberg. 1978. A kinetic model of cadmium metabolism in the human being. Environ. Res. 16:248-269.

Kjellstrom, T., K. Shiroishi, and P. E. Evrin. 1977. Urinary B_2-microglobulin excretion among people exposed to cadmium in the general environment. An epidemiological study in cooperation between Japan and Sweden. Environ. Res. 13:318-344.

Knowlton, P. H., W. H. Hoover, C. J. Sniffen, C. S. Thompson, and P. C. Belyea. 1976. Hydrolyzed leather scrap as a protein source for ruminants. J. Anim. Sci. 43:1095-1103.

Kobayashi, J. 1978. Pollution by cadmium and the itai-itai disease in Japan. pp. 199-260. In F. W. Oehme (ed.) Toxicity of Heavy Metals in the Environment. Marcel Dekker, Inc., New York.

Kojima, S., Y. Haga, T. Kurihara, T. Yamawaki, and T. Kjellstrom. 1979. A comparison between fecal cadmium and urinary B_2-microglobulin, total protein, and cadmium among Japanese farmers. Environ. Res. 14:436-451.

Kostial, K., D. Kello, M. Blanusa, T. Maljovic, and I. Rabar. 1979. Influence of some factors on cadmium pharmacokinetics and toxicity. Environ. Health Perspect. 28:89-95.

Kreula, M. S. 1947. Absorption of carotene from carrots in man and the use of the quantitative chromic oxide indicator method in absorption experiments. Biochem. J. 41:269-273.

Kubota, J. 1977. Molybdenum status of United States soils and plants. pp. 555-581. In W. R. Chappel and K. K. Peterson (eds.) Molybdenum in the Environment. Vol. 2. Marcel Dekker, Inc., New York.

Kubota, J., and W. H. Allaway. 1972. Geographic distribution of trace element problems. pp. 525-554. In J. J. Mortvedt, P. M. Giordano, and W. L. Lindsay (eds.) Micronutrients in Agriculture. Soil Sci. Soc. Am., Madison, WI.

Kubota, J., W. H. Allaway, D. L. Carter, E. E. Cary, and V. A. Lazar. 1967. Selenium in crops in the United States in relation to selenium-responsive diseases of animals. J. Agr. Food Chem. 15:448-453.

Kubota, J., and V. A. Lazar. 1958. Cobalt status of soils of the Southeastern United States: II. Ground-water podzols and six geographically associated soil groups. Soil Sci. 86:262-268.

Kubota, J., V. A. Lazar, and K. C. Beeson. 1960. The study of cobalt status of soils in Arkansas and Louisiana using the black gum as the indicator plant. Soil Sci. Soc. Am. Proc. 24:527-528.

Kubota, J., V. A. Lazar, L. W. Langan, and K. C. Beeson. 1961. The relationship of soils to molybdenum toxicity in cattle in Nevada. Soil Sci. Soc. Am. Proc. 25:227-232.

Kubota, J., E. R. Lemon, and W. H. Allaway. 1963. The effect of soil moisture content on the uptake of molybdenum, copper, and cobalt by alsike clover. Soil Sci. Soc. Am. Proc. 27:679-683.

Kubota, J., G. H. Oberly, and E. A. Naphan. 1980. Magnesium in grasses of three selected regions in the United States and its relation to grass tetany. Agron. J. 72:907-914.

Lahann, R. W. 1976. Molybdenum hazard in land disposal of sewage sludge. Water, Air, Soil Pollut. 6:3-

Lahouti, M., and P. J. Peterson. 1979. Chromium accumulation and distribution in crop plants. J. Sci. Food Agric. 30:136-142.

Lamand, M. 1979. Influence of silage contamination by soil upon trace elements availability in sheep. Ann. Rech. Vet. 10:571-573.

Landrigan, P. J., C. W. Heath, Jr., J. A. Liddle, and D. D. Bayse. 1978. Exposure to polychlorinated biphenyls in Bloomington, Indiana. Report EPI-77-35-2, Public Health Service-CDC-Atlanta.

Lauwerys, R., H. Roels, A. Bernard, and J. P. Buchet. 1980. Renal response to cadmium in a population living in a non-ferrous smelter area in Belgium. Int. Arch. Occup. Environ. Health 45:271-274.

Lee, C. Y., W. F. Shipe, Jr., L. W. Naylor, C. A. Bache, P. C. Wszolek, W. H. Gutenmann, and D. J. Lisk. 1980. The effect of a domestic sewage sludge amendment to soil on heavy metals, vitamins, and flavor in vegetables. Nutr. Rep. Int. 21:733-738.

Lee, D., and G. Matrone. 1969. Fe and Cu effect on serum ceruloplasmin activity of rats with Zn-induced Cu deficiency. Proc. Soc. Exp. Biol. Med. 130:1190-1194.

Leeper, G. W. 1978. Managing the Heavy Metals on the Land. Marcel Dekker, Inc., New York. 121pp.

Lepow, M. L., L. Bruckman, M. Gillette, S. Markowitz, R. Robino, and J. Kapish. 1975. Investigations into sources of lead in the environment of urban children. Environ. Res. 10:415-426.

L'Estrange, J. L. 1979. The performance and carcass fat characteristics of lambs fattened on concentrate diets. 4. Effects of barley fed whole, ground or pelleted and of a high level of zinc supplementation. Ir. J. Agr. Res. 18:173-182.

Levander, O. A. 1979. Lead toxicity and nutritional deficiencies. Environ. Health Perspect. 29:115-125.

Lexmond, T. M. 1981. A contribution to the establishment of safe copper levels in soil. pp. 162-183. In P. L'Hermite and J. Dehandtschutter (eds.) Copper in Animal Wastes and Sewage Sludge. Reidel Publ., Boston.

Lexmond, T. M., and F. A. M. De Haan. 1980. Toxicity of copper. pp. 410-419. In J. K. R. Gasser (ed.) Effluents from Livestock. Applied Science Publ., London.

Lewis, B. G., C. M. Johnson, and T. C. Broyer. 1974. Volatile selenium in higher plants: The production of dimethyl selenide in cabbage leaves by enzymatic cleavage of Se-methyl selenomethionine selenonium salt. Plant Soil 40:107-118.

Lewis, G. P., W. J. Jusko, L. L. Couglin, and S. Hartz. 1972. Cadmium accumulation in man: Influence of smoking, occupation, alcohol habit, and disease. J. Chron. Dis. 25:717-726.

Lichtenstein, E. P., G. R. Myrdal, and K. R. Schulz. 1964. Effect of formulation and mode of application of aldrin on the loss of aldrin and its epoxide from soils and their translocation into carrots. J. Econ. Entomol. 57:133-136.

Lichtenstein, E. P., G. R. Myrdal, and K. R. Schulz. 1965. Absorption of insecticidal residues from contaminated soils into five carrot varieties. J. Food Chem. 13:126-133.

Lichtenstein, E. P., and K. R. Schulz. 1965. Residues of aldrin and heptachlor in soils and their translocation into various crops. J. Agr. Food Chem. 13:57-63.

Lindsay, W. L. 1979. Chemical equilibria in soils. Wiley-Interscience, New York. 449pp.

Linne, C., and R. Martens. 1978. Examination of the risk of contamination by polycyclic aromatic hydrocarbons in the harvested crops of carrots and mushrooms after the application of composted municipal refuse (in German). Z. Pflanzenernaehr. Bodenkd. 141:265-274.

Lisk, D. J. 1972. Trace metals in soils, plants, and animals. Adv. Agron. 24:267-325.

Lloyd, C. A., R. L. Chaney, S. B. Hornick, and P. J. Mastradone. 1981. Labile cadmium in soils of long-term sludge utilization farms. Agron. Abstr. 1981:29.

Loneragen, J. F. 1975. The availability and absorption of trace elements in soil-plant systems and their relation to movement and concentrations of trace elements in plants. pp. 109-134. In D. J. D. Nicholas and A. R. Egan (eds.) Trace

Elements in Soil-Plant-Animals Systems. Academic Press, Inc., New York.

Lund, L. J., E. E. Betty, A. L. Page, and R. A. Elliott. 1981. Occurrence of naturally high cadmium levels in soils and its accumulation by vegetation. J. Environ. Qual. 10:551-556.

Lyon, G. L., R. R. Brooks, and P. J. Peterson. 1969. Chromium-51 distribution in tissues and extracts of Leptospermum scoparium J. R. et G. Forst. (Myrtaceae). Planta 88:282-287.

Lyon, G. L., R. R. Brooks, P. J. Peterson, and G. W. Butler. 1968. Trace elements in a New Zealand serpentine flora. Plant Soil 29:225-240.

Lyon, G. L., P. J. Peterson, and R. R. Brooks. 1969. Chromium-51 transport in the xylem sap of Leptospermum scoparium J. R. et G. Forst. (Myrtaceae). N. Z. J. Sci. 12:541-545.

Lyon, G. L., P. J. Peterson, R. R. Brooks, and G. W. Butler. 1971. Calcium, magnesium, and trace elements in a New Zealand serpentine flora. J. Ecol. 59:421-429.

MacLaren, A. P. C., W. G. Johnston, and R. C. Voss. 1964. Cobalt poisoning in cattle. Vet. Rec. 76:1148-1149.

Maclean, A. J. 1976. Cadmium in different plant species and its availability in soils as influenced by organic matter and additions of lime, P, Cd, and Zn. Can. J. Soil Sci. 56:129-138.

Mahaffey, K. R. 1978. Environmental exposure to lead. pp. 1-36. In J. O. Nriagu (ed.) The Biogeochemistry of Lead in the Environment, Part B. Elsevier/North Holland Biomedical Press. New York.

Mahaffey, K. R., and J. E. Vanderveen. 1979. Nutrient-toxicant interactions: Susceptible populations. Environ. Health Perspect. 29:81-87.

Mahler, R. J., F. T. Bingham, and A. L. Page. 1978. Cadmium-enriched sewage sludge application to acid and calcareous soils: Effect on yield and cadmium uptake by lettuce and chard. J. Environ. Qual. 7:274-281.

Majeta, V. A. and C. S. Clark. 1981. Health risks of organics in land application. J. Environ. Eng. Div. Proc. ASCE 107:339-357.

Matrone, G. 1974. Chemical parameters in trace-element antagonisms. pp. 91-103. In W. G. Hoekstra et al. (eds.) Trace Element Metabolism in Animals-2. University Park Press, Baltimore, MD.

Mayland, H. F., A. R. Florence, R. C. Rosenau, V. A. Lazar, and H. A. Turner. 1975. Soil ingestion by cattle on semiarid range as reflected by titanium analysis of feces. J. Range Mgmt. 28:448-452.

Mayland, H. F., D. L. Grunes, and V. A. Lazar. 1976. Grass tetany hazard of cereal forages based upon chemical composition. Agron. J. 68:665-667.

Mayland, H. F., G. E. Shewmaker, and R. C. Bull. 1977. Soil ingestion by cattle grazing crested wheatgrass. J. Range Mgmt. 30:264-265.

McClaslin, B. D., and V. L. Rodriguez. 1978. Gamma irradiated digested sewage sludge as micronutrient fertilizer on calcareous soil. Agron. Abstr. 1978:30-31.

McCaslin, B. D., and J. S. Sivinski. 1979. Dried gamma-irradiated sewage solids use on calcareous soils: crop yields and heavy metals uptake. 27pp. NTIS Rept. SAND-79-1538c.

McGhee, F., C. R. Creger, and J. R. Couch. 1965. Copper and iron toxicity. Poult. Sci. 44:310-312.

McGrath, D. 1981. Implications of applying copper-rich pig slurry to grassland -- Effects on plants and soils. pp. 144-153. In P. L'Hermite and J. Dehandtschutter (eds.) Copper in Animals Wastes and Sewage Sludge. Reidel Publ., Boston.

McKenzie, R. M. 1972. The manganese oxides in soils - a review. Z. Pflanzenern. Boden. 131:221-242.

McLellan, J. S., P. R. Flanagan, M. J. Chamberlain, and L. S. Valberg. 1978. Measurement of dietary cadmium absorption in humans. J. Toxicol. Environ. Health 4:131-138.

Menden, E. E., V. J. Elia, L. W. Michael, and H. G. Petering. 1972. Distribution of cadmium and nickel of tobacco during cigarette smoking. Environ. Sci. Technol. 6:830-832.

Mertz, D., T. Edward, D. Lee, and M. Zuber. 1981. Absorption of aflatoxin by lettuce seedlings grown in soil adulterated with aflatoxin B_1. J. Agr. Food Chem. 29:1168-1170.

Mertz, D., D. Lee, M. Zuber, and E. Lillehoj. 1980. Uptake and metabolism of aflatoxin by Zea mays. J. Agr. Food Chem. 28:963-966.

Mertz, W. 1969. Chromium occurrence and functions in biological systems. Physiol. Rev. 49:165-239.

Miller, J., and F. C. Boswell. 1979. Mineral content of selected tissues and feces of rats fed turnip greens grown on soil treated with sewage sludge. J. Agr. Food Chem. 27:1361-1365.

Millman, A. P. 1957. Biogeochemical investigations in areas of copper-tin mineralization in southwest England. Geochim. Cosmochim. Acta 12:85-93.

Mills, C. F. 1974. Trace-element interactions: Effects of dietary composition on the development of imbalance and toxicity. pp. 79-90. In W. G. Hoekstra (ed.) Trace Element Metabolism in Animals-2. University Park Press, Baltimore, MD.

Mills, C. F. 1978. Heavy metal toxicity and trace element imbalance in farm animals. World Cong. Anim. Feed., 3rd. 7:275-281.

Mills, C. F., I. Bremner, T. T. El-Gallad, A. C. Dalgarno, and B. W. Young. 1978. Mechanisms of the molybdenum/sulphur antagonism of copper utilization by ruminants. In M. Kirchgessner (ed.) Trace Element Metabolism in Man and Animals 3:150-158.

Mills, C. F., J. K. Campbell, I. Bremner, and J. Quarterman. 1980. The influence of dietary composition on the toxicity of cadmium, copper, zinc, and lead to animals. pp. 11-21. In Inorganic Pollution and Agriculture. Min. Agr. Fish. Food Reference Book 326. Her Majesty's Stationary Office, London.

Mills, C. F., and A. C. Dalgarno. 1972. Copper and zinc status of ewes and lambs receiving increased dietary concentrations of cadmium. Nature 239:171-173.

Mitchell, R. L., and J. W. S. Reith. 1966. The lead content of pasture herbage. J. Sci. Food Agric. 17:437-440.

Morrison, R. S., R. R. Brooks, and R. D. Reeves. 1980. Nickel uptake by Alyssum species. Plant Sci. Lett. 17:451-457.

Mortvedt, J. J., and P. M. Giordano. 1975. Response of corn to zinc and chromium in municipal wastes applied to soil. J. Environ. Qual. 4:170-174.

Moza, P., I. Schuenert, W. Klein, and F. Korte. 1979. Studies with 2,4',5-trichlorobiphenyl-^{14}C and 2,2',4,4',6-pentachlorobiphenyl-^{14}C in carrots, sugar beets and soil. J. Agr. Food Chem. 27:1120-1124.

Moza, P., I. Weisgerber, and W. Klein. 1976. Fate of 2,2'-dichlorobiphenyl-^{14}C in carrots, sugar beets and soil under outdoor conditions. J. Agr. Food Chem. 24:881.

Mukawa, A., K. Nogawa, and N. Hagino. 1980. Bone biopsy performed on women living in the cadmium polluted Jintzu River basin, Japan. J. Hyg. 35:761-773.

Mullens, G. L., D. C. Martens, W. P. Miller, E. T. Kornegay, and D. L. Hallock. 1982. Copper availability, form, and mobility in soils from three annual copper-enriched hog manure applications. J. Environ. Qual. 11:316-320.

Muller, H. 1976. Aufnahme von 3,4-benzpyren durch nahrungspflanzen aus kunstlich angereicherten substraten. Z. Pflanzenern. Bodenk. 139:685-695.

National Research Council. 1980a. Lead in the Human Environment. National Academy of Sciences, Washington, D. C. 525pp.

National Research Council. 1980b. Mineral Tolerance of Domestic Animals. National Academy of Sciences, Washington, D. C. 577pp.

Neathery, M. W., and W. J. Miller. 1975. Metabolism and toxicity of cadmium, mercury, and lead in animals: A review. J. Dairy Sci. 58:1767-1781.

Needleman, H. L. 1980. Lead and neuropsychological deficit: Finding a threshold. pp. 43-51. In H. L. Needleman (ed.) Low Level Lead Exposure: The Clinical Implications of Current Research. Raven Press, New York.

Needleman, H. L., C. E. Gunnoe, A. Leviton, R. Reed, H. Peresie, C. Maher, and P. Barrett. 1979. Deficits in psychologic and classroom performance of children with elevated lead levels. N. Engl. J. Med. 300:689-695.

Neudecker, C. 1978. Toxicological long-term animal feeding studies on carrots cultivated with composted garbage or sewage sludge (in German). Qual. Plant. 28:119-134.

Nogawa, K. 1978. Studies on Itai-itai disease and dose-response relationship of cadmium. pp. 213-221. In Proc. First International Cadmium Conference. Metals Bulletin, London.

Nogawa, K., and A. Ishizaki. 1979. A comparison between cadmium in rice and renal effects among inhabitants of the Jinzu River basin. Environ. Res. 18:410-420.

Nogawa, K., A. Ishizaki, M. Fukushima, I. Shibata, and N. Hagino. 1975. Studies on the women with acquired Franconi syndrome observed in the Ichi River basin polluted by cadmium. Is this itai-itai disease? Environ. Res. 10:280-307.

Nogawa, K., A. Ishizaki, and S. Kawano. 1978. Statistical observations of the dose-response relationships of cadmium based on epidemiological studies in the Kakehashi River basin. Environ. Res. 15:185-198.

Nogawa, K., E. Kobayashi, R. Honda, A. Ishizaki, S. Kwano, and H. Matsuda. 1980. Renal dysfunctions of inhabitants in a cadmium-polluted area. Environ. Res. 23:13-23.

Nolan, T., and W. J. M. Black. 1970. Effect of stocking rate on tooth wear in ewes. Ir. J. Agric. Res. 9:187-196.

O'Dell, G. D., W. J. Miller, W. A. King, S. L. Moore, and D. M. Blackmon. 1970. Ni toxicity in the young bovine. J. Nutr. 100:1447-1453.

Olsen, S. R. 1972. Micronutrient interactions. pp. 243-264. In J. J. Mortvedt, P. M. Giordano, and W. L. Lindsay (eds.) Micronutrients in Agriculture. Soil Sci. Soc. Am., Madison, WI.

Olson, O. E., and A. L. Moxon. 1939. The availability to crop plants of different forms of selenium in the soil. Soil Sci. 47:305-311.

Osuna, O., G. T. Edds, J. A. Popp, J. Monague, and K. E. Ferslew. 1979. Feeding trials of dried urban sludge and the equivalent cadmium level in swine. pp. 201-213. In Municipal Sludge Management. Impact of Industrial Toxic Materials on POTW Sludge. Information Transfer, Inc., Silver Spring, MD.

Ott, E. A., W. H. Smith, R. B. Harrington, and W. M. Beeson. 1966. Zinc toxicity in ruminants. II. Effect of high levels of dietary zinc on gains, feed consumption, and feed efficiency of beef cattle. J. Anim. Sci. 25:419-423.

Page, A. L. 1974. Fate and effects of trace elements in sewage sludge when applied to agricultural lands. A literature review study. U. S. Environ. Prot. Agency Rept. No. EPA-670/2-74-005. 108 pp.

Pahren, H. R., J. B. Lucas, J. A. Ryan, and G. K. Dotson. 1979. Health risks associated with land application of municipal sludge. J. Water Pollut. Contr. Fed. 51:2588-2601.

Parker, W. H., and T. H. Rose. 1955. Molybdenum poisoning (teart) due to aerial contamination of pastures. Vet. Rec. 67:276-279.

Peterson, P. J., M. A. S. Burton, M. Gregson, S. M. Nye, and E. K. Porter. 1976. Tin in plants and surface waters in Malaysian ecosystems. Trace Subst. Environ. Health 10:123-132.

Pinkerton, B. P., K. W. Brown, J. C. Thomas, and D. Zabcik. 1981. Plant uptake of cobalt from a petroleum sludge affected soil. Agron. Abstr. 1981:32.

Petrilli, F. L., and S. deFlora. 1977. Toxicity and mutagenicity of hexavalent chromium on Salmonella typhimurium. Appl. Environ. Microbiol. 33:805-809.

Polacco, J. C. 1977. Nitrogen metabolism in soybean tissue culture. II. Urea utilization and urease synthesis require Ni. Plant. Physiol. 59:827-830.

Poole, D. B. R. 1981. Implications of applying copper rich pig slurry to grassland -- Effects on the health of grazing sheep. pp. 273-286. In P. L'Hermite and J. Dehandtschutter (eds.) Copper in Animal Wastes and Sewage Sludge. Reidel Publ., Boston.

Porter, E. K., and P. J. Peterson. 1975. Arsenic accumulation by plants on mine waste (United Kingdom). Sci. Total Environ. 4:365-371.

Porter, E. K., and P. J. Peterson. 1977a. Arsenic tolerance in grasses growing on mine wastes. Environ. Pollut. 14:255-265.

Porter, E. K., and P. J. Peterson. 1977b. Biogeochemistry of arsenic on polluted sites in S. W. England. Trace Subst. Environ. Health. 11:89-99.

Preer, J. R., H. S. Sekhon, B. R. Stephens, and M. S. Collins. 1980. Factors affecting heavy metal content of garden vegetables. Environ. Pollut. B1:95-104.

Prince, A. L., F. E. Bear, E. G. Brennan, I. A. Leone, and R. H. Daines. 1949. Fluorine: its toxicity to plants and its control in soils. Soil Sci. 67:269-277.

Proctor, J., and S. R. J. Woodell. 1975. The ecology of serpentine soils. Adv. Ecol. Res. 9:255-366.

Raleigh, R. J., R. J. Kartchner, and L. R. Rittenhouse. 1980. Chromic oxide in range nutrition studies. Oregon State Univ. Agr. Exp. Sta. Bull. 641:1-41.

Rauser, W. E., and A. B. Samarakoon 1980. Vein loading in seedlings of Phaseolus vulgaris exposed to excess cobalt, nickel, and zinc. Plant Physiol. 65:578-583.

Reid, R. L., and D. J. Horvath. 1980. Soil chemistry and mineral problems in farm livestock: A review. Anim. Feed Sci. Technol. 5:95-167.

Reid, R. L., and G. A. Jung. 1974. Effects of elements other than nitrogen on the nutritive value of forage. pp. 395-435. In D. A. Mays (ed.) Forage Fertilization. Am. Soc. Agron., Madison, WI.

Rendig, V. V., and D. L. Grunes, (eds.). 1979. Grass Tetany. Am. Soc. Agron. Spec. Publ. No. 35, Madison, WI.

Rice, C., A. Fischbein, R. Lilis, L. Sarkozi, S. Kon, and I. J. Selikoff. 1978. Lead contamination in the homes of employees of secondary lead smelters. Environ. Res. 15:375-380.

Rocovich, S. E., and D. A. West. 1975. Arsenic tolerance in a population of the grass Andropogon scoparius Michx. Science 188:263-264.

Roels, H. A., J. P. Buchet, R. R. Lauwreys, P. Bruaux, F. Claeys-Thoreau, A. Lafontaine, and G. Verduyn. 1980. Exposure to lead by the oral and the pulmonary routes of children living in the vicinity of a primary lead smelter. Environ. Res. 22:81-94.

Roels, H. A., R. R. Lauwerys, J. P. Buchet, and A. Bernard. 1981a. Environmental exposure to cadmium and renal function of aged women in three areas of Belgium. Environ. Res. 24:117-130.

Roels, H. A., R. R. Lauwerys, J. P. Buchet, A. Bernard, O. C. Chettle, T. C. Harvey, and I. K. Al-Haddad. 1981b. In vivo measurement of liver and kidney cadmium in workers exposed to this metal: Its significance with respect to cadmium in blood and urine. Environ. Res. 26:217-240.

Romney, E. M., A. Wallace, and G. V. Alexander. 1975. Responses of bush bean and barley to tin applied to soil and to solution culture. Plant Soil 42:585-589.

Roth, J. A., E. F. Wallihan, and R. G. Sharpless. 1971. Uptake by oats and soybeans of copper and nickel added to a peat soil. Soil Sci. 112:338-342.

Ryan, J. A., L. D. Grant, J. B. Lucas, R. E. Marland, H. R., Pahren, W.A. Galke, and D. J. Ehreth. 1979. Cadmium health effects: Implications for environmental regulations. External review draft. 121 pp.

Ryan, J. A., H. R. Pahren, and J. B. Lucas. 1982. Controlling cadmium in the human food chain: A review and rationale based on health effects. Environ. Res. (In Press).

Saito, H., R. Shioji, Y. Hurukawa, K. Nagai, T. Arikawa, T. Saito, Y. Sasaki, T. Furuyama, and K. Yoshinaga. 1977. Cadmium-induced proximal tubular dysfunction in a cadmium-polluted area. Contr. Nephrol. 6:1-12.

Sayre, J. W., E. Charney, J. Vostal, and I. B. Pless. 1974. House and hand dust as a potential source of childhood lead exposure. Am. J. Dis. Child. 127:167-170.

Schurch, A. F., L. E. Lloyd, and E. W. Crampton. 1950. The use of chromic oxide as an index for determining the digestibility of a diet. J. Nutr. 41:629-636.

Schwarz, F. J., and M. Kirchgessner. 1978. Supplement of the feed ration with high amounts of copper for dairy cows (in German). Landwirtsch. Forsch. 31:317-326.

Schwartz, K. and J. E. Spallholz. 1978. Growth effects of small cadmium supplements in rats maintained under trace element-controlled conditions. pp. 105-109. In Proc. First International Cadmium Conference. Metal Bulletin, Inc., London.

Seckbach, J. 1968. Studies on the deposition of plant ferritin as influenced by iron supply to iron-deficient beans. J. Ultrastruct. Res. 22:413-423.

Seckbach, J. 1969. Iron content and ferritin in leaves of iron treated *Xanthium pennsylvanicum* plants. Plant Physiol. 44:816-820.

Shacklette, H. T., J. A. Erdman, T. F. Harms, and C. S. E. Papp. 1978. Trace elements in plant foodstuffs. pp. 25-68. In F. W. Oehme (ed.) Toxicity of Heavy Metals in the Environment. Marcel Dekker, Inc., New York.

Sharma, R. P., J. C. Street, M. P. Verma, and J. L. Shupe. 1979. Cadmium uptake from feed and its distribution to food products of livestock. Environ. Health Perspect. 28:59-66.

Sheaffer, C. C., A. M. Decker, R. L. Chaney, and L. W. Douglass. 1979. Soil temperature and sewage sludge effects on metals in crop tissue and soils. J. Environ. Qual. 8:455-459.

Shellshear, I. D., L. D. Jordan, D. J. Hogan, and F. T. Shannon. 1975. Environmental lead exposure in Christchurch children: Soil lead a potential hazard. N. Z. Med. J. 81:382-386.

Shewry, P. R., and P. J. Peterson. 1974. The uptake and transport of chromium by barley seedlings (Hordeum vulgare L.). J. Exp. Bot. 25:785-797.

Shewry, P. R., and P. J. Peterson. 1976. Distribution of chromium and nickel in plants and soil from serpentine and other sites. J. Ecol. 64:195-212.

Shigematsu, I., M. Minowa, T. Yoshida, and K. Miyamoto. 1979. Recent results of health examinations on the general population in cadmium-polluted and control areas in Japan. Environ. Health Perspect. 28:205-210.

Shrift, A., and T. K. Virupaksha. 1965. Seleno-amino acids in selenium-accumulating plants. Biochim. Biophys. Acta. 100:65-75.

Siegfried, R. 1975. The influence of refuse compost on the 3,4-benzpyrene content of carrots and cabbage (in German). Naturwissenschaften 62:300.

Siegfried, R., and H. Muller. 1978. The contamination with 3,4-benzpyrene of root and green vegetables grown in soil with different 3,4-benzpyrene concentrations (In German). Landwirtsch. Forsch. 31:133-140.

Silbergeld, E. K., and R. E. Hruska. 1980. Neurochemical investigations of low level lead exposure. pp. 135-157. In H. L. Needleman (ed.) Low Level Lead Exposure: The Clinical Implications of Current Research, Raven Press, New York.

Silva, S., and B. Beghi. 1979. Problems inherent in the use of organic fertilizers containing chromium. (In Italian) Ann. Della Fac. Agr. Univ. Cath. Sacro Cuore 19:31-47.

Simon, E. 1977. Cadmium tolerance in populations of Agrostis tenuis and Fetuca ovina. Nature 265:328-330.

Skeffington, R. A., P. R. Shewry, and P. J. Peterson. 1976. Chromium uptake and transport in barley seedlings (Hordeum vulgare L.). Planta 132:209-214.

Slater, J. P., and H. M. Reisenauer. 1979. Toxicity of Cr(III) and Cr(VI) added to soils. Agron. Abstr. 1979:38.

Sleper, D. A., G. B. Garner, C. J. Nelson, and J. L. Sebaugh. 1980. Mineral concentration of tall fescue genotypes grown under controlled conditions. Agron. J. 72:720-722.

Smith, G. S., H. E. Kiesling, J. M. Cadle, C. Staples, L. B. Bruce, and H. D. Sivinski. 1977. Recycling sewage solids as feedstuffs for livestock. pp. 119-127. In Proc. Third Nat. Conf. on Sludge Management, Disposal and Utilization. Information Transfer, Inc., Rockville, MD.

Smith, G. S., H. E. Kiesling, and E. E. Ray. 1979. Prospective use of sewage solids as feed for cattle. pp. 190-200. In Municipal Sludge Management Impact of Industrial Toxic Materials on POTW Sludge. Information Transfer, Inc., Silver Spring, MD.

Smith, G. S., H. E. Kiesling, E. E. Ray, D. M. Hallford, and C. H. Herbel. 1980. Sewage solids as supplemental feed for ruminants: bioassays of benefits and risks. (Abstract). pp. 357-358. In G. Bitton et al. (eds.) Sludge-Health Risks of Land Application. Ann Arbor Sci. Publ., Inc., Ann Arbor, MI.

Smith, G. S., H. E. Kiesling, and H. D. Sivinski. 1978. Nutrient usage and heavy metals uptake by sheep fed thermoradiated, undigested sewage sludge. pp. 239-254. In R. C. Loehr (ed.) Food, Fertilizer and Agricultural Residues. Ann Arbor Sci. Publ, Inc., Ann Arbor, MI.

Sommers, L. E. 1980. Toxic metals in agricultural crops. pp. 105-140. In G. Bitton et al. (eds.) Sludge -- Health Risks of Land Application. Ann Arbor Science Publ. Inc., Ann Arbor, MI.

Soon, Y. K. 1981. Solubility and sorption of cadmium in soils amended with sewage sludge. J. Soil Sci. 32:85-95.

Spence, J. A., N. F. Suttle, G. Wendham, T. El-Gallad, and I. Bremner. 1980. A sequential study of the skeletal abnormalities which develop in rats given a small dietary supplement of ammonium tetrathiomolybdate. J. Comp. Path. 90:139-153.

Spittler, T. M., and W. A. Feder. 1979. A study of soil contamination and plant lead uptake in Boston urban gardens. Commun. Soil Sci. Plant Anal. 10:1195-1210.

Sprague, R. T., L. W. Jacobs, M. J. Zabick, and J. Phillips. 1981. Concentration of organic and inorganic contaminants in Michigan sewage sludges. Agron. Abstr. 1981:35.

Standish, J. F., and C. B. Ammerman. 1971. Effect of excess dietary iron as ferrous sulfate and ferric citrate on tissue mineral composition of sheep. J. Anim. Sci. 33:481-484.

Standish, J. F., C. B. Ammerman, A. Z. Palmer, and C. F. Simpson. 1971. Influence of dietary iron and phosphorus on performance, tissue mineral composition and mineral absorption in steers. J. Anim. Sci. 33:171-178.

Standish, J. F., C. B. Ammerman, C. F. Simpson, F. C. Neal, and A. Z. Palmer. 1969. Influence of graded levels of dietary iron, as ferrous sulfate, on performance and tissue mineral composition of steers. J. Anim. Sci. 29:496-503.

Stara, J., W. Moore, M. Richards, N. Barkley, S. Neiheisel, and K. Bridbord. 1973. Environmentally bound lead. III. Effects of source on blood and tissue levels of rats. pp. 28-29. In EPA Environmental Health Effects Research Series A-670/1-73-036.

Sterritt, R. M., and J. N. Lester. 1981. Concentrations of heavy metals in forty sewage sludges in England. Water, Air, Soil Pollut. 14:125-131.

Stoewsand, G. S., W. H. Gutenmann, and D. J. Lisk. 1978. Wheat grown on fly ash: High selenium uptake and response when fed to Japanese quail. J. Agr. Food Chem. 26:757-759.

Stoszek, J. J., J. E. Oldfield, G. E. Carter, and P. H. Weswig. 1979. Effect of tall fescue and quackgrass on copper metabolism and weight gains of beef cattle. J. Anim. Sci. 48:893-899.

Street, J. J., W. L. Lindsay, and B. R. Sabey. 1977. Solubility and plant uptake of cadmium in soils amended with cadmium and sewage sludge. J. Environ. Qual. 6:72-77.

Strek, H. J., J. B. Weber, P. J. Shea, E. Mrozek, Jr., and M. R. Overcash. 1981. Reduction of polychlorinated biphenyl toxicity and uptake of carbon-14 activity by plants through the use of activated carbon. J. Agr. Food Chem. 29:288-293.

Stuedemann, J. A., S. R. Wilkinson, W. A. Jackson, R. L. Wilson, Jr., and J. B. Jones, Jr. 1974. Controlling grass tetany with a foliar applied MgO-bentonite-water slurry. J. Anim. Sci. 39:254.

Stuedemann, J. A., S. R. Wilkinson, D. J. Williams, H. Ciordia, J. V. Ernst, W. A. Jackson, and J. B. Jones, Jr. 1975. Long-term broiler litter fertilization of tall fescue pastures and health and performance of beef cows. pp. 264-268. In Proc. 3rd Int. Symp. Livestock Wastes, Am. Soc. Agr. Eng., St. Joseph, MI.

Stuedemann, J. A., S. R. Wilkinson, D. J. Williams, W. A. Jackson, and J. B. Jones, Jr. 1973. The association of fat necrosis in beef cattle with heavily fertilized fescue pastures. pp. 9-23. In Proc. Fescue Toxicity Conf. Univ. Missouri, Columbia, MO.

Suttle, N. F., B. J. Alloway, and I. Thornton. 1975. An effect of soil ingestion on the utilization of dietary copper by sheep. J. Agric. Sci. 84:249-254.

Suttle, N. F., and C. F. Mills. 1966. Studies of the toxicity of copper to pigs. I. Effects of oral supplement of zinc and iron salts on the development of copper toxicosis. Br. J. Nutr. 20:135-148.

Suzuki, M., N. Aizawa, G. Okano, and T. Takahashi. 1977. Translocation of polychlorobiphenyls in soil into plants: A study by a method of culture of soybean sprouts. Arch. Environ. Contam. Toxicol. 5:343-352.

Szandkowski, D., H. Schultz, K. H. Schaller, and G. Lehnert, 1969. On the ecological importance of the heavy metal content of cigarettes: Lead, cadmium, and nickel analyses of tobacco as well as its gas and particulate phases (in German) Arch. Hyg. Bakteriol. 153:1-8.

Takijima,, Y., and F. Katsumi. 1973. Cadmium contamination of soils and rice plants caused by zinc mining. 1. Production of high-cadmium rice on the paddy fields in lower reaches of the mine station. Soil Sci. Plant Nutr. 19:29-38.

Taylor, F. G., Jr. 1980. Chromated cooling tower drift and the terrestrial environment: A review. Nuclear Safety 21:495-508.

Ter Harr, G., and R. Aronow. 1974. New information on lead in dirt and dust as related to the childhood lead problem. Environ. Health Perspect. 7:83-89.

Thill, J. L., and J. R. George. 1975. Cation concentrations and K to Ca+Mg ratio of nine cool-season grasses and implications with hypomagnesaemia. Agron. J. 67:89-91.

Thornton, I., and D. G. Kinniburgh. 1978. Intake of lead, copper, and zinc by cattle from soil and pasture. In M. Kirchgessner (ed.) Trace Element Metabolism in Man and Animals 3:499.

Toepfer, E. W., W. Mertz, M. M. Polansky, E. E. Roginski, and W. R. Wolf. 1977. Preparation of chromium-containing material of glucose tolerance factor activity from brewer's yeast extracts and by synthesis. J. Agr. Food Chem. 25:162-166.

Towill, L. E., C. R. Shriner, J. S. Drury, A. S. Hammons, J. W. Holleman, and J. O. Pierce. 1978. Reviews of the environmental effects of pollutants: III. Chromium. EPA-600/1-78-023.

Trelease, S. F. 1945. Selenium in soils, plants, and animals. Soil Sci. 60:125-131.

Trelease, S. F., and H. M. Trelease. 1939. Physiological differentiation in *Astragalus* with reference to selenium. Am. J. Bot. 26:530-535.

Tsuchiya, K. (ed.) 1978. Cadmium Studies in Japan: A Review. Elsevier/ North-Holland Biomedical Press, New York. 376 pp.

Turner, M. A. and R. H. Rust. 1971. Effects of chromium on growth and mineral nutrition of soybeans. Soil Sci. Soc. Am. Proc. 35:755-758.

Tyler, G. 1976. Influence of vanadium on soil phosphatase activity. J. Environ. Qual. 5:216-217.

Underwood, E. J. 1977. Trace Elements in Human and Animal Nutrition. 4th Ed. Academic Press, New York. 545 pp.

Unwin, R. J. 1981. The application of copper in sewage sludge and pig manure to agricultural land in England and Wales. pp. 102-116. In P. L'Hermite and J. Dehandtschutter (eds.) Copper in Animal Wastes and Sewage Sludge. Reidel Publ., Boston.

U.S.D.A. 1978. Improving soils with organic wastes. Report to the Congress in response to Section 1461 of the Food and Agriculture Act of 1977 (PL 95-113). USGPO, Wash., D. C. 157 pp.

Van Campen, D. R., and R. M. Welch. 1980. Availability to rats of iron from spinach: Effects of oxalic acid. J. Nutr. 110:1618-1621.

Vergnano, O., and J. G. Hunter. 1952. Nickel and cobalt toxicities in oat plants. Ann. Bot. (London) 17:317-328.

Vlek, P. L. G., and W. L. Lindsay. 1977a. Thermodynamic stability and solubility of molybdenum minerals in soils. Soil Sci. Soc. Am. J. 41:42-46.

Vlek, P. L. G., and W. L. Lindsay. 1977b. Molybdenum contamination in Colorado pasture soils. pp. 619-650. In W. R. Chappell and K. K. Petersen (eds.) Molybdenum in the Environment, Vol. 2. Marcel Dekker, Inc., New York.

Wagner, K. H., and I. Siddiqi. 1971. Die Speicherung von 3,4-benzfluoranthen in Sommerweizen und Sommerrogen. Z. Pflanzenern. Bodenkd. 127:211-218.

Waldrup, P. W., C. M. Hillard, W. W. Abbott, and L. W. Luther. 1970. Hydrolyzed leather meal in broiler diets. Poult. Sci. 49:1259-1264.

Wallace, A., G. V. Alexander, and F. M. Chaudhry. 1977. Phytotoxicity of cobalt, vanadium, titanium, silver, and chromium. Commun. Soil Sci. Plant Anal. 8:751-756.

Wallace, A., and R. T. Mueller. 1973. Effects of chelated and non-chelated cobalt and copper on yields and microelement composition of bush beans grown on calcareous soil in a glasshouse. Soil Sci. Soc. Amer. Proc. 37:904-908.

Wallace, A., E. M. Romney, G. V. Alexander, S. M. Soufi, and P. M. Patel. 1977. Some interactions in plants among cadmium, other heavy metals, and chelating agents. Agron. J. 69:18-20.

Walsh, L. M., W. H. Erhardt, and H. D. Seibel. 1972a. Copper toxicity in snapbeans (Phaseolus vulgaris L.). J. Environ. Qual. 1:197-200.

Walsh, L. M., and D. R. Keeney. 1975. Behavior and phytotoxicity of inorganic arsenicals in soils. pp. 35-52. In E. A. Woolson (ed.) Arsenical Pesticides. ACS Symp. Ser. 7. American Chemical Society, Washington, D. C.

Walsh, L. M., D. R. Stevens, H. D. Seibel, and G. G. Weis. 1972b. Effects of high rates of zinc on several crops grown on an irrigated Plainfield sand. Commun. Soil Sci. Plant Anal. 3:187-195.

Walsh, L. M., M. E. Sumner, and R. B. Corey. 1976. Consideration of soils for accepting plant nutrients and potentially toxic nonessential elements. pp. 22-47. In Land Application of Waste Materials. Soil Conserv. Soc. Am., Ankeny, IA.

Warnock, R. E. 1970. Micronutrient uptake and mobility within corn plants (*Zea mays* L.) in relation to phosphorus-induced zinc deficiency. Soil Sci. Soc. Amer. Proc. 34:765-769.

Washko, P. W., and R. J. Cousins. 1977. Role of dietary calcium and calcium binding protein in cadmium toxicity in rats. J. Nutr. 107:920-923.

Webber, M. D., Y. K. Soon, T. E. Bates, and A. U. Hag. 1981. Copper toxicity resulting from land application of sewage sludge. pp. 117-135. In P. L'Hermite and J. Dehandtschutter (eds.) Copper in Animal Wastes and Sewage Sludge. Reidel Publ., Boston.

Weber, J. B., and E. Mrozek, Jr. 1979. Polychlorinated biphenyls: phytotoxicity, absorption, and translocation by plants, and inactivation by activated charcoal. Bull. Environ. Contam. Toxicol. 23:412-417.

Wedeen, R. P., D. K. Mallik, V. Batuman, and J. D. Bogden. 1978. Geophagic lead nephropathy: Case report. Environ. Res. 17:409-415.

Welch, R. M. 1981. The biological significance of nickel. J. Plant Nutr. 3:345-356.

Welch, R. M., and E. E. Cary. 1975. Concentrations of chromium, nickel, and vanadium in plant materials. J. Agr. Food Chem. 23:479-482.

Welch, R. M., and W. A. House. 1980. Absorption of radiocadmium and radioselenium by rats fed intrinsically and extrinsically labelled lettuce leaves. Nutr. Rept. Intern. 21:135-145.

Welch, R. M., W. A. House, and D. R. Van Campen. 1977. Effects of oxalic acid on availability of zinc from spinach leaves and zinc sulfate to rats. J. Nutr. 107:929-933.

Welch, R. M., W. A. House, and D. R. Van Campen. 1978. Availability of cadmium from lettuce leaves and cadmium sulfate to rats. Nutr. Rep. Intern. 17:35-42.

Welch, R. M., and E. W. D. Huffman, Jr. 1973. Vanadium and plant nutrition. The growth of lettuce (Lactuca sativa L.) and tomato (Lycopersicon esculentum Mill.) plants in nutrient solutions low in vanadium. Plant Physiol. 52:183-185.

Westcott, D. T., and D. Spincer. 1974. The cadmium, nickel, and lead content of tobacco and cigarette smoke. Beitrage zur Tabakforschung 7:217-221.

Whitby, L. G., and D. Lang. 1960. Experience with the chromic oxide method of fecal marking in metabolic balance investigations on humans. J. Clin. Invest. 39:854.

White, M. C., and R. L. Chaney. 1980. Zinc, cadmium, and manganese uptake by soybean from two zinc- and cadmium-amended coastal plain soils. Soil Sci. Soc. Am. J. 44:308-313.

WHO. 1972. Sixteenth report of the Joint FAO/WHO Expert Committee on Food Additives. WHO Tech. Rept. Ser. No. 505.

Wien, E. M., D. R. Van Campen, and J. M. Rivers. 1975. Factors affecting the concentration and bioavailability of iron in turnip greens to rats. J. Nutr. 105:459-466.

Wigham, H., M. H. Martin, and P. J. Coughtrey. 1980. Cadmium tolerance of Holcus lanatus L. collected from soils with a range of cadmium concentrations. Chemosphere 9:123:125.

Wild, H. 1970. Geobotanical anomalies in Rhodesia. 3. The vegetation of nickel-bearing soils. Kirkia 7, Suppl. 1-62.

Wild, H. 1974. Indigenous plants and chromium in Rhodesia. Kirkia 9:233-241.

Wilkinson, S. R., J. A. Stuedemann, J. B. Jones, Jr., W. A. Jackson, and J. W. Dobson. 1972. Environmental factors affecting magnesium concentrations and tetanigenicity of pastures. pp. 153-176. In J. B. Jones, Jr., M. C. Blount, and S. R. Wilkinson (eds.). Proc. Symp. Magnesium in the Environment: Soils, Crops, Animals, and Man. Taylor County Publ. Co., Reynolds, GA.

Williams, D. J., D. E. Tyler, and E. Papp. 1969. Abdominal fat necrosis as a herd problem in Georgia cattle. J. Am. Vet. Med. Assoc. 154:1017-1021.

Williams, P. H., J. S. Shenk, and D. E. Baker. 1978. Cadmium accumulation by meadow voles (Microtus pennsylvanicus) from crops grown on sludge-treated soil. J. Environ. Qual. 7:450-454.

Willoughby, R. A., E. MacDonald, B. J. McSherry, and G. Brown. 1972. Lead and zinc poisoning and the interaction between Pb and Zn poisoning in the foal. Can. J. Comp. Med. 36:348-359.

Willoughby, R. A., and W. Oyaert. 1973. Zinc poisoning in foals (in Dutch). Vlaams Diergeneeskd. Tijdschr. 42:134-143.

Woolson, E. A. 1973. Arsenic phytotoxicity and uptake in six vegetable crops. Weed Sci. 21:524-527.

Yamagata, N., and Y. Murakami. 1958. A cobalt-accumulator plant, *Clethra barbinervis* Sieb. et Zucc. Nature 181:1808-1809.

Yamagata, N., and I. Shigematsu. 1970. Cadmium pollution in perspective. Bull. Inst. Publ. Health 19:1-27.

Yoneyama, T. 1981. Detection of N-nitrosodimethylamine in soils amended with sludges. Soil Sci. Plant Nutr. 27:249-253.

Yost, K. J., L. J. Miles, and T. A. Parsons. 1980. A methodology for estimating dietary intake of environmental trace contaminants: Cadmium, a case study. Environ. Intern. 3:473-484.

Bioassay for Toxic Properties

L.J. Sikora and W.D. Burge

Mutagens

General Discussion--

The genetic basis for much human disease is well known. It can result from either gene mutation, chromosome rearrangement, or abnormal numbers of chromosomes (deSerres, 1976). The potential for an increased rate of genetic damage to humans and parallel increases in rates of carcinoma in modern industrial society has been reviewed by a number of authors (Ames, 1979; deSerres, 1976; Flamm, 1978; Fishbein, 1979). The problems in identifying chemical agents capable of producing genetic damage are great. As yet, methodology covering all aspects has not been devised, but methodology for identifying mutagenic activity is currently under intensive examination. This activity has been shown to be highly correlated with the carcinogenicity of substances. The carcinogenicity of the substances was originally determined as the result of epidemiological studies or by use of mice and rats or other experimental animals. The mutagenic activity has been revealed by use of a wide variety of assay systems ranging from bacteria, fungi, and insects to mammalian cells in culture (deSerres, 1976). Compared to assays utilizing mice or rats these procedures are rapid and inexpensive and have made it possible to conduct extensive screening of chemicals. Their application to environmental samples that might contain known or unknown chemicals capable of mutagenic activity is very recent.

The need to determine the fate of mutagenic chemicals in wastes applied to soils is apparent. EPA (1978) has devised a list of three groups of these recently developed bioassay systems for screening wastes and waste-treated soils for mutagens. Mutagens present in wastes applied to soils may be inactivated, but nonmutagenic chemicals have been converted to mutagens by bacterial intestinal flora (deSerres, 1976), in the sap of corn plants (deSerres, 1976), and in soils (Bartha and Pramer, 1976). Mutagens leaching into the environment have been shown to accumulate in shellfish (Parry and Al-Mossawi, 1979). Wastes containing mutagens that do not breakdown in soil or wastes containing substances convertible to nondegradable mutagens may present an environmental hazard that would preclude their application to soils. Wastes containing mutagens taken up by plants or chemicals converted to mutagens upon uptake by plants would be similarly excluded.

The chemicals responsible for the mutagenic activity in wastes/soils and plants should be identified. Accumulation of information on the role of possible mutagens in producing genetic damage and carcinomas may eventually demonstrate their involvement or eliminate them from consideration. Further, knowledge of their identity may be useful in epidemiological studies.

Use of fish and invertebrate bioassays.

The unknown toxicity of industrial waste discharges into waterways was the primary stimulus for the development of several bioassays. A bioassay is a biological measurement of the effects of a chemical or a group of chemicals with the purpose of determining the toxicity of the material on important indicator life forms. The term "indicator" is used here to mean a life form which reacts on the average similarly to a larger group of animals or plants, and thus is a good example of the toxic effects on that larger group. For example, a fathead minnow would be an indicator of a larger group of gamefish in a region.

Bioassays have been used to bridge the gap between the effects of a material on a single species, be it plant or animal, and a larger, more important group or groups of life forms which are directly in contact with the material in question. Bioassays can be used to detect effects of unknown substances as well as known substances in which case the bioassay is superior to chemical analyses. For chemical analyses as a toxicity determination, the specific component must be identified. Moreover, chemical assays for toxicity are valid only when results can be related to biological responses.

The benefit of bioassays to provide data on toxicity is obtained only when strict protocol for the bioassay is followed. The basic procedure for any bioassay for lethal concentration determination involves establishing the combination of concentration and exposure time which permits partial survival of the test organism. The percentage survival often chosen for measurement is 50% since this value is most accurate and permits bioassay data to be examined by probit analysis and manipulation by analytical/ mathematical techniques (Walden, 1976). The bioassay must be designed so that the effect of the toxicant is <u>not</u> indirect in that the addition of the test material to, for example, a fish tank results in changes of pH, decreases in O_2 content, or precipitation of essential nutrients. An incorrect conclusion on the toxicology of the test material may be formed under these conditions. Because time-concentration curves are vital in bioassays, the concentration of the toxicant in the test medium should ideally not change during the time period. In pulp and paper effluent bioassays, solution replacement is mandatory to maintain the concentration of the toxicants, which are otherwise depleted because of their instability, or because of their uptake by test fish (Walden, 1976).

The selection, care, and maintenance of test organisms are critical if the biological variation in the bioassay is to be minimal. The test organisms should react the same under control conditions in successive tests as well as within a reasonable, statistical variation within the same concentration-time bioassay. For instance, death of test fish during acclimation and maintenance prior to bioassay is indicative of stress, and bioassays should not be undertaken while these conditions occur (Walden, 1976).

The concentrations of a chemical to be tested should be chosen based upon data of concentrations of the chemical in the waste material and in the final disposal site.

Two types of bioassays have been routinely used for the determination of toxicity in aquatic systems. Acute toxicity bioassays are used to determine the lethal dosage of a waste material and sublethal bioassays are used to determine the dosage at which stress or poor performance of the test animal is observed. Warren and Doudoroff (1958) felt that acute bioassay results are, in themselves, valueless and must be related to the concentration producing no deleterious effects. The objective of sublethal investigations (or chronic toxicity tests) is to determine the effects of low concentrations of pollutants or chemicals on life forms from which the data can be used to determine threshold levels. Threshold levels are those concentrations or dilutions which show an effect and below which no effect is observed. The sublethal bioassay is valuable to industries such as the pulp and paper industry because these industries produce low toxicity, high volume materials which may have more chronic than lethal effects.

Bioassays have been used for several years but not until the last 20 to 30 years, when the environmental damage of waste disposal practices has been scrutinized, has the development of several bioassays using divergent life forms appeared. In Canada, bioassays have been an integral part of environmental protection efforts for several years and effluent control regulations are written based on bioassay data (Zanella and Berben, 1979). The Canadian regulations specify a test fish (rainbow trout) and a test methodology.

Union Carbide has evaluated the toxicity of 50 synthetic organic compounds of major commercial significance and indicated that brine shrimp provided a rapid and precise means of evaluating the acute toxicity of these petrochemicals in the seawater environment (Price, Waggy, and Conway, 1974).

Walden (1976) recently wrote a review on the bioassays used in the pulp and paper industry. Fish were the primary test organisms used in the acute studies and included salmon, trout, and goldfish. Both adult and juvenile forms were tested. Other bioassays included invertebrates such as shrimp, (fresh and seawater), daphnia, stoneflies, mayflies, and caddisflies.

Sublethal toxicity tests are most often performed with fish and include evaluations of loss of schooling, reluctance to eat, respiratory loss, abnormal gill movements, excessive mucous production, loss of equilibrium, and loss of weight. At times, biochemical studies are utilized in sublethal toxicity studies with some factor in carbohydrate metabolism monitored for changes.

The acute toxicity tests involve measurement of concentrations of the test material which will reduce the population of the test organism by 50% in a fixed exposure time. Exposure times routinely do not exceed 96 hr and are commonly 24, 48, or 96 hrs. The concentration of the test material which results in 50% die-off is referred to as LC_{50} or the lethal concentration permitting 50% survival. The units are those of the dilution factor of the test material with a diluent expressed as %(v/v). The same units are used in expressing the sublethal toxicity levels. Another means of expressing toxicity is the use of toxic units which relate the LC_{50} of a toxic constituent to its concentration in the test materials, and Chung, Meier,

$$\text{toxic unit (tu)} = \frac{\text{conc. of toxic constituent}}{LC_{50} - (96 \text{ hr})}$$

and Leach (1979) suggested that the toxicity of an effluent with a number of toxic constituents could be expressed as the sum of the various toxic units for each known constituent,

$$(96 \text{ hr})\ LC_{50} = \frac{100}{tu}$$

Several assumptions are made in considering the toxicity of an effluent by the summation of toxicities and these include non-synergistic effects of the toxic constituents, effects of constituents being the same (i.e., having the same toxic unit), their effects in different effluents, and that all toxic constituents in the effluent to be tested are known. Chung, Meier, and Leach (1979) found a reasonable correlation between the summation of toxic units of toxic constituents and bioassay results on the effluent. The value of such a determination is that the chemical analyses of an effluent can be accomplished in 1 1/2 hr versus the 96 hr necessary for the bioassay.

A diversity of organisms have been used for bioassays because researchers or industries have attempted to use an organism which is prevalent or is important in the region where the industry is located. Mollusks, such as oysters and clams, and other crustaceans, such as lobsters and crayfish, have been utilized in pulp and paper effluent bioassays (Walden, 1976). Compared to fish, limited work has been carried out on other aquatic organisms but the consensus of results from the studies is that the tolerance of invertebrates, notably insects and crustaceans, is comparable to or greater than that of fish (Walden and Howard, 1976).

More recent publications have estimated the importance of analyzing or bioassaying a combination of organisms to determine the effects of a toxic or new chemical on an ecosystem (Crow and Taub, 1979; Calamari, Galassi, and DaGasso 1979). In discussing the testing of a new chemical substance, Calamari, Galassi, and DaGasso (1979) suggested performing acute toxicity tests on organisms of at least three species at different trophic or food chain levels. The next step would be to assess the persistance of the molecule by means of biodegradability tests. The final analysis would involve a series of chronic assays selected on the basis of preliminary results of toxicity tests and persistance.

As hazardous wastes are permitted to be land applied, an animal and plant bioassay should be conducted to determine the effects of the amendment on these indicator life forms. Research data are needed to determine the types of bioassays which would be appropriate and the species to be utilized in the bioassay. As Calamari, Galassi, and DaGasso (1979) suggested, an assessment on several trophic levels such as primary producer, primary consumer, and secondary consumer should be made. Bioassays evaluating the effect on microorganisms should also be evaluated inasmuch as these organisms are responsible for 80 to 90% of the degradation of organic materials in soils. Bioassays are the mandatory bridge which spans the gap between the effects on ecosystems of hazardous waste application and the observed effects determined in a laboratory.

Assays for Plant Growth Inhibitors

Plant bioassays are generally based on measurements of seed germination, root, and hypocotyl elongation, or plant weight. Oats, soybean, mustard, sorghum, cucumber, corn, and numerous other plant species have been used as indicator plants in various bioassay tests (Craft, 1935; Espinoza, Adams, and Behrens, 1958; Swanson, 1946; Parker, 1966). The oat plant is probably the most frequently used indicator plant for herbicide bioassays because of its high sensitivity to many herbicide types and the ease with which it is grown (Horowitz, 1976).

Germination--

Many chemicals and waste materials may inhibit or delay the germination of various plant species. Bioassay tests utilizing the germination of seeds in Petri dishes or similar containers have been used to study bark extracts, various chemicals and volatile organic compounds (Still, Dirr, and Gartner 1976; Durrant, Draycott, and Payne, 1974; Mueller, 1965). Increasing germination inhibition of various plant species was noted in each study as the concentration of the test material increased. Similar results were also observed during a bioassay in which seeds were germinated in pots filled with mixtures of soil and sewage sludges (Wollen, Davis, and Jenner, 1978).

The effects of a chemical on seed germination have been shown to vary depending on seed size and species (Mueller, 1965). Recently, investigators have also shown that presoaking will often decrease the effect of a chemical or waste material on germination (Durrant, Draycott, and Payne, 1974).

Plant bioassays, however, are seldom based on germination tests alone, since sublethal concentrations of a chemical may severely inhibit the growth of plant roots or shoots (Horowitz, 1976).

Seedling Elongation--

Inhibition of root and hypocotyl elongation has been used as a rapid plant bioassay for herbicides (Swanson, 1946; Parker, 1966), bark extracts, various organic chemicals and elements (Still, Dirr, and Gartner, 1976; Cappaert, Verdonck, and DeBoodt, 1977). Most elongation studies have been conducted on filter paper, sand, agar, or some other media placed in Petri dishes. The seedlings are grown for a short prescribed period (one to five days) and the length of the shoot or root is measured.

Other elongation bioassays have used "ragdolls", solution cultures, or slanted glass trays for growing seedlings (Konzak, Polle, and Kittrick 1976; Ratsch, Grothans, and Tingey, 1979; Ramirez and Sikora, 1980). Investigation, however, has shown that seedlings grown by solution culture techniques may show phytotoxic effects at lower levels of chemical concentration than when they are grown on filter paper (Konzak, Polle, and Kittrick, 1976). Apparently, the roots are able to establish a beneficial gradient near the filter paper-root boundary which is not possible in the solution culture system. Some investigators have indicated that pregerminated seedlings or advancing seeds will result in less response variation by the indicator seedlings. However, presoaking seeds has also been found to reduce the relative response of seedlings to test chemicals (Durrant, Draycott, and Payne, 1974).

Experimental measurements obtained from seedling elongation studies are often expressed as a percent of the control length. But for comparison between various environmental conditions, or between a series of studies, similiar activity values are widely used. Values such as GR_{50} and EC_{50} (the dose or concentration which causes a growth reduction of 50% or which reduces elongation by 50%) or similar expressions are often used in plant bioassays (Behrens, 1970; Finney, 1964). Seedling diameter and relative curvature have also been used to study various chemicals (Imai and Siegel, 1973).

Plant Growth--

Bioassays in which plants are allowed to develop over several weeks or longer have been extensively used in herbicide investigations. Investigators have shown that many plants do not respond to herbicides, especially photosynthetic inhibiting types, while a substantial supply of nutrients is readily available to the seedling from the seed. Therefore, indicator plants are often grown until growth stages are reached at which photosynthesis becomes the primary energy source rather than the seed (Kratky and Warren, 1971). Furthermore, phytotoxicity due to plant accumulation of various elements may not be easily observed until the latter stages of plant development.

Plant growth bioassays are often conducted by growing an indicator plant species in a series of small pots to each of which a particular soil or medium and one of several concentrations of the test chemical have been added (Santelman, 1972). This is compared to the same plants grown in the same soil or media but which contains an unknown concentration of the chemical. Other plant growth bioassays which have utilized solution culture and hydroponic methods have been conducted to study aluminum toxicity and cationic herbicides respectively (Konzak, Polle, and Kittrick, 1976; Weber and Scott, 1966).

The determination of plant weight is probably the most commonly measured plant response used in plant growth bioassays. Usually, the plant tops alone are weighed since the separation of the roots from the growing media is often laborious and time consuming. Other commonly used measurements consist of leaf length, plant height, and petiole length of the first trifoliate leaf. These measurements are used to take into account modifications of growth habit (i.e., lodging and wilting). Height reduction in certain cases has been shown to be affected by certain chemicals more so than weight (Koren, Foy, and Ashton, 1968). Numerical scores have been utilized for visual estimation of relative plant injury intensity (Day, Jordan, and Hendrixson, 1961). However, the use of numerical scores is inherently subjective and therefore visual estimates do not result in a highly reproducible measurement. Reduction in photosynthesis, CO_2 exchange, water consumption, chlorophyll content and the recording of phytotoxic symptoms such as malformations of leaves, stem deformation, and chlorosis have been used to detect the presence of various herbicides and/or to determine the concentration present (Behrens, 1970).

Similar activity values, such as GR_{50} and ED_{50} for reporting the reduction in weight or height in plant growth bioassays, are found throughout the literature. The use of these values for assessing plant growth is discussed in detail by Finney (1964).

Summary and Research Needs

The application of hazardous wastes to land mandates that some type of plant bioassay should be established to evaluate general phytotoxicity of waste-amended soil before land application. Results from this bioassay may be used to determine the proper species to be used as plant cover or whether a harvestable crop could be established at waste disposal sites. Crop cover at these sites is necessary to stabilize the soil surface and to reduce surface water runoff.

The mutagenic activity of chemicals as identified by rapid, relatively inexpensive bioassays utilizing microbes and cell cultures has been correlated to a high degree with their carcinogenicity. To lower the genetic disease burden for man, other animals, and plants, the potential for land treatment to destroy mutagens in hazardous wastes warrants investigation. Also, the possible generation of mutagens in soils and plants from nonmutagenic substances in wastes during land treatment of hazardous wastes needs to be determined. In addition, the potential for movement of mutagens from the soil by leaching into groundwater or uptake by plants needs to be determined. The specific chemicals responsible for mutagenic activity in wastes/soils and plants need to be identified. This information may be of great value to epidemiological studies. Further research may either prove or disprove the carcinogenicity of these chemicals.

REFERENCES

Ames, B. N. 1979. The detection and hazards of environmental carcinogens/ mutagens. p. 1-11. In D. Schuetzle (ed.). Monitoring toxic substances. ACS Symposium Series 94. American Chemical Society, Washington, D.C.

Bartha, R., and D. Pramer. 1967. Pesticide transformation to aniline and azo compounds in soil. Science 156:1617-1618.

Behrens, R. 1970. Quantitative determination of triazine herbicides in soils by bioassay. Residue Rev. 32:355-369.

Calamari, D., S. Galassi, and R. DaGasso. 1979. A system of tests for the assessment of toxic effects on aquatic life. An experimental preliminary approach. Ecological and Environ. Safety. 3:75-89.

Cappaert, I. M. J., O. Verdonck, and M. DeBoodt. 1977. Degradation of bark and its value as a soil conditioner. pp. 123-129. In Proc. of the Symposium on Soil Org. Matter Studies. IAEA. Vienna, Austria.

Chung, L. T. K., H-P. Meier, and J. M. Leach. 1979. Can pulp mill effluent toxicity be estimated from chemical analyses. pp. 141-145. In Proc. Environ. Conf. of TAPPI.

Crafts, A. S. 1935. The toxicity of sodium arsenite and sodium chlorate in four California soils. Hilgardia 9:462-498.

Crow, M. E., and F. A. Taub. 1979. Designing a microcosm bioassay to detect ecosystem level effects. Int'l. J. Environ. Studies. 13:141-147.

Day, B. E., L. S. Jordan, and R. T. Hendrixson. 1961. The decomposition of amitrole in California soils. Weeds 9:443-456.

deSerres, F. J. 1976. The application of short-term tests for mutagenicity in the toxicological evaluation of environmental chemicals. p. 3-11. In F. J. deSerres, J. R. Fouts, J. R. Bend and R. M. Philpot. (ed.). In vitro metabolic activation in mutagenesis testing. Proc. of the Symposium on the Role of Metabolic Activation in Producing Mutagenic and Carcinogenic Environmental Chemicals, Research Triangle Park, N.C., Feb. 9-11, 1976.

Durrant, M. J., A. P. Draycott, and P. A. Payne. 1974. Some effects of sodium chloride on germination and seedling growth of sugar beets. Ann. Bot. 38:1045-1051.

E.P.A. 1978. Fed. Register. Dec. 18, 1978, Part IV: 58960.

Espinoza, W. G., R. S. Adams, Jr., and R. Behrens. 1958. Interaction effects of atrazine and CDAA, linuron, amiben or trifluralin on soybean growth. Agron. J. 60:183-186.

Finney, D. J. 1964. Statistical method in biological assay. Charles Griffin Co. Ltd. London.

Fishbein, L. 1979. Potential industrial carcinogens and mutagens. Elsevier Scientific Publishing, New York. 534 p.

Flamm, W. G. 1978. Genetic disease in humans versus mutagenicity test systems. p. 1-8. In W. G. Flamm, and M. A. Mehlman (ed.). Advances in modern toxicology, vol. 5, mutagenesis. John Wiley and Sons, New York.

Horowitz, M. 1976. Applications of bioassay techniques to herbicide investigations. Weed Res. 16:209-215.

Imai, I., and S. M. Siegel. 1973. A specific response to toxic cadmium levels in red kidney bean embryos. Physiol. Plant 29:118-180.

Konzak, C. F., E. Polle, and J. A. Kittrick. 1976. Screening several crops for aluminum tolerance. pp. 311-327. In M. J. Wright (ed.) Proc of Workshop on Plant Adaptation to Mineral Stress in Problem Soils, Beltsville, MD.

Koren, E., C. Foy, and F. M. Ashton. 1968. Phytotoxicity and persistence of four thiocarbamates in five soil types. Weed Sci. 16:172-175.

Kratky, B. A., and G. F. Warren. 1971. A rapid bioassay for photosynthetic and respiratory inhibitors. Weed Res. 19:658-661.

Mueller, W. H. 1965. Volatile materials produced by *Salvia leucophylla*: Effects on seedling growth and soil bacteria. *Botan. Gaz.* 126:195-200.

Parker, C. 1966. The importance of shoot entry in the action of herbicides applied to the soil. Weeds 14:117-121.

Parry, J. M., and M. A. J. Al-Mossawi. 1979. The detection of mutagenic chemicals in the tissue of the mussel *Mytilus edulis*. Environ. Pollution 19:175-186.

Price, K. S., G. T. Waggy, and R. A. Conway. 1974. Brine shrimp bioassay and seawater BOD of petrochemicals. Jour. Water Pollut. Control Fed. 46:63-77.

Ramirez, M. A., and L. J. Sikora. 1980. Development and evaluation of a plant bioassay for toxicity determination. Agron. Abst. p. 35.

Ratsch, H. C., L. C. Grothans, and D. T. Tingey. 1979. Root elongation protocol for complex effluents. *In* D. J. Brusick, R. R. Young, et al. I.E.R.L.-R.T.P. Procedures Manual: Level 1 Environmental Assessment Biological Test. US EPA contract #68022681. Indust. Environ. Res. Lab. Research Triangle, NC. (In press).

Santelman, P. W. 1972. Herbicide bioassay. pp. 91-101. *In* Research Methods in Weed Science. Weed Sci. Soc., USA.

Still, S. M., M. A. Dirr, and J. B. Gartner. 1976. Phytotoxic effects of several bark extracts on mung bean and cucumber growth. J. Amer. Soc. Hort. Sci. 101:34-37.

Swanson, C. P. 1946. A single bioassay method for the determination of low concentrations of 2,4-dichlorophenoxyacetic acid in aqueous solution. Bot. Gaz. 107:507-509.

Walden, C. C. 1976. The toxicity of pulp and paper mill effluents and corresponding measurement procedures. Water Res. 10:639-644.

Walden, C. C., and T. E. Howard. 1976. Toxicity research and regulation. pp. 93-99. *In* Proc. Environ. Conf. of TAPPI.

Warren, C. E., and P. Doudoroff. 1958. The development of methods for using bioassays in the control of pulp mill waste disposal. TAPPI 41:211-216.

Weber, J. B., and D. C. Scott. 1966. Availability of a cationic herbicide absorbed on clay minerals for cucumber seedlings. Science 152:1400-1402.

Wollen, E., R. D. Davis, and S. Jenner. 1978. Effects of sewage sludge on seed germination. Environ. Pollut. 17:195-205.

Zanella, E. F., and S. A. Berben. 1979. Evaluation of methodologies for the determination of acute toxicity in pulp and paper effluents. pp. 133-139. In Proc. Environ. Conf. of TAPPI.

Fate of Pathogens

W.D. Burge

Nature of Pathogenic Organisms

Pathogenic organisms in manufacturing wastes may have originated from a number of sources. Animals processed for food or other uses may at times have significant levels of infection that will result in pathogens being introduced into the wastes. Many manufacturers combine their plant sewage streams with their manufacturing wastes. In addition it may be advantageous at times to combine municipal sewage wastes with manufacturing wastes for land treatment. Although chlorination of wastewater treatment plant effluents and anaerobic or aerobic digestion of sewage sludges are capable of greatly reducing the number of pathogens in this material, the destruction of these organisms has been shown not to be complete (Foster and Engelbrecht, 1973).

The pathogenic organisms present in sewage and animal processing wastes can be placed into four groups of organisms: bacteria, viruses, intestinal worms, and protozoans. The kinds and numbers of these organisms in sewage and the diseases they cause have been discussed in a number of reviews (Foster and Engelbrecht, 1973; Burge and Marsh, 1978; Aiken et al., 1978). Some of the common bacterial pathogens that may be present are *Salmonella enteritidis*, *Salmonella typhi*, *Shigella* spp., *Vibrio comma*, *Mycobacterium tuberculosis*, and enteropathogenic *Escherichia coli*. The respective diseases caused by these organisms are salmonellosis, typhoid fever, shigellosis, cholera, tuberculosis and diarrhea. Common viruses isolated from sewage wastes include poliovirus, echovirus, coxsackievirus, reovirus, adenovirus, and rotavirus (Gerba, Wallis, and

Melnick, 1975). The viral agent for infectious hepatitis may also be present. The diseases caused by viruses vary from the common cold to fatal infections of the central nervous system, and many of the above viruses are capable of producing the entire spectrum of infections. The intestinal parasites most often found in sewage of temperate regions are the ova of Ascaris lumbricoides (round worm), Trichuris trichiura (whipworm), Enterobius vermiculus (pin worm), Hyenolepis spp. (dwarf tapeworm), and Taenia saginata (beef tapeworm), and cysts of the protozoan Entamoeba spp. (Hays, 1977). Most textbooks on parasitology mention that intestinal worm infections are usually subclinical and only become serious in people, usually children, weakened by malnutrition. Protozoan infections most often are subclinical but can result in diarrhea with extreme weakening of the host. Entamoeba histolytica is the most pathogenic of the protozoa, but in recent years Giardia lamblia, an organism formerly thought not to be pathogenic, has become the protozoan of greatest health significance (Aiken et al., 1978). Outbreaks of giardiasis have occurred recently in communities in California, Colorado, Oregon, and Pennsylvania that chlorinate but do not filter their drinking water obtained from surface sources such as streams, rivers, and lakes (Center for Disease Control, 1980). Lack of filtration apparently allows chlorine resistant Giardia cysts that may have originated from wildlife or sewage pollution into the drinking water distribution system.

The kinds of pathogens that might be expected in animal processing wastes are similar to those in municipal sewage wastes. According to Stauch (1976) who has reviewed the literature on animal wastes, salmonellae and leptospires provide the greatest hazard. In Germany a herd of dairy cattle was infected with salmonellosis from the feeding of meatmeal from a rendering plant. Salmonellosis in man is a foodborne disease. Leptospirosis infection in man is usually the result of contact with urine or water contaminated by urine of infected animals (Davis et al., 1973).

Survival

The enteric pathogenic bacteria and viruses present in sludge or sludge-amended soils do not have a special stage resistant to destruction such as that found in the spore-forming bacteria and fungi. They must reproduce or undergo inactivation. Viruses cannot reproduce external to cells of their hosts, and are inactivated at a rate that is directly related to temperature and other factors. Of the pathogenic enteric bacteria, as far as we know, only salmonellae are capable of growth in sludge/soil systems, but even these organisms usually face severe microbial competition and potential elimination.

Persistence of viruses is favored by low temperatures, alkaline pH, high levels of soil organic matter and moisture, and shielding from ultra-violet radiation (Lance, 1977). Presumably, these same conditions also favor bacterial survival. A number of reviews (Gerba, Wallis, and Melnick, 1975; Burge and Marsh, 1978) have listed the reported survival times for enteric pathogens in soil. Virus survival, even when favored by low temperatures in cold climates, probably seldom exceeds 3 to 4 months. Coxsackievirus type B3 seeded into sludge applied to soil of a lysimeter study in Denmark was inactivated at a rate of about one log unit per month (Damgaarde-Larsen, Jensen, and Nissen, 1977). In soil flooded with virus seeded sewage sludge and effluent, Larkin, Tierney, and Sullivan (1976) found that poliovirus type 1 survived for 96 days during winter months. The longest survival in the summer was 36 days on spray irrigated vegetables. Most pathogenic bacteria are also eliminated in a few months after application, but salmonellae may persist by regrowth in sludge treated soils. In sludge trenching studies at Beltsville (Walker et al., 1975), salmonellae could still be detected up to about 1 year, but this can be considered a maximum period for survival.

Protozoan cysts are susceptible to destruction by drying. Cysts of *Entamoeba histolytica* have been found to persist up to 8 days in soil but for less than 3 days when exposed on the surface of vegetables. *Ascaris* ova, however, have been shown to persist for as long as 7 years in soil (Parsons et al., 1975).

Movement

Manufacturing sludges from combined sewage and manufacturing waste streams and combined manufacturing wastes and municipal sludges applied to soils should be similar to municipal sludges in their ability to contain pathogens. When municipal sludges are applied to soil, the bacteria and viruses are intimately associated with the sludge. Cliver (1976) is of the opinion that the vigorous mechanical treatment needed to free viruses from sludge is evidence that they are associated with sloughed intestinal epithelium initially present in the fecal material. Movement of these organisms with aerosols, at least from filter-cake sludge, would seem unlikely. The greatest potential for movement would be with the sludge and soil as a result of erosion. For example, Walker et al. (1975) found that a tile drain intercepting subsurface and surface runoff from a watershed used in a sewage-sludge trenching study carried water contaminated with fecal coliforms, but a tile drain intercepting only groundwater was essentially free of these organisms. For the organisms to have moved into the subsurface drain they would first have to move from the sludge and then through the soil.

Movement of bacteria, viruses, helminthic ova and protozoan cysts with sludge particles, or movement of these organisms independent of the sludge material, would necessitate water movement through the sludge. Trenching and subsurface injection of sludge, and, to a great extent, even mixing of sludge into the soil surface leaves the sludge and soil as discrete entities. Compared to most soils, relatively large forces are required to move water through sludge. The water will tend to move around the sludge and through the soil until the sludge is altered considerably by biological processes.

The lack of water movement through sludge was demonstrated in a study conducted by Damgaarde-Larsen, Jensen, and Nissen (1977),who mixed sewage sludge containing coxsackievirus type B3 and tritiated water into the surfaces of sandy and clayey soils contained in lysimeters. After six months, only a small amount of the tritiated water was found in the leachate and in only that from the sandy soil. No virus was found in the leachate.

If pathogenic organisms should be washed from sludge and become associated with the liquid phase, removal by the soil from the liquid phase during infiltration and percolation constitutes the next barrier against migration to groundwater. Protozoan cysts and helminthic ova are too large to move through most soil-pore systems. In their review, McGauhey and Krone (1967) concluded that soils would provide an efficient medium for removal of the bacteria from infiltrating water and hypothesized that viruses also would be effectively removed. More recent work has shown that soils can effectively remove viruses from water but that under certain conditions both viral and bacterial movement may be extensive. Rapid infiltration systems for renovation of sewage or sewage effluents would be least effective at removing pathogenic organisms because extremely coarse soils must be utilized to ensure high flow rates. In one particular study where sewage was applied at extremely high rates to a system consisting of coarse unconsolidated alternating layers of silty sand and gravel, Schaub and Sorber (1977) found that fecal streptococcal bacteria, a tracer virus (f2 bacteriophage), and endemic enteric viruses flowed laterally some 183 m from the site of application to eventually contaminate groundwater.

In another study involving renovation of secondary sewage effluent in a dry river bed consisting of a loamy sand, low levels of fecal coliforms were detected in a well 45 m from the application site, but not in a well 91 m distant (Bouwer, Lance, and Riggs, 1974). However, viruses were not detected in sampling wells during a similar study (Gilbert et al., 1976). Why bacteria moved more readily than viruses is not clear. In column studies using soil from this same river bed, both viruses and bacteria were removed from infiltrating water at relatively shallow depths (Lance, Gerba, and Melnick, 1976).

Regulations for landfarming of hazardous wastes require fine grained soil types (OH, CH, MH, CL, and OL) for landfarms as defined by the United States Soil Classification System (Federal Register, Vol. 43, No. 243, p. 58990). These types of soils are agricultural soils. Most agricultural soils such as these would not be suitable for land treatment of sewage since the water infiltration rate would be too slow. Evidence reviewed by Lance (1977) indicates that most agricultural soils will effectively remove both bacteria and viruses from infiltrating water. Both of these groups of organisms can be removed by adsorption mechanisms, but because of their larger size, it is likely that bacteria are retained near the soil surface to a greater extent than viruses because of a physical clogging of the pores or voids. Viruses, on the other hand, are very small and move readily through most soil pores provided they are not adsorbed by soil particles.

Health Hazards

Humans--

Workers--The risks of contracting disease for workers landspreading wastes containing admixtures of sewage from employee waste streams or from municipal waste treatment plants would appear to be less than or, at least, no greater than that experienced by sewage farm, sewer maintenance, and wastewater treatment plant workers. Burge and Marsh (1978) concluded from their literature review that although available epidemiological data for these groups of workers are insufficient for definitive conclusions, they do not seem to indicate any acute hazards. A recent sero-epidemiologic study by Clark et al. (1980) helps support this view. Their study of more than 100 workers in the activated-sludge treatment plants located in three different cities failed to demonstrate any increased risk of infection or elevation of antibody titers to disease agents known to be in the wastewater treatment plant environment over that of control groups of workers. The cities were Cincinnati, OH; Chicago, IL; and Memphis, TN.

Community--Infectious disease hazards to a community, families, or individuals living near a hazardous waste land-farming site could come from pathogens physically transported from the site by particulate or liquid aerosols, by insects or small animals, by the clothes of resident workers or by surface flow or movement of water through the soil to ground water and eventually to potable water sources. Further, the pathogens could be transported by workers or animals infected at the site.

The threat presented by any of these routes from a land-farming site would seem less than that for communities surrounding wastewater treatment plants with the exception of the problem of surface runoff or movement through the soil. However, no studies have as yet demonstrated a clear danger to communities surrounding wastewater treatment sites. This is despite the fact that it has been thoroughly documented that liquid aerosols from wastewater treatment plants can carry pathogens exposing people residing nearby (Hickey and Reist, 1975).

In an attempt to determine health effects, epidemiological studies were made recently of urban populations surrounding wastewater treatment plants near Chicago, IL (Johnson et al., 1978), in Tecumseh, MI (Fannin et al., 1978), and in Skokie, IL (Carnow et al., 1979). Also, attendance at an elementary school was monitored for periods before and after construction of a wastewater treatment plant adjacent to the school grounds (Johnson et al., 1979). Evidence that the students were exposed to pathogen-containing aerosols was obtained by monitoring air concentrations close to aerosol sources for short periods using high volume aerosol samplers and extrapolating these data to the air concentrations students would encounter in classrooms and on the school playground. Calculations indicated that students would probably respire pathogens. In the Chicago, and Skokie, IL studies, serological techniques and health questionnaires were used to determine exposure. For the Tecumseh, MI study, only a health questionnaire was used. In all of these studies, there was no evidence of serious adverse effects on the health of the various populations; however, some relatively minor adverse effects were noted. For the Chicago, IL study, there was evidence of some increased skin disease, nausea, weakness, diarrhea, and breathing pain as compared with the control populations. Alpha- and gamma-hemolytic streptococcal throat isolations were higher for the aerosol-exposed populations, but this was regarded as having no health significance. For the Tecumseh, MI study, disease incidence was increased for people living within 60 m of the sewage plant, but this was thought to be related to low income and housing density, rather than to nearness to the plant. In the Skokie, Illinois study, the authors cautioned that, although there were no obvious adverse health effects, the population receiving the highest dosage levels was very small, making these conclusions somewhat tenuous.

The threat to people living near hazardous waste landfarming sites by runoff and leaching of pathogens would depend upon the nature of the site and its management. For sites with proper soils and grades to minimize the threat of erosion, proper management to prevent erosion should not be difficult, and application of water at rates that would produce pathogen movement through the profile would be difficult to achieve as previously mentioned.

Summary and Research Needs

Pathogens in industrial wastes may originate from the waste products of animals processed for food and other uses, and the combining of employee sanitation waste streams with manufacturing waste streams. The pathogenic organisms present will include bacteria, viruses, eggs of intestinal worms, and cysts of protozoans. Although standard methods of secondary treatment (sewage activation and anaerobic sludge digestion) will destroy many of these pathogens, some are capable of survival and may persist in chlorinated effluents and digested sludges produced for disposal. Primary wastes must be considered to be more hazardous than secondary wastes.

The health hazards to workers landfarming these wastes and to adjacent communities from infection by these pathogenic organisms would seem to be low, since studies of wastewater treatment plant workers and communities adjacent to wastewater treatment plants have failed to show any disease hazards in excess of those for controls.

The possibility of movement of pathogenic viruses or bacteria from the sites with surface runoff, or movement through the soil to eventually contaminate surface and ground waters will constitute the greatest potential disease hazard from the landfarming sites. Runoff hazards can be controlled by soil management practices to prevent erosion and capturing of runoff water for treatment or recycling if necessary. Movement through the profile can be controlled by utilizing soils whose characteristics are such that they will minimize the potential for movement. However, much needs to be known about movement of bacteria and viruses in soil so that the relationship between specific bacteria and viruses and the risk of their movement in certain soils under varying conditions can be evaluated. Basic studies on the factors influencing survival, adsorption, and movement of bacteria and viruses in soils are needed to change empirical approaches to more sound ones based upon well-grounded theory.

REFERENCES

Aiken, E. W., H. P. Pahren, W. Jakubowski, and J. B. Lukas. 1978. Health hazards associated with wastewater effluents and sludges: microbial considerations. In B. P. Sagik and C. S. Sorber (ed.) Risk assessment and health effects of land application of municipal wastewater and sludges. Center for Applied Research and Technology, The University of Texas, San Antonio, TX.

Bouwer, H., J. C. Lance, and M. S. Riggs. 1974. High-rate land treatment. II: Water quality and economic aspects of the Flushing Meadows project. J. Water Pollut. Control Fed. 46:844-859.

Burge, W. D., and P. B. Marsh. 1978. Infectious disease hazards of landspreading sewage wastes. J. Environ. Qual. 7:1-9.

Carnow, B., R. Northrop, R. Wadden, S. Rosenberg, J. Holden, A. Neal, L. Sheaff, and S. Meyer. 1979. Health effects of aerosols emitted from an activated sludge plant. Health Effects Research Laboratory, Office of Research and Development. U.S.E.P.A. Report 600/1-79-019.

Center for Disease Control. 1980. Waterborne giardiasis - California, Colorado, Oregon, Pennsylvania. Morbidity and Mortality Report 29:121-123.

Clark, C. S., G. L. VanMeer, C. C. Linnemann, H. S. Bjornson, P. S. Gartside, G. M. Schiff, S. E. Trimble, D. Alexander, E. J. Cleary, and J. P. Phair. 1980. Health effects of occupational exposure to wastewater. U.S.E.P.A. Symp. Wastewater Aerosols and Disease Proc. September 19-21, 1979.

Cliver, D. O. 1976. Surface application of municipal sludges. In L. B. Baldwin, J. M. Davidson, and J. F. Gerber (eds.), Virus aspects of applying municipal wastes to land. p. 77-83. Center for Environ. Programs, Institute of Food and Agricultural Sciences, Univ. of Florida, Gainesville, FL.

Damgaarde-Larsen, S., K. O. Jensen, and B. Nissen. 1977. Survival and movement of enterovirus in connection with land disposal of sludges. Water Res. 11:503-508.

Davis, B. D., R. Dulbecco, H. N. Eisen, H. S. Ginsberg, W. B. Wood, and M. McCarty. 1973. Microbiology. Harper and Row, Hagerstown, MD. 895 p.

Fannin, K. F., K. W. Cochran, H. Ross and A. S. Monto. 1978. Health effects of a wastewater treatment system. Health Effects Research Laboratory. Office of Research and Development, U.S.E.P.A. Report 600/1-78-062.

Foster, D. H. and R. S. Engelbrecht. 1973. Microbial hazards in disposing of wastewater on soil. p. 257-270. In W. E. Sopper and L. T. Kardos (eds.) Recycling treated municipal wastewater and sludge through forest and cropland. The Pennsylvania State Univ. Press, University Park, PA.

Gerba, C. P., C. Wallis, and J. L. Melnick, 1975. Fate of wastewater bacteria and viruses in soil. J. Irrig. Drain. Div., Amer. Soc. Civil Engr. 101:157-174.

Gilbert, R. G., R. C. Rice, H. Bouwer, C. P. Gerba, C. Wallis, and J. L. Melnick. 1976. Wastewater renovation and reuse: virus removal by soil filtration. Science. 192:1004-1005.

Hays, B. D. 1977. Potential for parasite disease transmission with application of sewage plant effluents and sludges. Water Res. 2:583-595.

Hickey, J. L. S. and P. C. Reist. 1975. Health significance of airborne microorganisms from wastewater treatment processes. II. Health significance and alternatives for action. J. Water Pollut. Control Fed. 47, 2758.

Johnson, D. E., D. E. Camann, H. J. Harding and C. A. Sorber. 1979. Environmental monitoring of a wastewater treatment plant. Health Effects Research Laboratory, Office of Research and Development, U.S.E.P.A. Report 600/1-79-027.

Johnson, D. E., D. E. Camann, J. W. Register, R. J. Prevost, J. B. Tillery, R. E. Thomas, J. M. Taylor, and J. M. Rosenfeld. 1978. Health implications of sewage facilities. Health Effects Research Laboratory. Office of Research and Development, U.S.E.P.A. Report 600/1-78-032.

Lance, J. C. 1977. Fate of pathogens in unsaturated and saturated soils. In 1977 National Conference on Composting Municipal Residues and Sludges. Silver Spring, MD. August 23-25, 1977. Sponsored by: Information Transfer Inc., and Hazardous Materials Control Research Institute.

Lance, J. C., C. P. Gerba, and J. L. Melnick. 1976. Virus movement in soil columns flooded with secondary sewage effluent. Appl. Environ. Microbiol. 32:520-526.

Larkin, E. P., J. L. Tierney, and R. Sullivan. 1976. Persistence of poliovirus in soil and on vegetables irrigated with sewage water. In L. B. Baldwin, J. M. Davison, and J. F. Gerba (eds.) Virus aspects of applying municipal wastes to land. Center for Environ. Programs, Institute of Food and Agricultural Sciences, Univ. Florida, Gainesville, FL.

McGauhey, P. H. and R. B. Krone. 1967. Soil mantle as a wastewater treatment system. Sanitary Engineering Research Laboratory, Rept. 67-11. Univ. California, Berkeley, CA.

Parsons, D., C. Brownlee, D. Wetler, A. Maurer, E. Haughton, L. Kornder, and M. Selzak. 1975. Health aspects of sewage effluents irrigation. Pollut. Control Branch, British Columbia Water Resour. Serv. Dept. Lands, Forests, and Water Resour., Victoria, British Columbia.

Schaub, S. A., and C. A. Sorber. 1977. Virus and bacteria removal from wastewater by rapid infiltration through soil. Appl. Environ. Microbiol. 33:609-619.

Stauch, D. 1976. Health hazards of agricultural, industrial and municipal wastes applied to land. pp. 317-342. In R. C. Loehr (ed.) Land as a waste management alternative. Ann Arbor Science Publishers, Inc., Ann Arbor, MI.

Walker, J. M., W. D. Burge, R. L. Chaney, E. Epstein, and J. D. Menzies. 1975. Trench incorporation of sewage sludge in marginal agricultural land. Environmental Protection Technology Series. EPA-600/2-75-034. 232 pp.

Composting of Chemical Industrial Wastes Prior to Land Application

G.B. Willson, L.J. Sikora, J.F. Parr

Introduction

Some chemical industrial wastes may be of such a highly toxic nature that their direct application to soil may cause extensive biocidal effects on the microflora, thereby impairing the soil's capacity to detoxify, degrade, and inactivate these chemicals. Shock loading effects from excessive applications of toxic organic chemicals to soils could last for weeks, months, or even years. Potential environmental pollution from other wastes may be difficult to manage in land treatment systems. Consequently, in developing effective and sustainable land treatment or landfarming systems for chemical wastes it will be necessary, in some cases, to consider methods for the initial pre-treatment or detoxification of certain waste chemicals. Composting may offer considerable potential for such accommodation.

The U.S. Department of Agriculture at Beltsville, Maryland in cooperation with The Maryland Environmental Service, recently developed an Aerated Pile Method for composting either undigested or digested sewage sludges. The method transforms sludge into usable compost in about 7 weeks, during which time the sludge is stabilized, odors are abated, and human pathogenic organisms are destroyed (Willson et al., 1980). The principal objective in composting sewage sludges is to produce an environmentally-safe, humus-like material that is free of malodors and pathogens, and can be used beneficially on land as a fertilizer and soil conditioner.

A number of definitions of composting have been proposed (Golueke, 1972; Gray, Sherman, and Biddlestone, 1971a,b; Finstein and Morris, 1975; Haug, 1980). However, the definition presented here, while similar to others, relates more appropriately to the composting method and technology developed at Beltsville. Thus, for the discussion which follows composting is defined as "the aerobic, thermophilic decomposition of organic wastes by mixed populations of indigenous microorganisms under controlled conditions which yields a partially-stabilized residual organic material that decomposes slowly when conditions again become favorable for microbiological activity" (Parr et al., 1978).

Factors Affecting the Composting Process

Composting is a microbiological process that depends on the growth and activities of a mixed population of bacteria, actinomycetes and fungi contained in the organic materials to be composted. When temperature, moisture and oxygen levels are favorable, these microorganisms will grow and aerobic decomposition proceeds. Nutrients such as carbon, nitrogen, phosphorus, and sulfur are released in this process and utilized by the microflora. As the activity continues, the temperature begins to increase from the heat generated through microbial oxidations and their respiratory functions. If the composting biomass is insulated, and assumes a geometrical form such as a pile or windrow, heat is retained for an extended period. Eventually, however, as the available carbon and other nutrients are depleted, microbial activity subsides, decomposition slows, and cooling occurs.

Poincelot (1975, 1977) discussed those factors which can most significantly affect the composting process and which must interact favorably to sustain rapid, effective, and efficient aerobic/thermophilic composting. The factors that affect composting are similar to those that influence the decomposition of residues and chemical wastes which were reviewed in an earlier section.

Temperature--

A typical time/temperature relationship for composting raw sewage sludge by the aerated pile method is shown in Figure 8. As microbial activity increases, temperatures increase rapidly (i.e., within several days) from the mesophilic into the thermophilic range which begins at about 40°C (104°F). At this point, the mesophilic microorganisms are inhibited by temperature while the thermophiles become active. Decomposition of the organic materials is most rapid in the thermophilic stage. Optimal temperatures for sewage sludge composting have been found to range from 60°C to 70°C. Higher temperatures can reduce the microbial

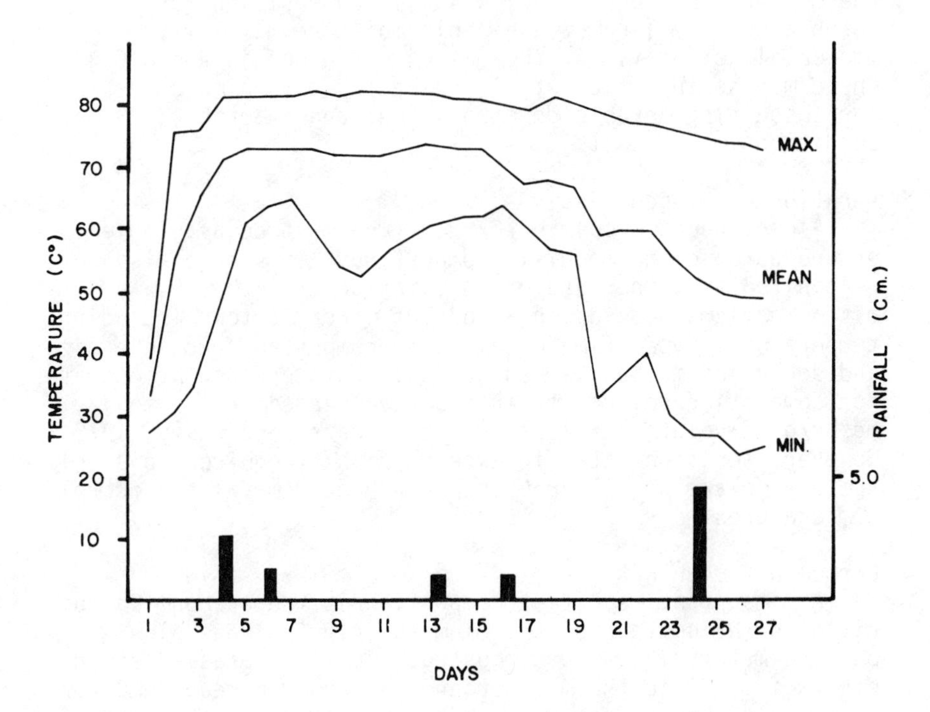

Figure 8. A typical time/temperature relationship for composting.

diversity that is essential for rapid, aerobic thermophilic composting. It has been hypothesized that much of the decomposition of organic substrates at high temperatures above 80°C may be principally due to chemical oxidations (Nell and Wiechers, 1978). Pathogen destruction has been shown to occur rapidly in the thermophilic stage according to specific time/temperature functions (Burge, Colacicco, and Cramer, 1979).

Moisture--

The optimal moisture content of the biomass for rapid aerobic/thermophilic composting is from 40 to 60% (w/w). If the moisture content is below 40%, decomposition may be aerobic but the lack of available moisture may limit microbial activity. If the moisture content is above 60%, there may be insufficient air space for oxygen movement and for sustaining aerobic decomposition, and anaerobic conditions may result.

Aeration or Oxygen Supply--

A continuous supply of oxygen (O_2) must be available to ensure aerobic/thermophilic composting. It should also be recognized that the rate of consumption of O_2 by microorganisms depends on a host of other factors including temperature, type of material being composted, particle size, and degree of mechanical agitation. Oxygen consumption rates for composting several materials have ranged from 2 to 13.7 mg/hr/g of volatile solids (Willson, Parr, and Casey, 1979). However, aeration rates in excess of that required to supply the oxygen are usually required for heat removal to control temperatures.

Carbon:Nitrogen Ratio--

An important aspect of composting is the carbon (C) and nitrogen (N) content of the biomass (e.g., sludge plus bulking material) to be composted, usually expressed as the C:N ratio. Since the microorganisms involved require C for growth and energy, and N for protein synthesis, the rate of decomposition is affected accordingly. Rapid and efficient composting is achieved with C:N ratios of between 25 and 35. Lower ratios can result in the loss of N as ammonia, while higher ratios can increase the amount of time required for composting. The availability of these nutrients to the microorganisms is perhaps more important than the actual ratio.

pH--

The optimum pH for rapid aerobic/thermophilic composting is somewhere between 6 and 7.5. Research at Beltsville has shown little effect of pH on the composting rate and product quality over a range of 5.5 to 11.0. However, composting was markedly suppressed, though temporarily so, at a pH level of 12 (Willson, Epstein, and Parr, 1976).

Particle Size Reduction (i.e., extent of grinding or shredding)--

In some cases, grinding or shredding of the starting materials can accelerate the rate of decomposition during composting by increasing (a) the surface area for microbial attack, and (b) the extent of contact between the sludge and the bulking material. Excessive grinding can reduce the porosity needed for air movement, which could lead to anaerobic conditions, and incomplete composting.

Other factors that can affect the composting process include the following.

Chemical Properties of Wastewater Sludges--

Wastewater sludges do vary considerably in their chemical composition, depending on the method of treatment used, and the kind and amount of industrial wastes discharged into the sanitary sewers. It follows then that the "compostability" of a particular sludge will depend on its chemical composition and the treatment process employed. Other chemical constituents of both sludges and bulking materials may, under some conditions, affect the composting process, depending largely on their concentration in the system. These include soluble salts, toxic organic chemicals such as pesticides, polychlorinated biphenyl's (PCB's), wood preservatives, sludge conditioning agents (lime, alum, and ferric chloride), and heavy metals.

Soluble Salts--

Sewage sludges contain varying amounts of soluble salts depending on their source, the type of wastewater treatment employed, and the extent of sludge dewatering. Where alum, lime, and ferric chloride ($FeCl_3$) are used as flocculating and conditioning agents during sludge treatment, potential problems can arise since the soluble salt content increases accordingly, and some plant species are highly sensitive to soluble salts. When prepared composts are used for agronomic and horticultural purposes, the danger of soluble salts is minimal because of dilution in soil after incorporation and mixing, and the extent of leaching after application.

Heavy Metals--

It is well known that certain heavy metals can cause adverse effects on the growth and metabolic processes of microorganisms. While a number of "high-metal" sewage sludges have been composted successfully at Beltsville with no apparent effect on the microflora, the metal concentrations seldom exceeded several thousand ppm. It is likely that the metal concentrations in some inorganic chemical industrial wastes will be substantially higher. There is little information on the effects of heavy metals and their interactions on the growth and activity of the composting microflora. The same can be said for the effects of toxic industrial organic chemicals.

Composting of Hazardous Industrial Wastes

There is growing interest in the feasibility of composting as a process for detoxifying, degrading, or inactivating hazardous wastes. Composting can provide a controlled system for containment of toxic constituents that may be subject to volatilization and leaching during biodegradation. While little research has been done in this area, most reports thus far are somewhat inconclusive and preliminary. Hill and McCarty (1967) reported the biodegradation of a number of chlorinated hydrocarbon pesticides during sewage treatment, though the effect of composting was not investigated. Rose and Mercer (1968) found that the insecticides diazinon, parathion, and dieldrin degraded rapidly when composted with cannery wastes. They also reported that DDT was relatively resistant to degradation. This is not surprising since it is now known that anaerobic conditions are necessary to support the initial reductive dechlorination of DDT to DDD. Such conditions did not occur in the system employed by Rose and Mercer. The results of composting petroleum refinery sludges were reported by Deever and White (1978) who found a significant reduction in the amount of toluene-hexane extractable grease and oil after composting. Potting mixes were prepared from the compost and satisfactory plant growth was obtained. Epstein and Alpert (1980) discuss other composting studies involving both crude and No. 6 oil, TNT, pulp and paper mill wastes, and pharmaceutical wastes. They suggest several classes of compounds that should be amenable to composting.

Combining Waste Streams for Composting Hazardous Industrial Wastes

As stated earlier, some waste industrial chemicals may be so highly toxic to the soil microflora that their direct application to soil is neither desirable nor feasible. Moreover, some chemical organic wastes may have adverse and undesirable chemical, physical, and/or microbiological properties that would either prevent or greatly limit the extent to which they could be composted alone. In such cases, it may be that selective combinations of these wastes with sewage sludges, municipal solid waste, or even agricultural wastes (i.e., crop residues and animal manures) may provide a readily compostable mixture. The combining of

certain chemical wastes with wastewater municipal s
a period of aerobic/thermophilic composting may pro
unique and desirable way of detoxifying, degrading,
inactivating toxic waste constituents prior to land
cation. The co-composting of toxic chemical wastes w
other municipal wastes, particularly sewage sludge, offers
certain advantages and useful inputs from the sludge
component. These include (a) sewage sludge contains an active indigenous population of microorganisms with a wide range in their degradative potential and physiological capabilities. This provides inoculum for some virtually sterile chemical wastes and enhances the probability of their degradation, (b) sewage sludge provides a unique buffering system for at least some neutralization of extremely acid or alkaline chemical wastes, (c) sewage sludge provides a readily available source of carbon, energy, and other nutrients to sustain a high level of biodegradative activity, and (d) sewage sludge mixed with a particular chemical waste lowers the effective concentration of toxic constituents through dilution and sorption mechanisms. High concentrations of some chemicals can inhibit microbial activity, and initial dilution may be necessary before active biodegradation can proceed.

Another approach might be that of applying varying rates of composted sewage sludge to land treatment/landfarming sites prior to the addition of chemical wastes. This would ensure an active indigenous population of microorganisms with the potential for rapid biodegradation of the toxic constituents. It would also provide the necessary nutrients to sustain microbial growth and activity, an essential buffering system, and sorption mechanisms for dilution and reduction of the effective concentration of toxic chemicals, thereby lowering the probability of biocidal effects.

The engineering, agronomic, economic, and environmental aspects of both of these approaches; i.e., co-composting and pre-conditioning of land treatment sites with composted or uncomposted sewage sludges, should be thoroughly explored by future research.

Design and Management Considerations

A number of different types of composting systems have been developed for composting municipal solid wastes. Some of these, described by Satriana (1974), may be adaptable for the co-composting of waste industrial chemicals and municipal solid wastes. The Aerated Pile Method developed at Beltsville for rapid aerobic/thermophilic composting of sewage sludges and cellulosic bulking materials could also be utilized for co-composting.

There are certain design and management considerations for effective composting of hazardous waste chemicals, by any of these systems, that are worthy of mention. These include (a) the degree of containment or enclosure (e.g., open air composting vs. closed vessel bioreactors), (b) degree of mechanization and automation, (c) type of material flow, and (d) the aeration system.

The degree of enclosure for composting systems ranges from little or none to completely enclosed facilities equipped with soil or compost filters for the exhaust gases. The probable need for close control of the composting process for detoxifying and degrading hazardous wastes would favor enclosed systems, although this may depend on the particular waste involved. In some cases, the exhaust gases may require filtering to prevent the loss of some volatile constituents to the atmosphere.

The degree of mechanization is mainly related to economic considerations and the volumes of materials to be handled. Worker safety, however, may impose some unique restrictions on the acceptable options. Automated control systems may be essential if close tolerances for operating conditions are necessary.

Materials flow patterns for composting systems can be classified as batch or continuous. The windrow and the aerated pile methods are examples of batch systems. Most of the enclosed systems are continuous slug flow; i.e., the materials flow continuously through the system with the oldest material exiting first. Recycling a portion of the compost may aid in conditioning the wastes for composting and may also provide a beneficial inoculum or adapted microflora to accelerate the biodegradation of specific wastes.

Aeration is probably the most important single consideration for managing the composting process. The availability or absence of oxygen determines whether the process is aerobic or anaerobic. Aeration is the principal means of removing water vapor and waste from the system. Regulation of the aeration rate controls the composting process rate since it is the only means of controlling the temperature. The aeration rate can also be varied to (a) control the rate of moisture removal, depending on the needs of the process and (b) provide alternating aerobic and anaerobic conditions to biodegrade waste chemicals having both oxidative and reductive steps in their degradation sequence.

Summary and Research Needs

Composting is a biological process that depends on the optimal growth and activity of a mixed population of mesophilic and thermophilic microorganisms. Rapid and effective aerobic/thermophilic composting is affected by a number of factors including temperature, moisture, C:N ratio, pH, particle size of materials. If these factors are not within optimum range the growth and activity of the microflora may be limited or even adversely affected and composting conditions may not be achieved. Composting can be controlled to select desired degradation processes and can be contained to prevent volatilization and leaching of hazardous constituents.

Composting of chemical industrial wastes or co-composting of these wastes with sewage sludges may offer an important means for initial detoxification, degradation, and inactivation of toxic constituents prior to their application to land treatment/landfarming systems. Such an approach could greatly minimize undesirable biocidal effects on the soil microflora, and potentially damaging environmental effects from direct land application. A study of the effectiveness of this approach as a management technique for the initial degradation of toxic chemical wastes should receive a high research priority.

Another approach to the land treatment of hazardous waste chemicals is that of amending or pre-conditioning a land treatment site with sewage sludge compost prior to the direct application of chemical wastes. The compost would provide an active indigenous microbial population for biodegradation, a highly effective and important buffering system, a means for diluting potentially biocidal concentrations of toxic chemicals, and carbon, energy, and nutrients to sustain a high level of biodegradative activity. The merits of this approach should also be fully explored by future research.

Research to evaluate composting methods and technologies for degradation of hazardous waste chemicals, and particularly under different conditions of aeration, should also be given a high priority. Effective composting technology, co-composting of chemical wastes prior to land application, and pre-conditioning of land treatment sites with compost prior to direct waste application could all markedly reduce the amount of land required for the accommodation of such wastes in land treatment/landfarming systems, and greatly minimize the potential environmental impact.

REFERENCES

Burge, W. D., D. Colacicco, and W. N. Cramer. 1979. Control of pathogens during sewage-sludge composting. p. 105-111. In Proc. of Natl. Municipal and Industrial Sludge Composting, Materials Handling. Information Transfer Inc., Silver Spring, MD.

Deever, W. R., and R. C. White. 1978. Composting petroleum refinery sludges. Texaco Inc., Port Arthur, TX. 24 p.

Epstein, E., and J. E. Alpert. 1980. Enhanced biodegradation of oil and hazardous residues. In Proc. Conf. on Oil and Hazardous Material Spills. Information Transfer, Inc., Silver Spring, MD. (In press)

Finstein, M. S., and M. L. Morris. 1975. Microbiology of municipal solid waste composting. Adv. Appl. Microbiol. 19:113-151.

Golueke, C. G. 1972. Composting: A study of the process and its principles. Rodale Press, Inc., Emmaus, PA. 110 p.

Gray, K. R., K. Sherman, and A. J. Biddlestone. 1971a. A review of composting - Part I. Process Biochem. 6(6):32-36.

Gray, K. R., K. Sherman, and A. J. Biddlestone. 1971b. A review of composting - Part 2. The practical process. Process Biochem. 6(10):22-28.

Haug, R. T. 1980. Compost engineering, principles and practice. Ann Arbor Sci. Publ. Inc., Ann Arbor, MI. 655 p.

Hill, D. W., and P. L. McCarty. 1967. Anaerobic degradation of selected chlorinated hydrocarbon pesticides. J. Water Poll. Cont. Fed. 39:1259-1277.

Nell, J. H. and S. G. Wiechers. 1978. High temperature composting. Water South Africa. 4:203-212.

Parr, J. F., G. B. Willson, R. L. Chaney, L. J. Sikora, and C. F. Tester. 1978. Effect of certain chemical and physical factors on the composting process and product quality. pp. 130-137. In Proc. of Nat'l Conf. on Design of Municipal Sludge Compost Facilities. Information Transfer, Inc., Silver Spring, MD.

Poincelot, R. P. 1977. The biochemistry of composting. pp. 33-39. In Proc. Nat'l Conf. on Composting of Municipal Residues and Sludges. Information Transfer, Inc., Silver Spring, MD.

Poincelot, R. P. 1975. The biochemistry and methodology of composting. Conn. Agric. Exp. Sta. Bull. No. 754. 18 pp.

Rose, W. W., and W. A. Mercer. 1968. Fate of pesticides in composted agricultural wastes. National Canners Association, Washington, D. C. 27 pp.

Satriana, M. J. 1974. Large Scale Composting. Noyes Data Corporation, Inc., Park Ridge, New Jersey. 269 pp.

Willson, G. B., E. Epstein, and J. F. Parr. 1976. Recent advances in compost technology. p. 167-172. In Proc. of the 3rd Natl. Conf. on Sludge Management, Disposal and Utilization. Information Transfer, Inc., Silver Spring, MD.

Willson, G. B., J. F. Parr, and D. C. Casey. 1979. Basic design information on aeration requirements for pile composting. pp. 88-99. In Proc. of Natl. Conf. Municipal and Industrial Sludge Composting - Materials Handling. Information Transfer, Inc., Silver Spring MD.

Willson, G. B., J. F. Parr, E. Epstein, P. B. Marsh, R. L. Chaney, D. Colacicco, W. D. Burge, L. J. Sikora, C. F. Tester, and S. B. Hornick. 1980. Manual for composting sewage by the Beltsville Aerated-Pile Method. Joint U.S. Dept. of Agriculture - U.S. Environmental Protection Agency publication. EPA-600/8-80-022. 64 pp.

PART II

RESULTS FROM LAND TREATMENT WITH SPECIFIC WASTES

Animal Wastes

S.B. Hornick

State universities, state agricultural experiment stations, USDA and cooperative regional research stations, and other agricultural researchers have thoroughly studied the benefits and problems associated with the landspreading of animal manures (Graber, 1973; Azevedo and Stout, 1974; Mathers et al., 1972; CAST, 1976; Southern Cooperative Series, 1979; Cornell University, 1969). This section will highlight the main concerns in the landspreading of animal wastes.

Of the 1.2 billion wet tons of manure produced by beef cattle, dairy cattle, horses, sheep, goats, swine, broilers, turkeys, and ducks (USDA, 1978), approximately 90% is used as a source of soil nutrients to fertilize and condition the soil for agricultural production. Beef cattle, dairy cattle, and horse excrement account for 89% of the total manure produced. In the past few decades, the demand for increased animal production for food purposes has removed many animals from the open range to confinement in feedlots, pens, and barns. This change from farming to agri-business has created a new problem: mainly, how to dispose of the manure in an environmentally safe manner. Approximately 35% of the total manure is produced in confinement (Graber, 1973).

The composition of manure is mainly dependent upon the feed ration. If a high roughage diet containing large amounts of lignin, hemicellulose, and other hard-to-digest materials is fed, the manure will contain large amounts of undigested feed, water, microbial cells, and wastes. Feedlot manure usually contains 85% water with the solid portion containing many organic compounds such as starches, sugars, proteins that contain N, and inorganic elements such as P and K which are essential plant nutrients. The range in composition for various animal excreta is shown in Table 10. The factors of major concern when landspreading an animal waste for its fertilizer value are the soluble salts in the inorganic fractions and the biodegradable materials of the organic fraction.

Soluble Salts

Since the soluble salt content of manure can be very high if animals are fed salt supplements, careful consideration must be made before the manure is applied to cropland. Whenever manure is applied in large quantities and limited amounts of irrigation or rainfall occur, the salts will not leach from the root zone and crop yields will be reduced.

Stewart and Meek (1977) showed that a waste containing 7.5% soluble salts could be applied to a soil at a 40 mt/ha rate to maintain a conductivity of 4 mmhos/cm or less if 10 cm of water moved through the profile annually. If higher application rates are used, then the plow depth should be increased so that the salts are distributed evenly over a greater soil mass. As discussed previously, soil conductivities of greater than 4 mmhos/cm will result in yield reduction.

Decomposition of Manure

The amount of organic decomposable or biodegradable materials in manure can be measured by the amount of oxygen which will be utilized by aerobic organisms that can degrade the organic material in an aerobic atmosphere at 20°C for five days. This measurement is the BOD_5. A more detailed discussion of BOD_5 is in an earlier chapter.

The BOD of the runoff from feedlot manure is very high and can easily deplete the O_2 supply in streams, rendering them anaerobic. In order to prevent the occurrence of this environmental problem, the feedlot runoff can be diluted and placed in a large, shallow oxidation pond or ditch where aerobic degradation by algae and bacteria can occur and odors will be kept at a minimum.

As degradation or decomposition occurs, the inorganic fraction or mineral portion increases, in that nitrogen is released as ammonium compounds which are later converted to nitrates. If manure is applied to land at rates greater than 110 mt/ha, the formation of nitrate will decrease because of ammonia formation (Mathers et al., 1972).

TABLE 10. ELEMENTAL COMPOSITION OF VARIOUS ANIMAL MANURES[a]

Element	Dry wt. basis	Type of animal manure					
		Hen litter	Dairy cow	Fattening cattle	Hog	Horse	Sheep
Ash	%	6.3	10.6	10.2	28.8	27.9	10.0
N	%	3.28	2.7	3.5	2.0	1.7	4.0
P	%	2.3	0.5	1.0	0.6	0.3	0.6
K	%	2.3	2.4	2.3	1.5	1.5	2.9
Ca	%	6.1	1.6	0.55	2.0	2.9	1.9
Mg	%	0.54	0.61	0.46	0.29	0.52	0.60
S	%	-	0.28	0.39	0.48	0.26	0.29
Na	%	0.48	-	-	-	-	-
Mn	ppm	251	56	23	72	37	32
Fe	ppm	340	222	182	1002	500	518
B	ppm	40	83	91	143	56	32
Cu	ppm	71	28	23	18	19	16
Zn	ppm	330	83	68	215	56	81
Mo	ppm	-	6	2	4	4	3
As	ppm	31	-	-	-	-	-
Al	ppm	730	-	-	-	-	-

a Azevedo and Stout (1974).

Nitrate accumulations in soils can be harmful if crops are used for animal feed. Nitrates do not accumulate in the grain but do accumulate in the forage. Therefore, caution should be taken in the feeding of silage grown on soils receiving greater than 100 mt/ha of manure each year.

Summary and Research Needs

Generally the use of animal manure for agricultural production is a safe practice as long as the N content and the soluble salt contents of the manure are monitored. Rapid incorporation of animal wastes into land reduces both odors and the loss of N through volatilization. Repeated application rates of less than 110 mt/ha should not lead to nitrate accumulation in the soil or groundwater. Irrigation or rainfall of 10 cm/yr or more should prevent salt damage to crops from application of 40 mt/ha of manure with a 7.5% soluble salt content.

Besides the N, P, and K fertilizer benefits obtained from manure applications, an increase in soil organic matter content and soil aggregation is noted (Unger and Stewart, 1974). In addition, bulk density and evaporation decrease as application rates increase. These changes in the physical properties of the soil help reduce soil erosion, which is a major threat to fertile agricultural soils.

Although much research on animal wastes has already been done, better management systems could be developed which optimize waste storage capacity while minimizing odor, nitrate, and handling problems. However, some tradeoffs concerning fertilizer value versus handling conveniences, etc., would have to be weighed. For instance, one possible management system which could stabilize animal waste and allow better storage and handling is composting. But during composting some N is lost through volatilization, reducing some of the N fertilizer value. Here, the economic importance of ease of storage, the aesthetic value of the absence of odor, and the freedom of worry over nitrate pollution must be evaluated. Salt or metal problems could be reduced by changing the animals' diet. However, research would be needed to maintain the same nutritional value of the diet.

Lastly, the mechanisms of nutrient and water availability in soils should be studied to better predict the ability of waste-amended soils to supply essential nutrients to plants under stress conditions such as drought. If some of these problems were further researched, the utilization of animal wastes could be maximized.

REFERENCES

Azevedo, J., and P. R. Stout. 1974. Farm animal manures: an overview of their role in the agricultural environment. Bull. No. 44., Calif. Agr. Exp. Sta., Berkely, CA.

CAST. 1976. Application of Sewage Sludge to Cropland. Report No. 64, EPA-430/9-76-013. U. S. Environ. Prot. Agen., Wash., D. C.

Cornell University. 1969. Animal waste management. Conference Jan., 1969. Syracuse, N.Y.

Graber, R. 1973. Agricultural animals and the environment. Oklahoma State University, Stillwater, OK.

Mathers, A. C., B. A. Stewart, J. D. Thomas, and B. J. Blair. 1972. Effects of cattle feedlot manure on crop and soil conditions. Tech. Rep. No. 11. Texas Agr. Exp. Sta., Bushland, TX.

Southern Cooperative Series. 1979. Animal waste treatment and recycling systems. Bull. No. 242. Univ. of Kentucky. Lexington, KY.

Stewart, B. A., and B. D. Meek. 1977. Soluble salt considerations with waste application. In L. F. Elliot and F. J. Stevenson (ed.) Soils for management of organic wastes and waste waters. Soil Sci. Soc. Am., Madison, WI.

Unger, P. W., and B. A. Stewart. 1974. Feedlot waste effects on soil conditions and water evaporation. Soil Sci. Soc. Amer. Proc. 38:954-957.

U.S.D.A. 1978. Improving soils with organic wastes. Report to the Congress in response to Section 1461 of the Food and Agriculture Act of 1977 (PL 95-113). USGPO, Wash., D. C. 157 pp.

Crop Residues and Food Processing Wastes

S.B. Hornick

When crops are grown for food and feed products, remaining parts of the plant can be returned to the land for recycling and maximum use of valuable plant nutrients.

More than 400 million tons of crop residue are produced in the U.S. each year. Three major crops, field corn, wheat, and soybean are responsible for 75% of this total residue. When utilized properly, crop residues can improve the organic matter content of the soil and reduce soil erosion. Crop residues are composed of organic materials such as carbohydrates, proteins, and cellulose, and substantial amounts of plant nutrients which become available as the residues decompose.

In some areas, crop residues can be excessive and must be carefully applied so that the soil's capacity to degrade them is not exceeded. If residues are applied in excess, the decomposition rate will slow and plant growth could be affected by the toxins and chemicals produced by microbes in the system. Also the excessive application of mature residues can lead to a high C:N ratio in the soil with immobilization of plant available N the result.

The processing of fresh fruit or food also produces a wide variety of organic wastes that are suitable for land spreading. Many of the solid wastes include fruit and vegetable peelings, spoiled or unacceptable fruit, whey and skim milk from dairy processes, and sludges from primary treatment of liquid wastes. The liquid wastes come from the washing and processing of the food and the cleaning of equipment after processing.

If poultry and meatpacking wastes are excluded, fruit and vegetables are the major contributors (89%) to the waste produced in the food processing industry (National Canners Assoc., 1973). The total yearly amount of solid waste produced is over 9.0 million dry tons (Hang, 1979). However, only part of the fruit and vegetable waste is available for land application. A major portion of the waste is used for animal feed with a smaller portion being utilized in the manufacture of by-products (USDA, 1978).

Residuals such as trimmings, peelings, culls, stems, and pits can be used for animal feeds if fed immediately and not stored and allowed to spoil. Citrus, corn, pineapple, and potato wastes are produced year round and are very dependable sources of animal feed. Because of its high vitamin content, citrus sludge can be dried and fed as an animal feed supplement (EPA, 1977).

Some of the by-products manufactured from fruit and vegetable residuals or solid wastes are molasses from citrus peels, vinegar fermented from the wastes, MSG, charcoal from pits and nuts, oils, and cosmetics (National Canners Assoc., 1973).

Because of the by-product manufacture and animal feed produced, it is mostly the wastewater (80 billion gallons from fruit and vegetables) rather than the sludge which comprises the major portion of land disposable waste in the food processing industry.

Unfortunately, not all processing plants have access to large areas of agricultural land. Since the wastes usually have high BOD_5's, limited available land may result in both water and BOD overloadings. Peeling and blanching operations usually account for the major portion of the BOD (80% or more) discharged from a cannery. Some of the ranges for BOD and pH from various processing wastes are as follows: apples, 240-19,000 mg/l, 4.1-8.2; sauerkraut, 300-41,000 mg/l, 3.6-6.8; and tomatoes, 454-1575 mg/l, 5.6-10.8, respectively (Hang, 1979). These ranges reflect differences in the type of raw products and processing methods used among processing plants. Some effluents are strongly alkaline because of lye-peeling liquors or alkaline cleaning compounds.

Average concentrations of heavy metals and pesticides are listed in Table 11. Both the metals and pesticide residues are low indicating essentially no threat of buildup or contamination in the soil environment.

TABLE 11. ELEMENTAL COMPOSITION OF FRUIT RESIDUALS[a]

Element	dry weight basis
N	0.97 %
P	0.14 %
K	0.09 %
S	0.13 %
Na	17.2 ppm
Ca	22.6 ppm
Mg	31.7 ppm
B	28.8 ppm
Cd	0.2 ppm
Cr	0.11[b] ppm
Cu	0.3 ppm
Fe	8.3 ppm
Pb	0.3 ppm
Mn	1.0 ppm
Hg	0.04[b] ppm
Zn	0.08 ppm
Chlorinated hydrocarbon	0.08[b] ppm
Organophosphate	0.13[b] ppm

a Noodharmcho and Flocker (1975).
b Rose, Katsuyama, and Steinberg (1972).

Summary and Research Needs

When applied to land, food processing wastes can supply small amounts of N, P, and K to fertilize crops grown on these sites. Although the addition of organics can benefit the physical properties of a soil, one must be careful that the high BOD of a processing sludge or wastewater does not result in unfavorable C:N ratio, especially when the bulk of the waste produced is limited to a four month period due to the seasonal nature of the crops. When water and BOD loadings are not excessive, fertilizer and soil condition benefits can be gained from the application of food processing wastes.

Research on the soil factors and microbial populations which would enable the soil to better assimilate these high BOD wastes should be conducted. In addition, crops which tolerate wet conditions and high C:N ratios can be screened and selected to grow on sites where food processing wastes are applied. Research in these areas would enable the maximum utilization of food processing wastes on land without threat to the natural soil processes.

REFERENCES

E.P.A. 1977. Pollution abatement in the fruit and vegetable industry: wastewater treatment. EPA-625/3-77-0007. U.S. Environ. Prot. Agen., Wash., D.C.

Hang, Y. D. 1979. Production of single-cell protein from food processing wastes. In J. H. Green and A. Kramer (ed.) Food processing waste management. AVI Publishing Co., Inc. Westport, CT.

National Canners Association. 1973. Solid waste management in the food processing industry. NTIS Publication No. PB-219 019.

Noodharmcho, A., and W. J. Flocker. 1975. Marginal land as an acceptor for cannery waste. J. Am. Soc. Hort. Sci. 100:682-684.

Rose, W. W., A. M. Katsuyama, and R. W. Steinberg. 1972. Ocean assimilation of food residuals. In Proc. 3rd Natl. Symposium on Food Processing Wastes.

U.S.D.A. 1978. Improving soils with organic wastes. Report to the Congress in response to Section 1461 of the Food and Agriculture Act of 1977 (PL 95-113). USGPO, Wash., D. C. 157 pp.

Leather Manufacturing Wastes

R.L. Chaney

Leather is stabilized collagen prepared from the hide of livestock. Hides are transformed into leather by two major methods: 1) chrome tanning, and 2) vegetable (polyphenol) tanning. Wastes from this industry were listed as hazardous wastes in May, 1980 (EPA, 1980a) because of Cr, Pb, and sulfide present in the wastes. Upon reevaluation, these wastes were removed from the listed hazardous wastes (EPA, 1980d) based on the presence of chromic rather than the chromate form of chromium in these wastes. The EP test was changed to consider chromate rather than total Cr (EPA, 1980c). Even this new test is likely to underestimate chromate due to chemical reduction during processing (Bloomfield and Pruden, 1980).

This chapter discusses sources and nature of wastes from the chrome tanning industry, results from feeding these wastes to livestock or applying them to land, and research needs for land treatment of tannery wastes. Much of the discussion of Cr is relevant to other waste streams declared hazardous because of Cr.

Sources and Nature of Wastes

EPA and its contractors have described tannery operations, waste streams and waste disposal practices (EPA, 1974; SCS Engineers, 1976; Bauer, Conrad, and Lofy, 1977). In the initial processing of raw-hides, flesh and hair are removed in mechanical and chemical operations which use lime and sodium sulfide. During treatment of the dehairing wastewater, manganous salts (Mn^{2+}) are added to catalyze oxidation of the sulfide remaining (Bailey and Humphreys, 1967). Unfortunately, this Mn could possibly contribute to the oxidation of sludge chromic to chromate (see below). Because of this chromic oxidation process, it seems unwise to mix Mn-treated wastewater or sludges from dehairing processes with Cr-rich sludges from waste tanning solution or tanned hide rinse water. Segregation of these waste streams would reduce the potential for chromate production from tannery sludges (Hauck, 1972; Pierce and Thorstenen, 1976; and Maire, 1977). The organic limed sludge which results from the treatment of dehairing wastewater is similar to limed raw municipal sewage sludge, but the tannery wastewater sludge is lower in heavy metals other than Cr and Mn (Table 12). This lime-organic waste sludge could be applied to cropland as a low grade limestone with low grade fertilizer value. If farmland is near the tannery, utilization of the liquid sludge on farmland should be less expensive than putting it in a landfill. Tannery "fleshing wastes" are often sold as a by-product for rendering or feed.

After flesh is removed from the hides, they are acidified and treated with the chrome tanning solution. This solution contains chromic sulfate at low pH. The Cr^{3+} reacts with the collagen and stabilizes it against deterioration. The waste chrome tanning solution is recycled directly in some tanneries; in others, the Cr^{3+} is precipitated by pH adjustment, collected, and reused; and in others, it is discharged to the wastewater treatment system or to publicly owned treatment works (POTW).

The tanned hide is rinsed, and then cut or split into a grain (outer) side used to make finished leather products, and a split side (inner, or flesh side) which is used to make rougher leather products.

The grain side is "shaved" to a uniform thickness, generating a coarse particulate "leather shavings" waste. Shaved tanned grain sides are then "buffed" or sanded to create a very smooth surface. This generates a fine particulate "leather buffing dust". Pieces of tanned leather are trimmed away to achieve the final leather shapes.

During other finishing steps dyes and colors are applied, the leather is lubricated, retanned with organic chemicals, and otherwise processed to make finished colored leather products (clothing, automobile and furniture coverings, etc.). Some waste dyes, organic solvents, etc., enter the waste stream at this point. Most tanneries have ceased use of Pb pigments which has sharply lowered Pb levels in their wastes.

Nature of Leather Manufacturing Wastes

EPA (1974), SCS Engineers (1976), and EPA (1980e) described

TABLE 12. COMPOSITION OF WASTES FROM A TANNERY

Waste Type	Cr	Mn	Cd	Zn	Pb	Cu	Ni	Fe	% Dry Matter
	mg/kg dry weight								
Fleshings	56.	50.1	0.24	18.7	2.0	6.7	6.1	1240	22
Shavings A	19800.	4.9	0.28	3.8	1.3	2.3	1.4	1670	43
B	21800.	5.9	0.13	3.8	0.8	3.3	2.0	1590	43
Buffing Dust	11900.	4.2	1.34	38.7	15.2	40.6	1.9	405	94
Wastewater Treatment Sludge	987.	2800.	2.59	42.7	6.1.	8.0	6.4	1260	52

R. L. Chaney and B. A. Coulombe. 1980. Wastes from a tannery were collected during a site visit. Samples were wet ashed (HNO_3 - H_2O_2) and analyzed for total metals using atomic absorption spectrophotometry.

the Cr, Pb, and sulfide levels in the many existing leather manufacturing wastes. Chromium ranged from 2,200 to 21,000 ppm wet wastes, and Pb from 40 to 724 ppm. During a site visit, researchers from our laboratory obtained samples of all solid wastes from a modern tannery except aeration lagoon solids and Cr recycle sludge. Analyses of these wastes samples are shown in Table 12. Segregation of Cr-rich wastewaters for separate treatment has kept the Cr level in waste streams other than Cr recycle sludge relatively low. Other heavy metals are generally low, as was reported by Thorstenen and Shah (1979).

Uses of Leather Wastes

Technology has been developed to utilize some tannery wastes. Leather trimming, shaving, and buffing wastes (collectively called leather tankage) are essentially pieces of Cr tanned leather, or fairly pure protein containing about 11% N and 1% Cr. Because of the protein-N, some of these wastes are now used as a feed ingredient or organic-N fertilizer.

Use as Feed--

In response to a petition, FDA developed rules for use of hydrolyzed leather meal as an approved feed ingredient for swine (Anonymous, 1967). It may contain no more than 2.75% Cr, and may not comprise more than 1% of the mixed feed.

Results of feeding trials of hydrolyzed leather meal have been reported by numerous researchers (Dilworth and Day, 1970; Gruhn and Luedke, 1972; Knowlton et al., 1976; Otevrelova and Mackova, 1977; Salimbaev, 1979; Waldrup et al., 1970). In general, they found that steaming or acid hydrolysis raises the protein equivalent value of leather tankage, and that the N is used in a manner similar to urea by sheep (i.e. as an N source rather than as a protein source). No toxicity has been reported. Reduced rates of liveweight gain may result if the hydrolyzed leather meal is used to replace too high a proportion of the total protein in a feed.

Use as a Fertilizer--

Because tannery wastes contain organic N in the form of the protein collagen, they have been tested for use as fertilizers. Leather tankage normally contains about 11% N. A process was developed to react leather tankage with urea and formaldehyde to produce a higher N analysis fertilizer. One such product, "Organiform", contains 22-24% N (Baker, 1972). The manufacturer of this product has reported N release curves for this product in comparison with other commerical fertilizers. The manufacturer claims it contains no more than 2% Cr, and low levels of other heavy metals.

Use as Soil Conditioner/Fertilizer--

Many studies have been conducted on different leather manufacturing solid wastes and Cr-rich wastewater treatment sludges. Because research on land application of sewage sludges has pointed out several experimental difficulties in studying wastes, it would be useful to reevaluate the literature on land application of leather wastes.

When high application rates of unstabilized organic wastes are land-applied, the excessive BOD can cause anaerobic soil conditions and increase soluble organic compounds in the soil. Some of these are phytotoxic. Also, soluble salts can be phytotoxic until leached from the surface soil. During anaerobic decomposition, the soluble Cr that is found is usually a Cr^{3+} chelate. Studies using soluble metal salt additions to soils can not be used to estimate the effects on plants or ground water of applying wastes containing precipitated metals. This last point is especially relevant to Cr, since Cr^{3+} compounds and complexes are kinetically quite inert (Allaway, 1977a). At pH 6 to 8, sludge Cr^{3+} precipitates equilibrate and a quite stable insoluble Cr sludge forms. Therefore, studies using soluble salts additions can lead to serious errors in estimating risks of Cr in leather manufacturing wastes.

Kick and Braun (1977) reported studies on a Cr-rich tanning wastewater recycle sludge containing only 0.08% N and 32% Cr (dry weight). Research on this unusual, low volume, inorganic Cr waste is not useful toward understanding the potential effects on the environment of bulk organic tannery wastes containing only 0.3-3.0% Cr.

A number of conclusions can be drawn from the whole body of research on Cr-rich leather wastes, including study of tannery combined wastewater treatment sludges (Mazur and Koc, 1976a,b, 1980; Koc, 1979, Volk, V.V., Orgeon State Univ., personal communication, 1980), POTW sludges produced largely from tannery wastewaters (Cunningham, Keeney, and Ryan, 1975; Dowdy et al., 1979, 1980; Dowdy and Larson, 1975; Dowdy and Ham, 1977; Grove and Ellis, 1980; Alther, 1975), and leather tankage (Andrzejewski, 1971; Kravchenko, 1979; Berkowitz et al., 1980; Amacher and Baker, 1980; Silva and Beghi, 1979, Shivas, 1980a,b).

1. The organic-N is apparently released more slowly than from other proteinaceous wastes because the Cr stabilizes the protein. Volk (ibid), based on field research, estimates that about 10% of tannery wastewater treatment sludge N becomes crop available during the year of application.

2. These wastes seldom caused any phytotoxicity, especially when rates were limited to the N requirement of crops. Crops were somewhat enriched in Cr during the first crop year.

Leaching Chromate from Leather Waste Amended Soils--

The critical remaining issue is leaching of Cr to groundwater. Under current federal regulations, chromate in groundwater may not exceed 50 ug/l (see EPA, 1980a,c). If the immobile Cr^{3+} applied to soils in tannery wastes could be oxidized to the mobile chromate by soil MnO_2 (Bartlett and James, 1979), it seems possible that this chromate could leach and contaminate groundwater (EPA, 1980e). However, Amacher and Baker (1980) found that a soluble Cr-fulvic acid extracted from ground leather was not significantly oxidized by MnO_2. When chromate is added to soils, it is rapidly reduced to form insoluble chromic (Cary, Allaway, and Olson, 1977b; Bartlett and Kimble, 1976b; Taylor, 1980), and/or adsorbed by the soil (Bloomfield and Pruden, 1980; Artiole and Fuller, 1979). Other research has shown that soil water at 1 m depth was not increased in Cr by land application of sewage sludge rich in tannery waste-borne Cr (Dowdy et al., 1979, 1980). Water leaching from pots treated with ground leather at N-fertilizer rates was also not enriched in Cr (Silva and Beghi, 1979).

Together, these studies support a model in which precipitated or chelated Cr is not readily oxidized by MnO_2, and that ground water will not be polluted by Cr from land-applied leather wastes if reasonable management is practiced regarding soil pH and waste application rate. This hypothesis has not been proven conclusively by the present data since the lower redox potentials which result from sludge application would more rapidly reduce any chromate formed back to chromic. Soils treated with leather wastes long ago should be evaluated since these are redox stabilized and would allow the testing of worst case field conditions. In any Cr studies, field-moist, never-dried soils must be studied to avoid the errors identified by Bartlett and James (1979).

A detailed discussion of plant uptake of Cr and animal tolerance of Cr is provided in other chapters in this book.

Summary and Research Needs

Most leather tanning in the U.S. uses chromium based processes. When tannery waste streams are segregated, only a few are high in Cr: the precipitated $Cr(OH)_3$ removed from used tanning solutions, and the leather trimmings, shavings, and buffing dust generated during transformation of tanned hides into finished leather. The precipitated Cr^{3+} waste has no practical use except to be recycled, and should not be considered for land application.

Leather wastes (tankage) have been converted to steamed leather meal, and are an approved swine feed ingredient. These wastes have been experimentally fed to swine, poultry, cattle, and sheep with no injury to the livestock. The protein in steamed leather meal is of low nutritional value (collagen); ruminants use the nitrogen in a manner similar to dietary urea rather than dietary protein.

Further, leather wastes (11%N) have also been treated by a urea-formaldehyde process to increase N content for use as a fertilizer. This is a commercial practice producing the 22-24% N product "Organiform."

Research has shown that land application of ground leather supplies slow release nitrogen. In these studies, chromium is somewhat increased in crops, at least during the first crop year. Chromium phytotoxicity did not result from the land application of precipitated Cr^{3+}. Some other leather manufacturing wastewater treatment sludges have caused crop problems, but not demonstrably due to Cr. Many problems occur in pot studies with wastes because soluble salts are in excess, or inadequate fertilizers are applied for pot studies, etc.

A small amount of research has been conducted on leaching of Cr from soils treated with sewage sludge, tannery sludge, or chromate. Chromate in soil water at 1 m depth has not exceeded the 50 ug/l groundwater standard even when large amounts of Cr^{3+} bearing wastes were added. Often groundwater Cr was unchanged by the surface applied Cr. Although soil MnO_2 can oxidize soluble Cr^{3+} to chromate, this appears to be environmentally irrelevant because Cr^{3+} has very low solubility at practical soil pH levels.

Research Needs--

Potential oxidation of chromic to chromate is important in environments rich in Cr^{3+}. More basic research is needed to assure that opportunities for environmental pollution by Cr are not ignored. The mechanisms of these reactions must be better understood. Chromium rich soils should be amended with MnO_2 to see whether environmentally stable chromic-chromium can be converted to chromate under some worst case conditions.

Land treatment sites freshly amended with tannery sludges will have lower oxidation potential than long-term well oxidized sites. These latter sites should be studied to see whether Cr^{3+} application can ever pollute groundwater with chromate. Economics of segregated treatment of tannery waste streams should also be explored.

REFERENCES

Allaway, W. H. 1977. Soil and plant aspects of the cycling of chromium, molybdenum, and selenium. Proc. Intern. Conf. Heavy Metals in the Environment 1:35-47.

Alther, E. W. 1975. Chromium-containing sludge and chromium uptake by forage plants. Leder 26:175-178.

Amacher, M. C., and D. E. Baker. 1980. Kinetics of Cr oxidation by hydrous manganese oxides. Agron. Abstr. 1980:136.

Andrzejewski, M. 1971. Effect of chromium fertilization on the yield of several plant species and on the chromium content of the soil (in Polish). Roczn. Nauk Rolnickzych A97:75-97.

Anonymous. 1967. Food additives permitted in the feed and drinking water of animals or for the treatment of food-producing animals. Fed. Reg. 32(157):11734.

Artiole, J., and W. H. Fuller. 1979. Effect of crushed limestone barriers on chromium attenuation in soils. J. Environ. Qual. 8:503-510.

Bailey, D. A., and F. E. Humphreys. 1967. The removal of sulphide from limeyard wastes by aeration. J. Soc. Leather Trades Chemists 51:154-172.

Baker, H. J., and Bro., Inc. 1972. Organiform LT. Product brochure. H. J. Baker Co., 360 Lexington Ave., New York, NY 10017

Bartlett, R. J., and B. James. 1979. Behavior of chromium in soils: III. Oxidation. J. Environ. Qual. 8:31-35.

Bartlett, R. J., and J. M. Kimble. 1976b. Behavior of chromium in soils: II. Hexavalent forms. J. Environ. Qual. 5:383-386.

Bauer, D. H., E. T. Conrad, and R. J. Lofy. 1977. Solid waste management practices in the leather tanning industry. J. Am. Leather Chem. Assoc. 72:270-280.

Berkowitz, J. B., S. E. Bysshe, B. E. Goodwin, J. C. Harris, D. B. Land, G. Leonardos, and S. Johnson. 1980. Field verification of land cultivation/ refuse farming. EPA-600/9-80-010. U.S. Environ. Prot. Agen., Wash., D.C. pp. 260-273. In O. Schultz (ed.) Proc. Sixth Ann. Res. Symp. on Disposal of Hazardous Waste.

Bloomfield, C., and G. Pruden. 1980. The behavior of Cr (VI) in soil under aerobic and anaerobic conditions. Environ. Pollut. 23:103-114.

Cary, E. E., W. H. Allaway, and O. E. Olson. 1977b. Control of chromium concentration in food plants. 2. Chemistry of chromium in soils and its availability to plants. J. Agr. Food Chem. 25:305-309.

Cunningham, J. D., D. R. Keeney, and J. A. Ryan. 1975. Yield and metal composition of corn and rye grown on sewage sludge-amended soil. J. Environ. Qual. 4:448-454.

Dilworth, B. C., and E. J. Day. 1970. Hydrolyzed leather meal in chick diets. Poult. Sci. 49:1090-1093.

Dowdy, R. H., and G. E. Ham. 1977. Soybean growth and elemental content as influenced by soil amendments of sewage sludge and heavy metals: Seedling studies. Agron. J. 69:300-303.

Dowdy, R. H., and W. E. Larson. 1975. The availability of sludge-borne metals to various vegetable crops. J. Environ. Qual. 4:278-282.

Dowdy, R. H., C. E. Clapp, D. R. Duncomb, and W. E. Larson. 1980. Water quality of snowmelt runoff from sloping land receiving annual sewage sludge applications. pp. 11-15. In Proc. Nat. Conf. Municipal and Industrial Sludge Utilization and Disposal. Information Transfer, Inc., Silver Spring, MD.

Dowdy, R. H., C. E. Clapp, W. E. Larson, and D. R. Duncomb. 1979. Runoff and soil water quality as influenced by five years of sludge applications on a terraced watershed. Agron. Abstr. 1979:28.

E.P.A. 1974. Development document for effluent limitations guidelines and new source performance standards for the leather tanning and finishing point source category. EPA-440/1-74-016-a. 158 pp.

E.P.A. 1980a. Hazardous Waste Management System: Identification and listing of hazardous waste. Federal Register 45(98):33084-33133.

E.P.A. 1980c. Hazardous waste management system: Identification and listing of hazardous wastes. Fed. Reg. 45(212):72029-72034.

E.P.A. 1980d. Hazardous waste management system: Identification and listing of hazardous wastes. Fed. Reg. 45(212):72037-72039.

E.P.A. 1980e. Listing background document: Leather tanning and finishing industry. Supports Appendix 7 in EPA (1980a). On file in Docket 3001 at EPA.

Grove, J. H., and B. G. Ellis. 1980. Extractable chromium as related to soil pH and applied chromium. Soil Sci. Soc. Am. J. 44:238-242.

Gruhn, K., and H. Luedke. 1972. Use of hydrolyzed chrome leather waste in pig feeding and the content of chromium in flesh, liver, kidneys, and hair (in German). Arch. Tierernahr. 22:113-124.

Hauck, R. A. 1972. Report on methods of chromium recovery and reuse from spent chrome tan liquor. J. Am. Leather Chem. Assoc. 67:722-430.

Kick, H., and B. Braun. 1977. The effect of chromium containing tannery sludges on the growth and the uptake of chromium by different crops (in German). Landwirtsch. Forsch. 30:160-173.

Knowlton, P. H., W. H. Hoover, C. J. Sniffen, C. S. Thompson, and P. C. Belyea. 1976. Hydrolyzed leather scrap as a protein source for ruminants. J. Anim. Sci. 43:1095-1103.

Koc, J. 1979. Effect of the temperature, moisture content in soil, and additions of fertilizers on the decomposition of tannery sludges (in Polish). Rocz. Glebozn. 30:73-83.

Kravchenko, S. N. 1979. Effect of leather waste on the yield and quality of some vegetable crops (in Russian). Nauchnye Trudy Ukr. S.-Kr. Akad. 228:62-65.

Maire, M. S. 1977. A comparison of tannery chrome recovery systems. J. Am. Leather Chem. Assoc. 72:404-418.

Mazur, T., and J. Koc. 1976a. The fertilizing value of tannery sludge. II. Effect of fertilization with tannery sludges on the crop (in Polish). Rocz. Glebozn. 27:113-122.

Mazur, T., and J. Koc. 1976b. The fertilizing value of tannery sludge. III. Effect of fertilization with tannery sludges on the chemical composition of plants (in Polish). Rocz. Glebozn. 27:123-135.

Mazur, T., and J. Koc. 1980. The fertilizing value of tannery sludges. pp. 328-338. In Handbook of organic waste conversions. Van Nostrand- Reinhold, New York.

Otevrelova, K., and M. Mackova. 1977. Tanning waste in feeds for fur animals. 1. Hydrolytic treatment of different types of waste to increase digestibility (in Czech.). Sbornik Vedeckych Praci Vyzkumneho Ustavu Vyzivy Zvirat, Pohorelice 11:41-48.

Pierce, R., and T. C. Thornstenen. 1976. Recycling of chrome tanning liquors. J. Am. Leather Chem. Assoc. 71:161-164.

Salimbaev, S. 1979. Use of tannery shavings for fattening pigs (in Russian). Zhivotnovodstvo 8:60-61.

SCS Engineers, Inc. 1976. Assessment of industrial hazardous waste practices - Leather tanning and finishing industry. NTIS # PB 261-018.

Shivas, S. A. J. 1980a. Factors affecting the oxidation state of chromium disposed in tannery wastes. J. Am. Leather Chem. Assoc. 75:42-48.

Shivas, S. A. J. 1980b. The effects of trivalent chromium from tannery wastes on plants. J. Am. Leather Chem. Assoc. 75:288-299.

Silva, S., and B. Beghi. 1979. Problems inherent in the use of organic fertilizers containing chromium. (In Italian) Ann. Della Fac. Agr. Univ. Cath. Sacro Cuore 19:31-47.

Taylor, F. G., Jr. 1980. Chromated cooling tower drift and the terrestrial environment: A review. Nuclear Safety 21:495-508.

Thorstenen, T. C., and M. Shah. 1979. Technical and economic aspects of tannery sludge as a fertilizer. J. Am. Leather Chem. Assoc. 74:14-23.

Waldrup, P. W., C. M. Hillard, W. W. Abbott, and L. W. Luther. 1970. Hydrolyzed leather meal in broiler diets. Poult. Sci. 49:1259-1264.

Munitions and Explosives Wastes

R.H. Fisher and J.M. Taylor

Introduction

The explosives and munitions industry has 586 commercial and 35 government-owned contractor-operated (GOCO) facilities for the production, blending, loading, and packing of explosives and related products (TRW Group, 1975). As of 1980, commercial production, as derived from reports on explosives consumption, was estimated to be 1.9 million metric tons/year (Bureau of Mines, 1981). According to Bureau of Mines' annual estimates, commercial production has shown a steady increase during the 1970's, primarily as a result of increased consumption by the mining and construction industries. Data on production of explosives and propellants for military consumption during the 1970's are not readily available. However, in 1975 the TRW Group estimated military production for 1973 at 0.4 million metric tons/year. The generation of waste sludges for disposal from the manufacture of explosives and propellants by government and commercial facilities was 21,500 dry tons in 1977 (EPA, 1980). GOCO facilities produced 92 percent of these wastes. Of all 1977 wastes, 93 percent were disposed of by spreading on concrete pads and open uncontrolled burning, with greater than 90 percent being disposed of on site.

Manufacturing Processes and Wastewater Characteristics

The explosive compound that is produced in the largest quantity is 2,4,6-trinitrotoluene (TNT). Other explosives and propellants, in order of their production, are cellulose nitrate (nitrocellulose), cyclotrimethylene trinitramine

(RDX), nitroglycerin, cyclotetramethylene tetranitramine (HMX), and smaller amounts of specialized explosive and initiator compounds. Other materials such as waxes, plastics, and metal powders are used as oxidizers or catalysts. Different amounts of the explosives and propellants are blended together for specific needs.

Both manufacturing processes and loading and packing operations generate wastewaters, the latter tending to have lower concentrations of mixtures of various explosive compounds. Patterson et al. (1976a) and Patterson, Shapira, and Brown (1976) made a comprehensive survey of wastewater production from the military munitions industry. Little information has been published on the treatment of wastewater from commercial explosives manufacturing.

TNT is made by the step-wise nitration of toluene with nitric and sulfuric acids, then purified by the addition of a concentrated solution of sodium sulfite to extract isomers and impurities. Wastewater from this process, red water, is highly acidic (pH 2), and contains approximately 17% organic compounds including sulfonated nitrotoluene compounds, complex dye-like compounds, and various nitrotoluene compounds, as well as sodium sulfate, sodium sulfite, sodium nitrate, and sodium nitrite. The red water accounts for only 4% of the total wastewater flow from TNT production, but it is the most concentrated waste stream.

Clean-up water from TNT manufacturing or from loading and packing operations is comprised mainly of TNT and a few closely related compounds. This pink water can contain up to 120 mg/l TNT. After exposure to sunlight or upon neutralization with lime, an intensification of color in the red or pink water occurs that is accompanied by an increased toxicity and chemical stability (Nay, Randall, and King, 1972). Photochemical or chemical reactions of TNT and other compounds in the waste result in the formation of a variety of azo and azoxy oligomers with three or four aromatic rings per molecule and many smaller molecules also. These compounds are similar in structure to some azo dyes (Burlinson et al., 1979). Before 1968, TNT manufacture involved a batch process which is still being used in some less-modern plants. More modern plants have installed a counter-current, continuous flow process which results in the recycling of larger amounts of wastewater, eliminating some of the pollution problems.

Another wastewater common to the manufacture of most explosives and propellants originates from the production of nitric and sulfuric acids. These are produced on site and the total quantity produced is several times greater than the acid production by the inorganic chemicals industry. Waste water from acid production is generally neutralized with lime or soda and causes no serious treatment problems. Cooling waters are also generated in high volumes but generally have low concentrations of pollutants.

Cellulose nitrate is produced by the nitration of cotton linters or raw wood pulp with nitric acid and sulfuric acid. Wastewater which contains fine particles of cellulose nitrate is partially recycled; it also contains acid production wastes (Wendt and Kaplan, 1976). RDX and HMX are produced by the nitration of hexamine in the presence of nitric acid, acetic acid, ammonium nitrate, and other chemicals. Organic solvents are used to recrystallize the RDX and HMX to purify it. Various amounts of all of these chemicals occur in wastewaters. Nitroglycerin is made by nitration of glycerin by nitric and sulfuric acids. Wastewaters generated are similar to those previously mentioned in cellulose nitrate production.

The production of pelletized propellants, such as ball powder, involves the treatment of cellulose nitrate with benzene, ethyl acetate, and dinitrotoluene to obtain a compound with desired burning properties for shells and rocket motors. These organic compounds are found in the wastewater along with cellulose nitrate and sodium sulfate.

Initiators are highly reactive compounds usually used to detonate a larger mass of less reactive explosive. They are produced in small quantities from a variety of starting materials. These initiators include lead azide, lead styphanate, trinitroresorcinol, and many other compounds. Process wastewaters are treated with sodium hydroxide and heat to remove residual explosive compounds, but a considerable amount of lead is present in the wastewater (EPA, 1980).

Physical, Chemical, and Biological Treatment of Munitions Wastewater

An extensive survey of wastewater treatment methods in the munitions manufacturing industry was made by Patterson et al. (1976b). Early studies by Schott, Ruchhoft, and Megregian (1943) and Ruchhoft, LeBosquet, and Meckler (1945) showed that treatment practices such as neutralization, chemical precipitation, ozonation, and electrolytic reduction were not effective in treating red or pink water. Chlorination and bromination were effective in partial color removal although little further work has been done in this area. Activated carbon has been shown to be effective in removing TNT and related colored compounds by adsorption (Schott,

Ruchhoft, and Megregian, 1943; Patterson, Shapira, and Brown, 1976; Patterson et al., 1976b). The activated carbon could reduce TNT in pink water from 178 to 3.7 mg/l (98% removal) and could also be used to adsorb the various other related nitrobodies, although to a lesser extent. However, the effluent from the activated carbon is still colored. The activated carbon cannot be regenerated but instead is incinerated, which causes an air pollution problem and leads to high operating costs.

Presently the only practical method for treating red water is to evaporate it in rotary kilns to 35-40% solids. The resultant sludge is either sold to the paper industry for its sodium sulfate content or incinerated. The highly soluble ash is then disposed of in a landfill. Major air pollution problems exist with this method, since the emissions from one facility alone were estimated to be 50 tons/yr particulates, 5 tons/yr SO_x, and 648 tons/yr NO_x (Patterson, Shapira, and Brown, 1976).

An alternative method is to add the red water to the fuel oil used by the plant for heating. The TNT and nitrobodies are solubilized in the oil phase and incinerated during the normal burning of the fuel (Patterson et al., 1976b). The use of reverse osmosis to concentrate red water has also been tried but was unsuccessful as a result of membrane failure. Current research into new membrane materials may eventually make this process feasible. Other preliminary research has been done on treatment by UV light-catalyzed ozonation and reduction by the Tampella process and other processes (Patterson et al., 1976b). Ruchhoft, LeBosquet, and Meckler (1945) reported on the adsorption of TNT wastes in soil columns but ultimate disposal of the TNT-contaminated soil was not mentioned. Andren et al. (1975) reported that a polymeric adsorbent resin could be used to remove TNT, RDX, and various nitrobodies from pink water. The resin could be regenerated by acetone and the resulting acetone-nitrobody mixture could be distilled leaving a concentrated sludge that could be incinerated. Okamoto et al. (1977) reported on the removal of TNT from pink water by addition of diamine compounds and a cationic surfactant. The resulting complex was not soluble in water and could be precipitated, but again there is no mention of ultimate disposal.

TNT wastes and/or TNT are toxic to many microorganisms and will reduce the efficiency of biological wastewater treatment methods if present at concentrations of 2 mg/l TNT or higher (Ruchhoft, LeBosquet, and Meckler, 1945; Nay, Randall, and King, 1974). Lower concentrations of the more highly colored nitrobody complexes were found to be toxic to fish.

Studies have shown the rapid removal of TNT by the activated sludge process (Carpenter et al., 1978; Hoffsommer et al., 1978). The 97% removal of 10-50 mg/l TNT occurred in 3 to 5 days when activated sludge had supplemental C sources added to increase TNT degradation. ^{14}C-TNT was utilized in the study and only 0.4% $^{14}CO_2$ was found to be evolved, showing very little if any complete mineralization by microorganisms. The remaining radioactivity was approximately equally divided between the sludge floc, probably by adsorption, and the supernatant. A variety of nitrogen-containing compounds were found to be formed from the ^{14}C-TNT. Additional laboratory work by Traxler (1974) and Traxler, Wood, and Delaney (1975) with ^{14}C-TNT as the sole C source showed similar rates of TNT removal and $^{14}CO_2$ production in pure culture studies. The microbial biomass contained 6.1% of the labelled C. Carpenter et al. (1978) suggested that the radioactive fraction associated with the biomass was incorporated into lipid or protein fractions as cross-linked polyamide polymers which appeared resistant to further degradation.

Wastes from cellulose nitrate production have been successfully treated by a combination of chemical and biological methods (Wendt and Kaplan, 1976). The cellulose nitrate wastewaters were treated with 3% sodium hydroxide and heat resulting in the production of organic acids, ammonia, cyanide, nitrite, nitrate, and other unidentified compounds although this mixture was still colored. After neutralization, this waste could be biologically treated.

Wang et al. (1978) found that the addition of polyelectrolytes and bentonite resulted in the precipitation of nitrocellulose fines and enabled it to be centrifuged or filtered. Glover and Hoffsommer (1979) have reported on the photolysis of RDX by UV light in water or ethanol. Decomposition products included formaldehyde, nitrous oxide, formamide, ammonia, nitrite, and nitrate. A combination of UV light and ozone resulted in the production of cyanic acid, cyanuric acid, and CO_2. The potential hazards of some of these degradation products make this an unlikely treatment method at present.

The treatment of wastewater generated from torpedo refueling operations has been studied by Kessick, Characklis, and Elvey (1977). The fuel contains propylene glycol dinitrate (PGDN) as a propellent, di-n-butyl sebecate as a stabilizer, and 2-nitrodiphenylamine, which is biodegradable. The fuel was initially non-biodegradable and toxic. Alkaline hydrolysis of the PGDN resulted in the formation of propylene glycol, which is biodegradable, and nitrate. The hydrolyzed effluent was then passed through activated carbon and discharged to sanitary sewers.

Microbial Degradation of Explosives and Propellants

Recently, there has been a great deal of research concerning the mechanisms involved in the degradation of TNT. McCormick, Feeherry, and Levinson (1976) reported on the reduction of the nitro groups of TNT to amino groups by enzymes of the obligate anaerobe *Veillonella alkalescens*, and also *Escherichia coli* and a *Pseudomonas* sp. While all nitro groups could be reduced, there was a preferential reduction of the 4-nitro group. Intermediate hydroxylamino compounds were also found. A wide variety of mono-, di-, and trinitroaromatic compounds were also examined and were found to vary greatly in the reduction of their nitro groups. Depending upon the starting compound, a large number of monoaminodinitrotoluene and diaminomononitrotoluene compounds were found.

Previous work with dinitrophenols and dinitrocresols showed that the nitro group could be removed as nitrate, and that the aromatic ring could undergo a hydroxylation with ultimate cleavage of the ring by 'meta' fission. This has not been seen with nitrotoluene compounds, probably because of the different electron withdrawing powers of the sidechains. The hydroxylamino intermediates can also undergo a nonenzymatic reaction resulting in azoxy compounds. Naumova, Amerkhanova, and Shaikhatdinov (1979) reported that *Ps. denitrificans* isolated from TNT-contaminated soil could degrade TNT rapidly at 200 mg/l if other nutrient sources were present. There was a preferential reduction of the 2-nitro group and hydrolysis of nitro to nitrate accounted for only 3% of nitro group loss.

Klausmeier et al. (1976) determined that enzymes from *Bacillus subtilis* and *Ps. aeruginosa* could reduce nitro groups and that the presence of a "TNT reductase" was probably fairly widespread across a broad range of microorganisms. They also found that microorganisms have a large variation in their tolerance to TNT that may be a result of higher levels of this enzyme present in microorganisms adapted to growth in the presence of TNT. The nitrotoluene compound must have at least two nitro groups to be reduced effectively and a specific arrangement of the nitro groups on the ring is not necessary. Polar substituents like hydroxyl groups also reduced "TNT reductase" activity. These findings are in agreement with McCormick, Feeherry, and Levinson (1976).

Won et al. (1974) reported the degradation of TNT by Pseudomonas-like organisms isolated from TNT-contaminated soil. There was some O_2 uptake by cells when TNT was the sole C source, but it is not known whether TNT was actually metabolized. In the presence of added nutrients, 99% removal of 100 ug/ml TNT was possible in 24 hours. The TNT was degraded to several aminonitrotoluene compounds and azoxy-toluene compounds, the latter probably formed non-enzymatically from hydroxylaminotoluene intermediates. After depletion of TNT after 24 hours, the azoxy compounds eventually disappeared but the nature of the final decomposition products is not clear.

Parrish (1977) made a survey of 190 fungal species in 98 genera and found that 183 of them could transform TNT to other compounds, but only 5 species could transform 2,4-dinitrotoluene.

Osmon and Klausmeier (1973) and Klausmeier, Osmon, and Walls (1974) found that bacteria isolated from soils near TNT manufacturing and loading and packing operations had a greater tolerance to a high concentration of TNT (100 mg/l) than did isolates from other soils. This was probably a result of the adaptation of the population to grow in the presence of TNT. The presence of 1% TNT in the soil was found to inhibit bacterial growth.

Traxler (1974) and Traxler, Wood, and Delaney (1975) reported the isolation of gram-negative bacteria that could grow with TNT as the sole C and/or N source and that heterotrophic CO_2 fixation occurred since added $NaH^{14}CO_3$ was subsequently found incorporated into cell biomass. Degradation of TNT by rabbit liver and kidney cells resulted in the formation of many nitrotoluene and azo compounds which were found in the urine (Channon, Mills, and Williams (1944). Burlinson (1980) investigated the fate of TNT in natural streams and found that it was subject to biotransformation and photo decomposition with the formation of a number of compounds similar to those in _in vitro_ studies.

Soli (1973) found that nitro groups of RDX were reduced to amino groups during growth by anaerobic purple photosynthetic bacteria although no actual RDX metabolism was found. Osmon and Klausmeier (1973) found that neither RDX nor ammonium picrate could serve as a sole C source for pseudomonads.

Wyman et al. (1979) reported on the microbial degradation of 2,4,6-trinitrophenol (picric acid) by _Ps. aeruginosa_. Although ammonium picrate is no longer produced as an explosive, large quantities still exist in storage facilities or in obsolete munitions. The picric acid was reduced to

2-amino-4,6-dinitrophenol (picramic acid), showing preferential reduction of the 2-nitro group.

Osmon and Andrews (1978) investigated the biodegradation of TNT in soil and in compost. They found that degradation processes in the soil are too slow to treat 1% TNT wastes practically, even with added nutrients. Although the TNT was totally removed, the presence of transformed products showed incomplete degradation. In a field study, no leaching of TNT was observed but no analyses were reported to determine possible leaching of transformation products. Bench-scale composting was also done, using a variety of starting materials such as paper, garbage, sewage sludge, and grass clippings. Composting was successful in removing 99% of TNT in 53 days when applied at a level of 10% TNT. When ^{14}C-TNT was composted in a laboratory-scale method, many of the normal degradation products were absent. TNT mineralization produced 1.08% of label as $^{14}CO_2$ and 0.4% was found to be associated with cell mass. A water-soluble filtrate contained 44% of the radioactive C in an unidentified form. Almost 25% of the label was not recovered, perhaps as a result of handling losses or volatilization. Although these workers stated that complete mineralization of TNT to CO_2, N_2, and H_2O can occur, they provided no documentation to support this claim.

Nitrocellulose compounds were found to be extremely resistant to microbial breakdown by cultures of cellulolytic bacteria or fungi (Siu, 1951). Nitroglycerin (glyceryl trinitrate) was noted to be reduced in rat liver cells with the production of glyceryl di- and mononitrates. Glycerol or labelled $^{14}CO_2$ from labelled nitroglycerin could not be found, showing the incomplete breakdown of this substance (Needleman et al., 1971).

Toxicity of Nitroaromatic Compounds

TNT levels greater than 50 mg/l were observed to prevent the growth of many fungi, actinomycetes, and gram-positive bacteria, but growth was not prevented at 20 mg/l. Gram-negative bacteria were more resistant and some isolates could grow in the presence of 100 mg/l TNT (Klausmeier, Osmon, and Walls, 1974).

Studies by Osmon and Klausmeier (1973) and the U.S. Army (Anonymous, 1971) showed that TNT was toxic to bluegills at 2-3 mg/l. Reports of toxic effects in munitions plant workers have been discussed by McCormick, Feeherry and Levinson (1976) and include damage to liver, spleen, and kidney cells.

Wyman et al. (1979) reported that a strain of *Ps. aeruginosa* could reduce picric acid to picramic acid. The picramic acid was observed to be both a frame shift mutagen and a base substitution mutagen by the Ames test. The picric acid could also be reduced to picramic acid by the rat liver cell homogenate used in the test. This reaction also occurs in human liver, spleen, and kidney cells. Christensen, Luginbyhl, and Carroll (1975) have suggested that certain nitroamine transformation products may be toxic and more recent studies by the U.S. Army[1] have also considered this possibility.

Won, DiSalvo, and Ng (1976) noted that TNT was toxic to green algae, copepods, and oyster larvae at 2.5 mg/l, 5 mg/l, and 5 mg/l, respectively. TNT was also found to be a frame shift mutagen by the Ames test at 0.5 to 10 ug/ml. Transformation products of the TNT did not cause a mutagenic reaction according to this report.

Summary and Research Needs

Physical treatments, such as sorption by activated carbon or ion exchange resins or coagulation and filtration, are effective in removing TNT and other related compounds present in munitions manufacturing wastewater. The absence of proper ultimate disposal methods for waste sludges from these physical treatment methods produces waste sludges that are unlikely candidates for direct land disposal. Pretreatment to degrade and detoxify complex organic ring compounds should be investigated for use prior to exploration of the land disposal alternative. The use of landfills, lagoons, and incineration are all either prohibited, uneconomical, or undesirable as a result of pollution problems.

The use of chemicals as a pretreatment method has been shown to be successful with PGDN and cellulose nitrate. When sodium hydroxide and heat were applied to these wastes, the resulting products could be treated biologically by activated sludge methods. The potential for land application of chemically treated munitions wastes should be investigated.

Preliminary research is being conducted on the photo- and photochemical decomposition of TNT, RDX, and related compounds. Some of the breakdown products from these pretreatment methods are biodegradable, but more research needs to be done to determine the extent to which the pretreatment will

---

[1] Personal communication. Jesse Barkley, U. S. Army Medical Bioengineering Research and Development Laboratory. Environmental Protection Div., Ft. Detrick, MD, Dec. 1980.

be successful in increasing biodegradability. Once biodegradability has been established, the land treatment of munitions manufacturing wastes can be investigated.

The use of soil to dispose of TNT was observed to be too slow to effectively treat 1% TNT in soil. However, bench-scale composting of up to 10% TNT showed that almost complete removal of TNT occurred in 55 days. Many of the toxic transformation products formed in activated sludge and soil were not found in the composted TNT. Further research needs to be done in the area of identifying the unknown volatile and soluble compounds formed during TNT composting and to determine their biodegradability and potential toxicity. Composted wastes of TNT and other munitions must be examined for the leachability of undecomposed chemicals and transformation products prior to land treatment (Kaplan and Kaplan, 1982).

Safety considerations will obviously be a paramount concern in any investigations and applications of research results relating to munitions wastes.

REFERENCES

Andren, R. K., J. M. Nystrom, R. P. McDonnell, and B. W. Stevens. 1975. p. 816-825. In Proc. 30th Indus. Waste Conf. Purdue Univ., Lafayette, IN. Ann Arbor Sci. Publ., Inc., Ann Arbor, MI.

Anonymous. 1971. U. S. Army Environmental Hygiene Agency Sanitary Engr. Sp. Study No. 24-007-70/71. Evaluation of toxicity of selected TNT wastes on fish, Phase 1 - Acute toxicity of alpha-TNT to bluegills. U. S. Army Environmental Hygiene Agency, Edgewood Arsenal, MD.

Bureau of Mines. 1981. Explosives Annual. U.S. Department of Interior, Mineral Industry Surveys. (October 1981) Washington, D.C. 12 p.

Burlinson, N. E. 1980. Fate of TNT in an aquatic environment: photodecomposition vs. biotransformation. NSWC TR 79-445. Naval Surface Weapons Center, Silver Spring, MD. 22p.

Burlinson, N. E., M. E. Sitzmann, P. J. Glover, and L. A. Kaplan. 1979. Photochemistry of TNT and related nitroaromatics: part III. NSWC/WOL TR 78-198. Naval Surface Weapons Center, Silver Spring, MD. 53p.

Carpenter, D. F., N. G. McCormick, J. H. Cornell, and A. M. Kaplan. 1978. Microbial transformation of ^{14}C-labelled 2,4,6-trinitrotoluene in an activated-sludge. Appl. Environ. Microbiol. 35:949-954.

Channon, H. J., G. T. Mills, and R. T. Williams. 1944. The metabolism of 2,4,6-trinitrotoluene (α-TNT). Biochem. J. 38:70-85.

Christensen, H. E., T. T. Luginbyhl, and B. S. Carroll (ed.). 1975. NIOSH registry of toxic effects of chemical substances. U.S. Department of Health, Education, and Welfare. U.S. Government Printing Office, Washington, D. C.

E.P.A. 1980. Explosives. p. 636-669. In Background document. Resource conservation and recovery act. Subtitle C-hazardous waste management. Environmental Protection Agency, Office of Solid Waste, Washington, D. C.

Glover, D. J., and J. C. Hoffsommer. 1979. Photolysis of RDX. Identification and reactions of products. NSWC TR 79-349. Naval Surface Weapons Center, Silver Spring, MD. 25p.

Hoffsommer, J. C., L. A. Kaplan, D. J. Glover, D. A. Dubose, C. Dickinson, H. Goya, E. G. Kayser, C. L. Groves, and M. E. Sitzmann. 1978. Biodegradability of TNT: a three-year pilot plant study. NSWC/WOL TR 77-136. Naval Surface Weapons Center, Silver Spring, MD. 65p.

Kaplan, D. L., and A. M. Kaplan. 1982. Composting Industrial Waste - Biochemical Consideration. BioCycle 23(3):42-44.

Kessick, M. A., W. G. Characklis, and W. Elvey. 1977. Treatment of wastewater from torpedo refueling facilities. pp. 442-449. In Proc. 32nd Indus. Waste Conf. Purdue Univ., Lafayette, IN. Ann Arbor Sci. Publ., Inc., Ann Arbor, MI.

Klausmeier, R. E., J. A. Appleton, E. S. DuPre, and K. Tenbarge. 1975. The enzymology of trinitrotoluene reduction. pp. 779-805. In J. M. Sharpley and A. M. Kaplan (ed.) Proc. Third Intl. Biodegradation Symp. Applied Science Publishers, Ltd, London.

Klausmeier, R. E., J. L. Osmon, and D. R. Walls. 1974. The effect of trinitrotoluene on microorganisms. Dev. Ind. Microbiol. 15:309-317.

McCormick, N. G., F. E. Feeherry, and H. Levinson. 1976. Microbial transformation of 2,4,6-trinitrotoluene and nitroaromatic compounds. Appl. Environ. Microbiol. 31:949-958.

Naumova, R. P., N. N. Amerkhanova, and V. A. Shaikhutdinov. 1979. Study of the first stage of the conversion of trinitrotoluene under the action of Pseudomonas denitrificans. Appl. Biochem. Microbiol. 15:33-38.

Nay, M. W., Jr., C. W. Randall, and P. H. King. 1972. Factors affecting color development during the treatment of TNT waste. pp. 983-993. In Proc. 27th Indus. Waste Conf. Purdue Univ., Lafayette, IN. Ann Arbor Publ., Inc., Ann Arbor, MI.

Nay, M. W., Jr., C. W. Randall, and P. H. King. 1974. Biological treatability of trinitrotoluene manufacturing wastewater. J. Water Pollut. Control. Fed. 46:485-497.

Needleman, P., D. J. Blehm, A. B. Harkey, E. M. Johnson, Jr., and S. Lang. 1971. The metabolic pathway in the degradation of glyceryl trinitrate. J. Pharmacol. Exp. Therap. 179:347-353.

Okamoto, Y., E. J. Chou, J. Wang, and M. Roth. 1977. The removal of 2,4,6-trinitrotoluene (TNT) from aqueous solution with surfactants. pp. 249-253. In Proc. 1977 Natl. Conf. Treatment and Disposal Ind. Wastewaters and Residues. Information Transfer, Inc., Rockville, MD.

Osmon, J. L., and C. C. Andrews. 1978. The biodegradation of TNT in enhanced soil and compost systems. Tech. Rep. ARLCD-TR-77032. U. S. Army Armament Research and Development Command. Large Caliber Weapon Systems, Dover, NJ. 81p.

Osmon, J. L., and R. E. Klausmeier. 1973. The microbial degradation of explosives. Dev. Ind. Microbiol. 14:247-252.

Parrish, F. W. 1977. Fungal transformation of 2,4-dinitrotoluene and 2,4,6-trinitrotoluene. Appl. Environ. Microbiol. 34:232-233.

Patterson, J. W., N. I. Shapira, and J. Brown. 1976. Pollution abatement in the military explosives industry. pp. 385-394. In Proc. 31st Indus. Waste Conf. Purdue Univ., Lafayette, IN. Ann Arbor Sci. Publ., Inc., Ann Arbor, MI.

Patterson, J. W., N. I. Shapira, J. Brown, W. Duckert, and J. Polson. 1976a. State-of-the-art: military explosives and propellents production industry. Vol. I - The military explosives and propellents industry. EPA- 600/2-76-213a, U.S. Environmental Protection Agency, Cincinnati, OH. 96 pp.

Patterson, J. W., N. I. Shapira, J. Brown, W. Duckert, and J. Polson. 1976b. State-of-the-art: military explosives and propellents production industry. Vol. III. - Wastewater treatment. EPA-600/2-76-213c, U.S. Environmental Protection Agency, Cincinnati, OH. 169 pp.

Ruchhoft, C. C., M. LeBosquet, Jr., and W. G. Meckler. 1945. TNT wastes from shell-loading plants. Ind. Eng. Chem. 37:937-943.

Schott, S., C. C. Ruchhoft, and S. Megregian. 1943. TNT wastes. Ind. Eng. Chem. 35:1122-1127.

Siu, R. G. H. 1951. Microbial decomposition of cellulose with special reference to cotton textiles. Reinhold Publishing Corp., New York. 531 p.

Soli, G. 1973. Microbial degradation of cyclonite (RDX). AD-762 751. NTIS, Springfield, VA. 4 p.

Traxler, R. W. 1974. Biodegradation of alpha-TNT and its production isomers. Tech. Rep. 75-113-FSL. United States Army Natick Development Center. Natick, MA. 23 p.

Traxler, R. W., E. Wood, and J. M. Delaney. 1975. Bacterial degradation of alpha-TNT. Dev. Ind. Microbiol. 16:71-76.

TRW Systems Group 1975. Assessment of Industrial Hazardous Waste Practices, Organic Chemicals, Pesticides, and Explosives Industries. PB 251-307. NTIS, Springfield, Va.

Wang, L., M. Pressman, W. Shuster, R. Shade, F. Bilgen, and T. Lynch. 1978. Treatment of nitrocellulose-manufacturing wastewater. pp. 27-35. In 3rd Ann. Conf. Treatment and Disposal Indus. Wastewaters and Residues. Information Transfer, Inc., Rockville, MD.

Wendt, T. M., and A. M. Kaplan. 1976. A chemical-biological treatment process for cellulose nitrate disposal. J. Water Pollut. Control Fed. 48:660-667.

Won, W. D., L. H. DiSalvo, and J. Ng. 1976. Toxicity and mutagenicity of 2,4,6-trinitrotoluene and its microbial metabolites. Appl. Environ. Microbiol. 31:576-580.

Won, W. D., R. J. Heckly, D. J. Glover, and J. C. Hoffsommer. 1974. Metabolic disposition of 2,4,6-trinitrotoluene. Appl. Microbiol. 27:513-516.

Wyman, J. F., H. E. Gerard, W. D. Won, and J. H. Quay. 1979. Conversion of 2,4,6-trinitrophenol to a mutagen by *Pseudomonas aeruginosa*. Appl. Environ. Microbiol. 37:222-226.

Pesticide and Organic Chemical Manufacturing Wastes

D.D. Kaufman

Pentachlorophenol, Chlorophenols, and Dioxins

Chlorinated phenols are manufactured in large amounts and have widely varying applications. They are used as insecticides, fungicides, herbicides, antiseptics, and disinfectants; their effectiveness generally increases with the degree of chlorination. Other uses are as starting materials for production of certain pesticides, dyes, and pigments (Doedens, 1963). The annual production of chlorinated phenols and their derivatives is hard to estimate but approximately 150,000 tons has been deemed reasonable (Nilsson et al. 1978). The most widely used methods for preparing chlorinated phenols are direct chlorination, and alkaline hydrolysis of the appropriate chlorobenzene, the particular method used depending upon the isomer desired (Doedens, 1963; Melnikov, 1971).

The isomers manufactured via direct chlorination include 2,4-dichlorophenol, 2,4,6-trichlorophenol, 2,3,4,6-tetrachlorophenol, and pentachlorophenol. 2,4-Dichlorophenol is most often used as the starting material in the production of the herbicide (2,4-dichlorophenoxy)acetic acid. The most important use of 2,4,6-trichlorophenol, 2,3,4,6-tetrachlorophenol, and pentachlorophenol is as wood preservatives, both as phenols and the corresponding sodium or potassium salts. In addition, they are used as bactericides; pentachlorophenol is used in slime control in pulp and paper mills.

The chlorophenols manufactured via hydrolysis of chlorobenzenes include 2,4,5-trichlorophenol and pentachlorophenol. The hydrolysis of chlorobenzenes can give rise to the unwanted formation of chlorinated dibenzo-*p*-dioxins as by-products (Milnes, 1971), an undesirable side reaction favored by high temperatures.

The mass death of chickens, several industrial accidents where a large number of employees were seriously poisoned by exposure (Goldman, 1972; May, 1973; Baughman, 1974; Hay, 1976), and their detection in the environment have focused interest on the dioxin content of commercial chlorophenol formulations as well as industrial wastes. In earlier work only neutral impurities such as the chlorinated dioxins, dibenzofurans, and diphenylethers were taken into account. However, since chlorophenols were shown to contain chlorinated phenoxyphenols and these can undergo thermal and photochemical ring closure to dioxins (Jensen and Renberg, 1972; Rappe and Nilsson, 1972; Nilsson and Renberg, 1974; Nilsson et al., 1974; Levin, Rappe, and Nilsson, 1976), most scientists have also taken the phenoxyphenols into account.

Chlorinated dibenzo-*p*-dioxins and dibenzofurans have been known to chemists for many years. 2,8-Dichlorodibenzo-*p*-dioxin was first reported in 1941 (Ueo, 1941); the 2,3,7,8-tetrachlorodibenzo-*p*-dioxin (TCDD) and octachlorodibenzo-*p*-dioxin (OCDD) analogs followed in 1957 (Sanderman, Stockman, and Castor, 1957). Extensive research by Gilman and Dietrich (1957) and by Pohland and Yang (1972) provided data on many others. Similarly, simple chlorinated dibenzofurans have been reported since the early 1930's (Gilman et al., 1934), while experience with the more highly chlorinated ones is comparatively recent (Kimmig and Schulz, 1957; Vos, 1970). Major concern arose only when the toxic and teratogenic properties of TCDD became apparent in widely-distributed pesticides such as (2,4,5-trichlorophenoxy)acetic acid. Subsequent analyses (Vos, 1970; Firestone, 1972; Woolson, Thomas, and Ensor, 1972) demonstrated that a variety of chlorinated dibenzo-*p*-dioxins and dibenzofurans can occur as impurities from the manufacture of many industrial and agricultural chemicals based on chlorophenols and certain chlorinated aromatic hydrocarbons. To cite but a single example, production statistics (U.S. Tariff Commission, 1970) suggest that at least 25,000 tons of 2,4,5-trichlorophenol and its derivatives are manufactured in the United States each year, almost all of which contains low but detectable levels of TCDD (Crosby, Moilanen, and Wong, 1973). The potential occurrence of such compounds in wastes of chlorophenol-producing industries would thus seem considerable.

Little information is available on the disposal of pentachlorophenol manufacturing wastes. Based on available information regarding impurities in pentachlorophenol itself, one can reasonably expect that the bulk of the wastes are unreacted phenols or chlorophenols in addition to the polymeric compounds mentioned in preceding paragraphs. Most chlorophenol formulations contain an assortment of chlorophenols: in pentachlorophenol, the principal contaminant is tetrachlorophenol. In one pentachlorophenol formulation manufactured in Sweden, the pentachlorophenol can actually be regarded as an impurity since the main constituent (80%) is tetrachlorophenol (Nilsson et al., 1978). In addition to other chlorophenols, technical chlorophenols have been shown to contain a variety of dimeric impurities. Of most concern are the dioxins, due to the inherent toxicity of certain dioxins in particular. Chlorinated phenoxyphenols which can act as precursors to dioxins are also present at 1-5%. Other dimeric impurities found are diphenylethers, dibenzofurans, and dihydroxybiphenyls. Although considerable information is known about the degradation and fate of most chlorophenols in soil and other segments of the environment, very little is known about the degradation of the chlorinated biphenyls, dibenzofurans, dioxins, diphenylethers, and phenoxyphenols.

A remarkable level of scientific attention has been devoted to the chlorinated dibenzodioxins and dibenzofurans. A limited number of articles represents the four decades preceding 1970, but several recent conferences alone have provided more than 100 scientific research papers. Although the earlier papers were primarily concerned with the health effects, more recent concerns have been primarily associated with the possible environmental occurrence, effects, and fate. While modernized production methods have reduced the level of contamination by dioxins, etc., there is some evidence that these chemicals may also be formed during degradation of chlorophenols in the environment.

Despite the suggestion (Buu-Hoi et al., 1971) that phenol derivatives such as (2,4,5-trichlorophenoxy)acetic acid might be decomposed to dibenzodioxins by heat, further investigation has failed to demonstrate the conversion (Langer, 1973; Johnson, 1971). One very real possibility which exists, however, is the current production of pentachlorophenol and wood products treated with pentachlorophenol.

The burning of treated lumber provides sufficient heat and concentration to convert pentachlorophenol into dioxins and predioxins. This concept was examined by Crosby, Moilanen, and Wong (1973) who burned woodchips and trapped the volatiles emanating therefrom. The OCDD level in the wood initially was approximately 1 ng/g, while that in the smoke emanating from the woodchips was enriched in hepta- and hexachloro-p-dioxin. Photochemical formation has also been observed during the photolysis of pentachlorophenol. Octa- but not tetrachlorodibenzo-p-dioxins were formed.

Repeated attempts to detect TCDD after the irradiation of (2,4,5-trichlorophenoxy)acetic acid, 2,4,5-trichlorophenol, or sodium 2,4,5-trichlorophenate solution were unsuccessful (Crosby, Moilanen, and Wong, 1973). The expected photo-nucleophilic formation of phenols from other aromatic halides such as chlorobiphenyls has been demonstrated with both simple models (Crosby and Moilanen, 1971; Safe and Hutzinger, 1971; Ruzo, Zabik, and Schuetz, 1972; Hutzinger, Safe, and Zitko, 1972) and with highly complex polychlorinated biphenyl (PCB) mixtures (Arochlors) (Ruzo, Zakik, and Schuetz, 1972). Irradiation of an aqueous suspension of 4,4-di-chlorobiphenyl provided 4-chloro-4-hydroxybiphenyl and 4-chlorobiphenyl. However, as in the classical examples of synthesis (Zahn and Schimmelschmidt, 1940; Case and Schock, 1943), appropriately substituted chlorinated biphenyls also might be expected to form chlorodibenzofurans.

Failure to detect TCDD as a product of 2,4,5-trichloro-phenol photolysis can be explained on the basis of the extreme instability of the lower chlorinated dioxins to light (Crosby et al., 1971). In sunlight, 2,7-dichloro-, 2,3,7-trichloro-, and 2,3,7,8-tetrachlorodibenzo-p-dioxins were entirely decomposed in a few hours, whereas OCDD was more stable under the experimental conditions.

Although improved manufacturing and surveillance methods have dramatically reduced dioxin and dibenzofuran levels among industrial chemicals (Woolson, Thomas, and Ensor, 1972), there is evidence that under idealized conditions, at least, both types of compounds can be generated from commonly used phenols and chlorinated biphenyls under environmental conditions. The driving force can be either the ultraviolet component of sunlight impinging upon thin films, surfaces, or solutions, or heat such as might be encountered in burning wood.

However, environmental persistence or accumulation theoretically may be balanced by the simultaneous destructive action of light or heat. Photolysis can be rather rapid under idealized conditons, although a hydrogen donor appears to be required, and OCDD is rather stable compared to the highly toxic TCDD. Photosensitization must be an important consideration, as demonstrated by the effect of dichloro-benzophenone on the photoreduction of chlorinated dibenzo-furans. Such sensitizers have been shown to be expected under field conditions (Plimmer and Kingebiel, 1971; Ivie and Casida, 1971; Ross and Crosby, 1972).

Considerable research has been conducted on the degradation of phenolic compounds in some environmental situations (Kozak et al., 1979; Bevenue and Beckman, 1967; Doedens, 1963). Nearly all are known to be subject to degradation under certain environmental conditions. 2-Chlorophenol and 2,4-dichlorophenol are amenable to metabolism by bacteria found in soil and aquatic environments. Phenol-adapted bacteria isolated from soil rapidly metabolize 2-chlorophenol to produce 3-chlorocatechol. Microbial breakdown proceeds further to include ring cleavage.

Soil microorganisms capable of growing on and degrading (2,4-dichlorophenoxy)acetic acid are also adapted to degrade 2-chlorophenol and 2,4-dichlorophenol. Degradation by these microorganisms proceeds through the formation of 3,5-dichlorocatechol to eventually form dicarboxylic acids, acetate, and chloride ion.

The presence of these microorganisms is extremely important for the biological treatment of waste generated from the manufacture of chlorophenols and 2,4-D.

Effective microbial degradation of trichlorophenol and tetrachlorophenol has been demonstrated in activated sludge, lagoon effluent, and enrichment cultures. Thus, effective waste treatment of chlorophenol- containing wastewaters occurs when appropriate bacterial populations are present. Microorganisms capable of metabolizing trichlorophenols and tetrachlorophenols have not been isolated from soils, nor have degradation pathways been elucidated. However, pentachlorophenol, which is the most refractory chlorophenol compound, is susceptible to microbial degradation in soil, water, pentachlorophenol-treated wood, and sewage sludge. Certain soil bacteria and fungi are able to metabolize pentachlorophenol to pentachloroanisole. Other microbial isolates are able to detoxify pentachlorophenol quantitatively with release of Cl^-, quantitative disappearance of the substrate, almost quantitative oxygen uptake, and elimination of significant portions of the molecule as $^{14}CO_2$. The degradation of pentachlorophenol via the formation of tetra- and trichlorophenols, suggests that these compounds are also reasonably degradable (Kozak et al., 1979). Thus, the ubiquitous nature and presence of chlorophenols in wastes should not preclude consideration of land treatment for these wastes.

In the case of polychlorinated dibenzo-*p*-dioxins, the question of TCDD bioactivity is indisputable, as it is known to be one of the most toxic compounds known to occur as a chemical impurity (Higginbotham et al., 1968; Courtney et al., 1970; Sparschu, Dunn, and Rowe, 1971). Its chemical stability is also questionable. Thus, the central question of its hazard to, and its persistence in, the environment should be investigated. Published data on the environmental fate of chlorodibenzo-p-dioxins are confusing at present. Zitko (1972) could not find residues of dioxins in several aquatic animals at detection limits of 0.01-0.04 ug/g. Isensee and Jones (1971) studied absorption and translocation of TCDD and its 2,7-dichloro-analog in root and foliage, and concluded that dioxins are neither readily absorbed from the soil, nor translocated into foliage. Matsumura and Benezet (1973) attempted to measure the degree of bioaccumulation of TCDD in relation to several well established pesticides in several model ecosystems. They concluded that TCDD was not likely to accumulate in as many biological systems as DDT. This was probably because of TCDD's low solubility in water and lipids as well as its low partition coefficient in lipids.

Investigations by Kearney, Woolson, and Ellington (1972) indicated that the average TCDD remaining after weathering in soil for one year was of the order of 50-60% of that applied at all concentrations tested (1-100 ppm), whereas actual field survey (Woolson, Young, and Hunter, 1972) of the test areas with Agent Orange (1962-1970) revealed that much less TCDD residues were remaining in soil than was expected based on Kearney, Woolson, and Ellington's (1972) observations. Thus, it would seem reasonable that any source of waste with chlorophenols or dioxins present in them should be considered subject for further study.

Summary and Research Needs

Wastes from the organic chemicals manufacturing industries are poorly characterized. Because such a wide range of compounds may be present in these wastes, the many factors should be considered for each waste. If the compounds are biodegradable, then land treatment would be useful. Many of the industrial wastes will be activated sludge from treatment of process, cleaning, and sanitary wastewater, and the potentially toxic organic chemicals will be relatively dilute.

The fate of wastes from production of chlorophenols is considered in detail. Byproducts and undesired side reaction products which are removed in preparing the product include the dioxins. Dioxins may also result from burning these wastes, and even during degradation of the wastes.

However, dioxins are biodegradable in soil. Land treatment appears to offer considerable utility in disposal of these chlorophenol manufacturing wastes.

Research Needs--

Wastes from organic chemical manufacturing should be better characterized. So little is presently known that it is difficult to plan land treatment for even one factory. Segregation of waste streams may alleviate contamination of bulk wastes with non-biodegradable toxic organic compounds.

Many of the fundamental questions which require research for these industrial wastes should be studied first with pure compounds. When known toxic (hazardous) compounds are present in a waste, research on land treatment of that waste must address fate of these specific measurable compounds. Composting or other pretreatment of wastes, appropriate fertilization, soil type effects, etc., all remain to be evaluated.

REFERENCES

Baughman, R. W. 1974. Tetrachlorodibenzo-p-dioxins in the environment: High resolution mass spectrometry and the picogram level. Thesis, Harvard Univ., Cambridge, Mass.

Bevenue, A., and H. Beckman. 1967. Pentachlorophenol: A discussion of its properties and its occurrence as a residue in human and animal tissues. Residue Reviews 19:83-134.

Buu-Hoi, N. P., G. Saint-Ruf, P. Bigot, and M. Mangane. 1971. Preparation, properties, and identification of "dioxin" (2,3,7,8-tetrachlorodibenzo-dioxin) in the pyrolyzate of defoliants based on 2,4,5-trichlorophenoxyacetic acid and its esters and contaminated vegetation. Compt. Rend. Acad. Sci. Paris 273D:708.

Case, F. H., and R. V. Schock, Jr. 1943. Nitration of halobiphenyls (II) Di- and tetranitro derivatives of 2,2'-dichlorobiphenyls. J. Amer. Chem. Soc. 65:2086.

Courtney, K. D., D. W. Gaylor, M. D. Hogan, H. L. Falk, R. R. Bates, and J. Mitchell. 1970. Teratogenic evaluation of 2,4,5-T. Science 168:864-866.

Crosby, D. G., and K. W. Moilanen. 1971. Annual Report, Food Protection and Toxicology Center. Univ. Calif., Davis, CA.

Crosby, D. G., K. W. Moilanen, and A. S. Wong. 1973. Environmental generation and degradation of dibenzodioxins and dibenzofurans. Environ. Health Perspect. 5:259-266.

Crosby, D. G., A. S. Wong, J. R. Plimmer, and E. A. Woolson. 1971. Photodecomposition of chlorinated dibenzo-p-dioxins. Science 173:748.

Doedens, J. D. 1963. Chlorophenols. pp. 325-338. In Kirk-Othmer Encyclopedia of Chemical Technology, 2nd Ed., Vol. 5. John Wiley and Sons, Interscience Publishers, New York.

Firestone, D. 1972. Determination of polychloro-dibenzo-p-dioxins and related compounds in commercial chlorophenols. J. Assoc. Offic. Anal. Chem. 55:85-92.

Gilman, H., and J. J. Dietrich. 1957. Halogen derivatives of dibenzo-p-dioxin. J. Amer. Chem. Soc. 79:1439.

Gilman, H., G. E. Brown, W. G. Bywater, and W. H. Kirkpatrick. 1934. Dibenzofurans III. Nuclear substitution. J. Amer. Chem. Soc. 56:2473.

Goldman, P. 1972. Enzymology of carbon-halogen bonds. pp. 147-165. In Degradation of Synthetic Organic Molecules in the Biosphere. National Academy of Science, Washington, D.C.

Goldman, P. J. 1972. Schwerste akute Chlorakne durch trichlorophenol-zersetzungsprodukte. Arbeitsmed. Sozialmed. Arbeitshyg. 7:12.

Hay, A. 1976. Toxic cloud over Seveso. Nature, 262:636.

Higginbotham, G. R., A. Huang, D. Firestone, J. Verrett, J. Ress, and A. D. Campbell. 1968. Chemical and toxicological evaluations of isolated and synthetic chloro derivatives of dibenzo-p-dioxins. Nature 220:702.

Hutzinger, O., S. Safe, and V. Zitko. 1972. Photochemical degradation of chlorobiphenyls (PCB's). Environ. Health Perspect. No. 1:15.

Isensee, A. R., and G. E. Jones. 1971. Absorption and translocation of root and foliage applied 2,4-dichlorophenol, 2,7-dichlorodibenzo-p-dioxin, and 2,3,7,8-tetrachloro-dibenzo-p-dioxin. J. Agr. Food Chem. 19:1210.

Ivie, G. W., and J. E. Casida. 1971. Sensitized photodecomposition and photosensitizer activity of pesticide chemicals exposed to sunlight on silica gel chromatoplates. J. Agr. Food Chem. 19:405-409.

Jensen, S., and L. Renberg. 1972. Contaminants in penta-chlorophenol chlorinated dioxins and predioxins. Ambio. 1:62.

Johnson, J. E. 1971. Safety in the development of herbicides. Proc. Calif. Weed Conf. 23:43.

Kearney, P. C., E. A. Woolson, and C. P. Ellington, Jr. 1972. Persistence and metabolism of chlorodioxins in soils. Environ. Sci. Technol. 6:1017-1019.

Kimmig, J., and K. H. Schulz. 1957. Berufliche Anke (sog. Chlorakne) durch chlorierte aromatische zyklische ather. Dermatologica 115:540.

Kozak, V. P., G. V. Simsiman, G. Chesters, D. Stensby, and J. Harkin. 1979. Reviews of the environmental effects of pollutants: XI. Chlorophenols. ORNL/EIS-128. U.S. Department of Commerce, National Technical Information Service, Springfield, Va. 491pp.

Langer, H. G. 1973. The formation of dibenzodioxins and other condensation products from chlorinated phenols and derivatives. Environ. Health. Perspect. No. 5:3.

Levin, J. O., C. Rappe, and C. A. Nilsson. 1976. Use of chlorophenols as fungicides in sawmills. Scandinavian J. Work Environ. Health. 2:71-81.

Matsumura, F., and H. J. Benezet. 1973. Studies on the bioaccumulation and microbial degradation of 2,3,7,8-tetrachlorodibenzo-p-dioxin. Environ. Health Perspect. 5:253-258.

May, G. 1973. Chloracne from the accidental production of tetrachlorobenzoidioxin. British J. Ind. Medicine 30:276.

Melnikov, N. N. 1971. Chemistry of Pesticides. Springer-Verlag, New York.

Milnes, M. H. 1971. Formation of 2,3,7,8-tetrachlorodioxin by thermal decomposition of sodium 2,4,5-trichlorophenate. Nature. 232:395.

Nilsson, C. A., A. Norstrom, K. Andersson, and C. Rappe. 1978. Impurities in commercial products related to pentachlorophenol. pp. 313-324. In K. Ranga Rao (ed.) Pentachlorophenol. Plenum Press, New York.

Nilsson, C. A., K. Andersson, C. Rappe, and S. O. Westermark. 1974. Chromatographic evidence for the formation of chlorodioxins from chloro-2-phenoxyphenols. J. Chromatogr. 96:137-147.

Nilsson, C. A., and L. Renberg. 1974. Further studies on impurities in chlorophenols. J. Chromatogr. 89:325-333.

Plimmer, J. R., and U. I. Klingebiel. 1971. Riboflavin photosensitized oxidation of 2,4-dichlorophenol: assessment of possible chlorinated dioxin formation. Science 174:407-408.

Pohland, A. E., and G. C. Yang. 1972. Preparation and characterization of chlorinated dibenzo-p-dioxins. J. Agr. Food Chem. 20:1093-1099.

Rappe, C., and C. A. Nilsson. 1972. An artifact in the gas chromatographic determination of impurities in pentachlorophenol. J. Chromatography. 67:247-253.

Ross, R. D., and D. G. Crosby. 1972. The photolysis of ethylene thiourea. J. Agr. Food Chem. 21:335-337.

Ruzo, L. O., M. J. Zabik, and R. D. Schuetz. 1972. Polychlorinated biphenyls: photolysis of 3,4,3',4'-tetrachlorobiphenyl and 4,4'-dichlorobiphenyl in solution. Bull. Environ. Contam. Toxicol. 8:217.

Safe, S., and O. Hutzinger. 1971. Polychlorinated biphenyls: photolysis of 2,4,6,2',4',6'-hexachlorobiphenyl. Nature 232:641.

Sanderman, W., H. Stockman, and R. Caston. 1957. Pyrolysis of pentachlorophenol. Chem. Ber. 90:960.

Sparschu, G. L., F. L. Dunn, and V. K. Rowe. 1971. Study on the teratogenicity of 2,3,7,8-tetrachlorodibenzo-p-dioxin in the rat. Food Cosmet. Toxicol. 9:405.

Ueo, S. 1941. 2,6-dichlorodiphenylene dioxide. Bull. Chem. Soc. Japan 16:177.

Vos, J. G. 1970. Identification and toxicological evaluation of chlorinated dibenzofuran and chlorinated naphthalene in two commercial polychlorinated biphenyls. Food Cosmet. Toxicol. 8:625.

U. S. Tariff Commmission. 1970. Synthetic organic chemicals: Production and sales. TC Publ. 479. Washington, D. C.

Woolson, E. A., R. F. Thomas, and P. D. Ensor. 1972. Survey of polychlorodibenzo-p-dioxin content of selected pesticides. J. Agr. Food Chem. 20:351-354.

Woolson, E. A., A. L. Young, and J. H. Hunter. 1972. Chemical analysis for dioxin and defoliant residues in soil from tested area C-52A, Eglin Air Force Base, Florida. Weed Sci. Soc. Amer. Abstr. No. 173.

Zahn, K., and K. Schimmelschmidt. 1940. Hydroxybiphenylene oxides. U. S. Pat. 2,172,572 (Sept. 12, 1940). Chem. Abstr. 34:1032(1940).

Zitko, V. 1972. Absence of chlorinated dibenzodioxins and dibenzofurans from aquatic animals. Bull. Environ. Contam. Toxicol. 7:105.

Petroleum Wastes

S.B. Hornick, R.H. Fisher, and P.A. Paolini

The six major groups of operations and processes in a petroleum refinery are (1) storage of crude oil intermediates and final products; (2) fractionation such as distillative separation and vacuum fractionation; (3) decomposition such as thermal cracking, catalytic cracking, and hydrocracking; (4) hydrocarbon rebuilding and rearrangement such as polymerization, alkylation, reforming, and isomerization; (5) extraction such as solvent refining and solvent dewaxing; and (6) product finishing such as drying and sweetening, lube oil finishing, blending, and packing (FWPCA, 1967).

The above processes lend specific characteristics to a refinery waste stream. The major constituents of a refinery waste that determine its utilization for landspreading are pH, BOD, phenol content, sulfide content, oil content, and heavy metal content.

The chemical constituents of typical refinery wastes vary depending upon the petroleum product being produced. Table 13 shows a representative composition of 12 API refinery wastes (Overcash and Pal, 1979).

Most refinery wastewaters are alkaline mainly from the cracking process. Solvent refining contributes to a lesser degree. Acidic wastes are produced in alkylation and polymerization, but large volumes of cooling and wash waters usually dilute very acid or very caustic discharges. High BOD results from the soluble hydrocarbons and sulfides which are produced on the cracking and solvent refining processes. Phenolic compounds are the results of catalytic cracking, crude oil fractionation, and product treating. The

TABLE 13. COMPOSITION OF 12 API REFINERY WASTES (OVERCASH AND PAL, 1979)

	Minimum	Maximum	Average
Sulfides mg/ℓ	1.3	38	8.8
Phenol mg/ℓ	7.6	61	27
BOD mg/ℓ	97	280	160
COD mg/ℓ	140	640	320
pH	7.1	9.5	8.4
Oil mg/ℓ	23	130	57

decomposition of polycyclic aromatics such as anthracene and phenanthrene produce phenols in catalytic cracking (FWPCA, 1967). Chlorinated phenols generally cause taste and odor problems in drinking water. Sulfides are produced from crude desalting, crude distillation, and cracking processes.

Although product recovery is maximized, the refining of crude oil produces effluents and sludges with substantial amounts of oil and grease. Crude oil refineries are probably the largest contributors of oily waste. The sources of these wastes are crude oil tank bottoms, slop tank bottoms, API separator bottoms, biotreatment solids, coker blowdown, and water impoundment pond bottoms (Overcash and Pal, 1979; Huddleston and Cresswell, 1976). Table 14 shows various petroleum products and wastes and their hydrocarbon composition.

TABLE 14. HYDROCARBON TYPES CONTAINED IN VARIOUS OIL PRODUCTS AND WASTE OILS

	Oil Composition - % by Weight		
Material	Paraffins	Aromatics	Asphaltic
Arabian heavy crude[a]	46	40	14
No. 6 fuel[a]	37	46	17
Coastal mix[a]	54	33	13
Used crankcase oil[a]	71	18	11
Crude oil tank bottom[b]	36	56	8

a Cross and Lawson
b Lewis (1976)

In addition to refining wastes, the production of petrochemicals, i.e., synthetic detergent bases, solvents, fuel additives, plastics and resins, and synthetic fiber bases, produces waste with a wide range of characteristics, some of which are high dissolved solids; high BOD; compounds inhibitory to biological treatment; heavy metal contaminants from catalysts used; and due to product revision, wide variation in organic constituents. Some of the products generated in the petrochemical industry will be discussed in greater detail in the organic chemicals section.

Landfarming of petroleum wastes has proven to be a successful alternative to incineration when energy conservation is considered (Knowlton and Rucker, 1978). Different types of refinery wastes have been successfully farmed, e.g., crude and distillate tank cleanings (20-50% oil), refinery and chemical plant separator cleanings (10-20% oil), cleanings from sewers, desalters and spills (10% oil), filter clays from jet and white oil (8% oil), and biological sludge from wastewater treatment (Knowlton and Rucker, 1978). Understandably, addition of oil waste to a soil will change its chemical, biological, and physical makeup.

Hydrophobic effects of oil applications to soil tend to be short-lived. There is an initial impedance of water movement due to the nature of the oil and to the accumulation of hydrophobic mucilagenous substances generated by increased microbial growth. The oil flow then becomes dependent on gravitational and surface tension forces (Overcash and Pal, 1979).

Long-term effects of applied oil on soil physical properties tend to be beneficial. Aggregation, soil porosity, water holding capacity, and retention, all increase remarkably. There is an initial decrease of water infiltration and movement until the oil is decomposed to organic matter. As would be expected, bulk density is also markedly decreased.

The metal and elemental composition of oily wastes which have been successfully applied to land are shown in Table 15. The wastes differ greatly in the amount of organic matter (%C) they will contribute to the soil and the amount of N, P, and K, essential plant nutrients, which they will provide. Chromium is a heavy metal that is usually present in a large amount in refinery wastes. However, when added to the soil, Cr is quickly reduced to insoluble oxides and hydroxides and is unavailable to plants and unable to leach to the groundwater. If lands are limed to pH 6.5, the addition of alkaline oil waste will maintain the soil pH near neutral. Thus,

TABLE 15. ANALYSIS OF REFINING WASTES (BERKOWITZ ET AL., 1980)

	Refining Waste	
	Holding Pond Sludge	API Separation Bottoms
Element	--------ug/g dry weight, unless noted---------	
C	28.2 %	46.4 %
N	2.08%	0.23%
As	105	8
B	39	4
Ca	1.5%	8600
Cd	9.6	2
Cr	1.3 %	1600
Cu	310	160
K	7900	1900
Mg	5400	3400
Mo	22	10
Na	2.1 %	3000
Ni	19	44
P	3.1 %	750
Pb	89	93
Se	150	3
V	39	26
Zn	6500	1100

other metals harmful to the food chain such as Cd and Pb or harmful to plants such as Zn and Cu, and Na will precipitate as hydroxides. In addition the added organic material will increase the CEC and increase the adsorption of metals in the soil (Phung and Ross, 1978). The major concern with metals in oily wastes should be their interactions with microorganisms that can degrade oily and other potentially harmful constituents in the wastes which can inhibit normal soil reactions.

The general make-up of crude oil is rich in hydrocarbons resulting in a very high C:N ratio. Each oil differs in amounts of alkanes (paraffins), cycloalkanes, and aromatics depending upon its source. For example, an Arabian heavy crude contains more aromatics than the coastal crudes, but less paraffins (Raymond, Hudson, and Jameson, 1975).

After application of oily waste, the initial biological response is one of decreased microbial activity. This may be due to the same initial decrease in mineral N resulting from N immobilization by hydrocarbon-metabolizing microbes using up all available N. Given time, the microorganisms evolve to adjust to the high C:N ratio increasing the total microbial population (Overcash and Pal, 1979). Twenty-eight genera of bacteria, thirty genera of filamentous fungi, and twelve genera of yeasts are known at this time to decompose oil (Zobell, 1973).

The oil-loading rate on soil is an important factor in determining the length of time between the initial lag in microbial growth and the subsequent increase in the population. Degradation of oil is, therefore, dependent on which hydrocarbons are present and the ability of the microorganisms to evolve at that loading site (Overcash and Pal, 1979). Aromatic compounds are the least degradable by microbes and straight chain paraffins are the easiest to degrade (Evans, Deuel and Brown, 1980). These undergo oxidation to form alcohols, aldehydes, and acids. The most degradable alkanes or hydrocarbons are those with molecular weights in the C_6 to C_{28} range (Zobell, 1973).

There are several soil factors influencing the biodegradation of oil. An optimum soil moisture between 50 and 70% of the soil water-holding capacity aids decomposition (Dibble and Bartha, 1979). Adequate oxygen or "oxidized zones" are necessary for co-oxidation, which is the oxidation of a compound to an end product that might be of use to an entirely different organism (Evans, Deuel, and Brown, 1980; Exxon Research, 1976). It is estimated that 3-4 g of O_2 are required per gram of C oxidized in paraffin oil (Overcash and Pal, 1979). A pH preferably from 6 to 8 assures oil degradation (Knowlton and Rucker, 1978) and reduces significant heavy metals movement (Powell, 1979). Soil temperature also affects movement of oil. Cold weather is conducive to lateral spreading, while warm temperatures improve vertical movements (Overcash and Pal, 1979; Raymond, Hudson, and Jameson, 1975; Exxon Research, 1976). Temperature also influences the reaction rates of degradation, with warmer temperatures being more conducive for decomposition (Overcash and Pal, 1979; Exxon Research, 1976). Soil nutrient levels can greatly affect oil decomposition and reduction. Supplemental N and P increase oil decomposition by a factor of 2 (Overcash and Pal, 1979).

In general, intermittent landfarming may permit vegetative growth with no excessive metal uptake (Knowlton and Rucker, 1978). Adverse effects on plants only occur after fresh application of oil. Upon decomposition, the growth of plants is improved. A further look at germination and total crop production is needed (Overcash and Pal, 1979).

Microorganisms that play an important role in oil degradation have been able to utilize heavy metals, even at toxic levels, as energy sources or electron acceptors in their respiratory processes. These reactions may involve precipitation, adsorption, or volatilization of the metals and thereby make the environment more favorable for other microbial species (Ehrich, 1978).

There are two types of mechanisms that microorganisms use to rid their environment of metal cations. The first involves nonspecific binding of the metal cation onto the surface of the microbial cell. The second method of metal uptake by a microorganism is metabolism dependent. What the organism doesn't require can be precipitated intracellularly and stored. Microorganisms are also capable of producing organic compounds such as citric acid and oxalic acid which act as binding or chelating agents. Production of H_2S by microbes is of great importance in that heavy metals form insoluble sulfides. Both bacteria and yeast have exhibited H_2S production and have created a more tolerant environment for more sensitive organisms.

In an earlier discussion, aromatic compounds were found to decompose more slowly than unbranched alkanes. Huddleston and Cresswell (1976) found that resin-asphaltenes in general were extremely slow to decompose. Kincannon (1972) found that naphthenic acids and polyaromatic oils either decomposed very slowly or not at all. The following detailed discussion on the degradation of organic compounds shows the need for investigation into the persistence of these constituents.

Petroleum Degradation by Microorganisms

The degradation of hydrocarbon and phenolic compounds found in petroleum has been the subject of a considerable amount of research in recent years. Extensive reviews on the microbial degradation of aliphatic and aromatic hydrocarbons have been compiled by Van der Linden and Thijsse (1965), McKenna and Kallio (1965), Davis (1967), and Humphrey (1967) as well as others. Reviews have also been written on specific fractions such as aliphatic hydrocarbons (Klug and Markovetz, 1971; and McKenna, 1972) and aromatic compounds in general (Chapman, 1972; Dagley, 1971; and Gibson, 1971, 1972). The generalizations on the microbial degradation of petroleum hydrocarbons that follow are based on the information presented by these authors and others.

A large number of species of bacteria, actinomycetes, yeasts, and filamentous fungi have been shown to degrade hydrocarbons or aromatic compounds that are found in petroleum. These microorganisms have been found to vary greatly in their substrate preference and the regulation and type of catabolic pathways for the degradation of hydrocarbon and aromatic compounds. In addition to the actual usage of a hydrocarbon as an energy source, there is a process called co-metabolism or co-oxidation that occurs with many microorganisms. This concept was first proposed by Leadbetter and Foster (1959) and has been the subject of reviews by Horvath (1972) and Raymond, Jamison, and Hudson (1971). During co-oxidation, a substrate that cannot be utilized as a sole carbon source is oxidized concurrently with a metabolizable substrate and is transformed into a different compound which is still unusable to that microorganism. This probably occurs because of the broad substrate specificity of the initial enzyme involved. Although the compound formed during co-oxidation is not degradable by the microorganism capable of co-oxidation, it can usually be degraded by another microorganism at a later time. Co-oxidation is probably the cause of the accumulation of a variety of compounds by microorganisms when grown on hydrocarbons. Abbott and Gledhill (1971) compiled an extensive review of compounds produced by co-oxidation and also biosynthetic pathways during growth on hydrocarbons. The diversity of catabolic pathways for the degradation of hydrocarbons between different species and even strains of microorganisms makes it difficult to summarize all of the varied mechanisms and reactions that can occur. Instead, the major degradative pathways will be looked at in general with a mention of minor or alternate pathways.

Degradation of n-Alkanes--

The degradation of n-alkanes by microorganisms is similar to the degradation of fatty acids. The terminal methyl group is enzymatically oxidized by incorporation of molecular oxygen by a monooxygenase producing a primary alcohol with further oxidation to an acid group, although involvement of a dioxygenase is also postulated. Once the fatty acid is produced, it is degraded into two-carbon units via the beta-oxidation pathway. Dunlap and Perry (1967) have shown that the fatty acids produced from certain chain length alkanes may be directly incorporated into membrane lipids instead of going through the beta-oxidation pathway. Another pathway for n-alkane degradation that is encountered less often is the oxidation of both terminal carbons to form a dioic acid with subsequent beta-oxidation. Subterminal oxidation of the 2-carbon atom is seen mainly in C_3-C_6 alkanes, although it does occur in longer chain alkanes also. This is thought to be due to a free radical equilibrium between the 1- and 2-carbons and results in a 2-ketone or a secondary alcohol. Depending upon the microorganism involved, the 2-ketones and secondary alcohols formed by subterminal oxidation may or may not be metabolized (Markovetz, 1971). A dehydrogenation of the n-alkane may also occur yielding an alkene which is then converted to an alcohol although there is little evidence for this theory. Some microorganisms have been shown to have both terminal and subterminal oxidation, each having very different rates of activity.

The different chain lengths of n-alkanes are degraded to different extents by microorganisms and both the physical properties of the alkane chain and the enzymatic capability of the microorganism are involved. C_2-C_6 alkanes are inhibitory to some microorganisms possibly because their size allows them to penetrate into cell membranes. This is also seen with cycloalkanes of similar size. This could be the reason for the "toxicity" of short chain alkanes. However, this has only been observed with a few microorganisms. The vapor phase of these short chain alkanes is less "toxic" than the liquid phase. The "toxicity" of the short chain alkanes is also related to temperature since a higher temperature will increase the amount of alkane in the vapor phase and decrease the concentration of the liquid alkane.

At chain lengths greater than C_6 the degradability generally increases until about C_{11}-C_{12}. As the alkane chain length increases, the molecule becomes less soluble in water. However at chain lengths of C_{11}-C_{12} and above, the liquid n-alkanes are "accommodated" in water at a higher concentration than would be extrapolated from the solubility of a series of lower alkanes. This is thought to be due to a change in the structure of the water molecules surrounding the alkanes. During the microbial utilization of long chain liquid alkanes, microbial attachment to droplets of alkane is seen with "transport" of these long chain alkanes through the cell membrane. This is also thought to occur with solid long chain alkanes although in both cases the actual mechanism is unknown.

Perry (1968) compared growth on long and short chain alkanes by some bacteria. He found that although the initial oxygenase had a broad specificity and would oxidize C_1-C_8 alkanes, cells grown on C_4-C_8 alkanes did not oxidize the shorter chain alkanes to a significant extent. Therefore, entirely separate transport mechanisms may exist for gaseous alkanes, short chain liquid alkanes, "accommodated" intermediate chain liquid alkanes, and long chain solid alkanes.

Degradation of Alkenes--

Unsaturated 1-alkenes are generally oxidized at the saturated end of the molecule by the same mechanism as used for alkanes. Some microorganisms, such as the yeast _Candida lipolytica_, attack at the double bond and convert the alkene into an alkane-1,2-diol. Other minor pathways have been shown to proceed via an epoxide which eventually is converted into a fatty acid. For similar amounts of degradation, the chain length of 1-alkenes must be longer than the corresponding alkane. Many microorganisms will not grow on 1-alkenes less than C_{12}. Since 2-alkenes are more readily attacked than 1-alkenes, the presence of a terminal methyl group at each end of the molecule might be more important than just one as in 1-alkenes. There may be some steric interference by the double bond in 1-alkenes towards oxidation of the saturated terminal methyl group particularly in short chain alkenes (Jurtshuk and Cardini, 1971; Buswell and Jurtshuk, 1971). Recently Watkinson and Somerville (1976) reported that butadiene, which contains two double bonds between carbons, could be oxidized and then decarboxylated to acetate. This shows that the presence of double bonds at both terminal ends of a molecule does not preclude its biodegradability as was previously thought.

Degradation of Branched and Cyclic Alkanes--

Isoalkanes (branched alkanes) are degradable by a large number of microorganisms. The occurrence of a small number of methyl or ethyl side chains does not drastically decrease the degradability but complex branched chains and especially terminal branching are harder to degrade. Although the oxidation of 2-methylalkanes has been shown to occur at either terminal carbon, there is a preferential oxidation at the unbranched end of the molecule, again showing possible steric interference of the side chain on the enzyme. The position of the side chain also has an effect on the degradability since it has been shown that 1-phenylalkanes were more degradable than interior substituted ones and that the further the side group was into the molecule, the slower the degradation.

Cycloalkanes are much more resistant to degradation than straight chain alkanes. This is probably due to the absence of an exposed terminal methyl group for the initial oxidation. Co-oxidation appears to be very important in the degradation of cycloalkanes since only a few species of bacteria have been shown to use cyclohexane as a sole carbon source. Co-oxidation changes the cycloalkane to a cycloalkanone. This oxidation can be carried out by a variety of oxygenases such as peroxidase, polyphenoloxidase, and others (Beam and Perry, 1973, 1974). A commensalistic symbiosis is postulated with the cycloalkanones produced by co-oxidation being used as a sole carbon source by a wide range of microorganisms (Donoghue et al, 1976). Alkyl side chains on cycloalkanes are degraded by the normal alkane oxidation mechanism before degradation of the cycloalkane itself depending on the size of the side chain.

Degradation of Aromatic Compounds--

The microbial degradation of aromatic compounds has probably seen more research than any other class of hydrocarbons. The degradation of aromatic compounds can be divided into two groups of reactions. The first group is the modification of the side groups of the ring to produce one or two basic molecules which are then cleaved by the second group of enzymes and further degraded to molecules utilizable by the cell. The most common ring cleavage mechanism is the 'ortho' cleavage pathway in which 1,2-dihydricphenols, such as catechol or protocatechuic acid that have two hydroxyl groups situated on adjacent carbons of the ring, are the basic compounds capable of undergoing ring cleavage. The ring cleavage occurs between these carbons and is caused by a dioxygenase which inserts molecular oxygen into the ring to form a dicarboxylic acid. A series of enzymatic reactions then occurs with the final products being low molecular weight organic acids and aldehydes that are readily incorporated into the tricarboxylic acid cycle. Depending on the initial structure of the aromatic compound immediately prior to ring cleavage, a variety of different intermediate and end products can be formed.

The other major pathway for ring cleavage of 1,2-dihydricphenols is the 'meta' cleavage mechanism where the aromatic ring is cleaved by a dioxygenase between one of the carbons bearing a hydroxyl group and an adjacent carbon that lacks a hydroxyl group. An entirely different enzymatic pathway is present with the initial ring breakage product being a keto acid or an aldehydo-acid. There are also other less commonly found ring cleavage mechanisms such as those for 1,4-dihydricphenols and trihydricphenols.

The preliminary reactions that are required to modify the wide range of aromatic compounds into the few basic compounds that can undergo ring cleavage include hydroxylation, demethylation and decarboxylation and are generally enzymatic reactions. Depending upon the microorganisms involved, a certain compound could be altered in more than one way. When an odd-numbered carbon alkyl chain occurs as a side group on an aromatic ring, it can be degraded via the beta-oxidation pathway which eventually results in benzoic acid which can be degraded by many bacteria. Even-numbered carbon alkyl side chains are also degraded via beta-oxidation, but the phenylacetic acid which is the end product of the beta-oxidation is not degradable by most microorganisms capable of degrading the alkyl side chain. 'Ortho' and 'meta' cleavage both occur in some microorganisms but are usually not alternative degradation pathways for a single compound. Instead, the 'meta' cleavage pathway will degrade alkyl substituted phenolic compounds more readily than non-alkyl substituted phenolics, although there are exceptions. Regulation of ring cleavage and intermediate enzyme formation has been intensively studied with the identification of inducer and repressor compounds in many different microorganims under a variety of conditions.

Polycyclic aromatic hydrocarbons such as anthracene and phenanthrene are also degraded by the 'ortho' cleavage pathway. Initial reactions are very similar to those used for simpler aromatic compounds with an initial hydroxylation by a mono- or dioxygenase to form a 1,2-dihydroxy substituted polycyclic aromatic hydrocarbon. The reduction in microbial degradability of polycyclic aromatic hydrocarbons with increasing number of rings is mainly due to their lower solubility in water.

When an alkyl chain is present on a polycyclic aromatic hydrocarbon, it is removed by beta-oxidation if it is larger than an ethyl group. Ring cleavage usually can occur when methyl side chains are present, however, when certain locations on the ring are substituted, the resulting compound is very resistant to degradation (McKenna and Heath, 1976; Gibson, 1976). Another component of tars and oils that can be microbially degraded is the sulfur-containing thiophen ring, which can be used as a sole carbon source by many organisms (Trudgill, 1976).

Although aerobic conditions are generally required for the degradation of hydrocarbon and aromatic compounds, there are some reactions that can take place under anaerobic conditions. Shelton and Hunter (1975) reported that petroleum can be microbially degraded anaerobically by the reduction of sulfates and nitrates. An alkane dehydrogenase is proposed to be the initial enzyme involved with production of an alkene as the first intermediate compound although there is still doubt about the existence of this reaction. Rosenfeld (1947) showed that fatty acids were produced from the alkanes and could be fermented under anaerobic conditions. Alkanes shorter than C_9 can be degraded anaerobically while some of the longer chain alkanes may be tranformed into napthalenes and other polycyclic aromatic hydrocarbons. Anaerobic degradation of aromatic compounds has been shown with both the aromatic ring and side chains of substituted aromatic compounds being used as carbon sources for many anaerobic microorganisms (Balba and Evans, 1977).

Factors Influencing the Degradation of Petroleum in Soil--

The degradation of petroleum hydrocarbons and associated compounds in soils is affected by many factors. The most important of these may be the enzymatic capabilities of the microorganisms to degrade the various compounds. Hegeman (1972) theorized that the observed recalcitrance of many compounds in vitro is due to the lack of properly designed experiments under the appropriate conditions which are conducive to degradation. He also theorized that recalcitrance could also be due to insufficient time to evolve enzymatic pathways to degrade certain chemical compounds. The inability to degrade a certain substrate may just be related to the inability to induce certain enzymes or to transport the compound into the cell. Alteration of the enzymatic capability to degrade hydrocarbons by genetic manipulation has been reported by Friello, Mylroie, and Chakrabarty (1976). They constructed a strain of *Pseudomonas* that contains multiple transmissible plasmids containing the genetic information coding for the pathways to degrade octane, napthalene, salicylate, and other compounds. Normally only one plasmid containing the genes to degrade a hydrocarbon would be present in a cell. The ability to degrade a wide range of hydrocarbon and aromatic compounds should give this strain of *Pseudomonas* a selective advantage over other bacteria in degrading a mixture of petroleum hydrocarbons. Soli and Bens (1973) showed that when different groups of bacteria that show preferential attack on certain classes of hydrocarbons are grown together, there is little synergistic effect on hydrocarbon degradation noted. However, both of these studies were done using marine bacteria and there may be different bacterial interactions occurring in terrestrial environments.

Other factors relating to the degradation of hydrocarbons in soils were reported by Verstraete et al. (1976). They include the adsorption and volatilization of various petroleum fractions in soils and the mineral composition of the soil. Sometimes the addition of nitrogen or phosphorus to the soil will increase the amount of hydrocarbon degradation. Adaptation of the microorganism population is not considered a major factor in long-term degradation of petroleum in soils although the addition of soil containing previously adapted microorganisms may improve initial degradation rates. Other important factors are aeration and moisture level in soils. La Riviere (1955) pointed out that the production of surface active agents by some bacteria would aid in the dispersal of increased amounts of hydrocarbon into the aqueous phase for improved growth. The absence of adequate environmental conditions and not of suitable microorganisms is probably more important in determining the rapidity and extent of petroleum degradation in soil. Since Coty (1967) determined that some hydrocarbon-degrading bacteria also have the ability of fix atmospheric nitrogen, there may be an increased nitrogen content of the soil after addition of petroleum hydrocarbons.

Summary and Research Needs

Currently some petroleum wastes are land applied with subsequent crop cultivation. In these instances, application rates must be monitored so that the microbial population is not overwhelmed by the oil, metal, or organic compound fractions and overloading results.

If the petroleum wastes could be somewhat degraded or stabilized by a pretreatment process such as composting, application rates could be increased without environmental problems resulting. In addition to pretreatment, there are several other research areas that require further investigation in order to improve the feasibility of landfarming petroleum wastes. Optimum levels for sludge oil content, loading rates, soil pH, nutrient requirements, and mixing frequencies need to be determined. The long term effects of soluble salts and heavy metals on microbial growth and oil degradation rate need to be investigated. The actual oil degradation products and the economics of soil incorporation programs also need to be researched. The possible reclamation of disturbed land by oily waste should also be considered, as should the persistence of various organic constituents and their degradation products.

REFERENCES

Abbott, B. J., and W. E. Gledhill. 1971. The extra-cellular accumulation of metabolic products by hydrocarbon-degrading microorganisms. Adv. Appl. Microbiol. 14:249-388.

Balba, M. T., and W. C. Evans. 1977. The methanogenic fermentation of aromatic substrates. Biochem. Soc. Trans. 5:302-304.

Beam, H. W., and J. J. Perry. 1973. Co-metabolism as a factor in microbial degradation of cycloparaffinic hydrocarbons. Arch. Microbiol. 91:87-90.

Beam, H. W., and J. J. Perry. 1974. Microbial degradation of cycloparaffinic hydrocarbons via co-metabolism and commensalism. J. Gen. Microbiol. 82:163-169.

Berkowitz, J. B., S. E. Bysshe, B. E. Goodwin, J. C. Harris, D. B. Land, G. Leonardos, and S. Johnson. 1980. Field verification of land cultivation/ refuse farming. EPA-600/9-80-010. U.S. Environ. Prot. Agen., Wash., D.C. pp. 260-273. In O. Schultz (ed.) Proc. Sixth Ann. Res. Symp. on Disposal of Hazardous Waste.

Buswell, J. A., and P. Jurtshuk. 1969. Microbial oxidation of hydrocarbons measured by oxygraphy. Arch. Mikrobiol. 64:215-222.

Chapman, P. J. 1972. An outline of reaction sequences used for the bacterial degradation of phenolic compounds. p. 17-55. Degradation of synthetic organic molecules in the biosphere. Natural, pesticidal, and various other man-made compounds. Natl. Acad. Sci., Washington, D.C.

Coty, V. F. 1967. Atmospheric nitrogen fixation by hydrocarbon-oxidizing bacteria. Biotechnol. Bioeng. 9:25-32.

Cross, F. L., and J. R. Lawson. A new petroleum refinery. Am. Ins. of Chem. Eng. Symp. Series 70(136):812.

Dagley, S. 1971. Catabolism of aromatic compounds by microorganisms. Adv. Microbial Physiol. 6:1-46.

Davis, J. B. 1967. Petroleum microbiology. Elsevier Publishing Company, New York. 604 p.

Dibble, J.T., and R. Bartha. 1979. Effect of environmental parameters on the biodegradation of oil sludge. Appl. Environ. Microbiol. 37:721-739.

Donoghue, N. A., M. Griffin, D. B. Norris, and P. W. Trudgill. 1976. The microbial metabolism of cyclohexane and related compounds. p. 43-56. In J. M. Sharpley and A. M. Kaplan (ed.) Proc. Third Intl. Biodegradation Symp. Applied Science Publishers Ltd, London.

Dunlap, K. R., and J. J. Perry. 1967. Effect of substrate on the fatty acid composition of hydrocarbon-utilizing microorganisms. J. Bacteriol. 94:1919-1923.

Ehrich, H. L. 1978. How microbes cope with heavy metals, arsenic and antimony in their environment. p. 381-408. In D. J. Kushner (ed.) Microgial life in extreme environments. Academic Press, New York.

Evans, G. B., Jr., L. E. Deuel, Jr. and R. W. Brown. 1980. Mobility of water soluble org. constituents of API separator waste in soils. EPA Grant No. R805474010 Report.

Exxon Research. 1976. Program to optimize the parameters associated with the microbial degradation of petroleum sludges in soil. Cincinnati, OH.

Federal Water Pollution Control Administration (FWPCA). 1967. Petroleum refining industry wastewater profile. Contract 14-12-100.

Friello, D. A., J. R. Mylroie, and A. M. Chakrabarty. 1976. Use of genetically engineered multi-plasmid microorganisms for rapid degradation of fuel hydrocarbons. p. 205-214. In J. M. Sharpley and A. M. Kaplan (ed.) Proc. Third Intl. Biodegradation Symp. Applied Science Publishers Ltd, London.

Gibson, D. T. 1971. The microbial oxidation of aromatic hydrocarbons. CRC Crit. Rev. Microbiol. 1:199-233.

Gibson, D. T. 1972. Initial reactions in the degradation of aromatic hydrocarbons. p. 116-136. In Degradation of synthetic organic molecules in the biosphere. Natural, pesticidal, and various other man-made compounds. Natl. Acad. Sci., Washington, D.C.

Gibson, D. T. 1976. Microbial degradation of polycyclic aromatic hydrocarbons. p. 57-66. In J. M. Sharpley and A. M. Kaplan (ed.) Proc. Third Intl. Biodegradation Symp. Applied Science Publishers Ltd, London.

Hegeman, G. D. 1972. The evolution of metabolic pathways in bacteria. p. 56-72. In Degradation of synthetic organic molecules in the biosphere. Natural, pesticidal, and various other man-made compounds. Natl. Acad. Sci., Washington, D.C.

Horvath, R. S. 1972. Microbial co-metabolism and the degradation of organic compounds in nature. Bacteriol. Rev. 36:146-155.

Huddleston, R. L., and L. W. Cresswell. 1976. The disposal of oily wastes by land farming. The Management of Petroleum Refinery Wastewaters Forum. Tulsa, OK.

Humphrey, A. E. 1967. A critical review of hydrocarbon fermentations and their industrial utilization. Biotechnol. Bioeng. 9:3-24.

Jurtshuk, P., and G. E. Cardini. 1971. The mechanism of hydrocarbon oxidation by a *Corynebacterium* species. CRC Crit. Rev. Microbiol. 1:239-289.

Kincannon, C. B. 1972. Oily waste disposal by soil cultivation process. EPA-R2-72-110. U.S. Environ. Prot. Agen., Wash., D.C.

Klug, M. I., and A. J. Markovetz. 1971. Utilization of aliphatic hydrocarbons by micro-organisms. Adv. Microbial Physiol. 5:1-43.

Knowlton, H. E., and J. E. Rucker. 1978. An overview of petroleum industry use of landfarming. 71st Annual Meeting of American Institute of Chemical Engineers, Miami Beach, FL.

Leadbetter, E. R., and J. W. Foster. 1959. Oxidation products formed from gaseous alkanes by the bacterium *Pseudomonas methanica*. Arch. Biochem. Biophys. 82:491-492.

La Riviere, J. W. M. 1955. The production of surface active agents and its possible significance in oil recovery. Antonie van Leeuwenhoek. 21:9-27.

Lewis, R. A. 1976. Sludge farming of refinery wastes as practiced at Exxon's Bayway Refinery and Chemical Plant. Natl. Conf. Disposal of Residues on Land., St. Louis, MO.

Markovetz, A. J. 1971. Subterminal oxidation of aliphatic hydrocarbons by microorganisms. CRC Crit. Rev. Microbiol. 1:225-237.

McKenna, E. J., and R. F. Kallio. 1965. The biology of hydrocarbons. Ann. Rev. Microbiol. 19:183-208.

McKenna, E. J. 1972. Microbial metabolism of normal and branched chain alkanes. pp. 73-97. _In_ Degradation of synthetic organic molecules in the biosphere. Natural, pesticidal, and various other man-made compounds. Natl. Acad. Sci., Washington, D.C.

McKenna, E. J., and R. D. Heath. 1976. Biodegradation of polynuclear aromatic hydrocarbon pollutants by soil and water microorganisms. UILU-WRC-76-0113, University of Illinois at Urbana Water Resources Center, Urbana, IL. 25 p.

Overcash, M. R. and D. Pal. 1979. Design of land treatment systems for industrial wastes - theory and practice. Ann Arbor Science Publishers, Inc., Ann Arbor, MI. 684 pp.

Perry, J. J. 1968. Substrate specificity in hydrocarbon utilizing microorganisms. Antonie van Leeuwenhoek. 34:27-36.

Phung, H., and D. E. Ross. 1978. Soil incorporation of petroleum waste. Water Vol. 75.

Powell, R. W. 1979. Leaching of refinery and petrochemical plant wastewater treatment sludges. 11th Mid-Atlantic Indus. Wastes Conf. Penn State Univ., University Park, PA.

Raymond, R. L., J. O. Hudson, and V. W. Jameson. 1975. Final Report on Cleanup of Oil in Soil by Biodegradation. Project OS21.3,21.4, Committee on Environmental Affairs. Amer. Petrol. Inst.

Raymond, R. L., V. W. Jamison, and J. O. Hudson. 1971. Hydrocarbon cooxidation in microbial systems. Lipids. 6:453-457.

Rosenfeld, W. D. 1947. Anaerobic oxidation of hydrocarbons by sulfate-reducing bacteria. J. Bacteriol. 54:664-665.

Shelton, T. B., and J. V. Hunter. 1975. Anaerobic decomposition of oil in bottom sediments. J. Water Pollut. Control Fed. 47:2256-2270.

Soli, G., and E. M. Bens. 1973. Selective substrate utilization by marine hydrocarbonoclastic bacteria. Biotechnol. Bioeng. 15:285-297.

Trudgill, P. W. 1976. Microbial degradation of alicyclic and heterocyclic compounds. Biochem. Soc. Trans. 4:458-462.

Van der Linden, A. C., and G. J. E. Thijsse. 1965. The mechanisms of microbial oxidations of petroleum hydrocarbons. Adv. Enzymol. 27:469-546.

Verstraete, W., R. Vanloocke, R. De Borger, and A. Verlinde. 1976. Modelling of the breakdown and the mobilization of hydrocarbons in unsaturated soil layers. pp. 99-112. In J. M. Sharpley and A. M. Kaplan (ed.) Proc. Third Intl. Biodegradation Symp. Applied Science Publishers Ltd, London.

Watkinson, R. J., and H. J. Somerville. 1976. The microbial utilization of butadiene. pp. 35-42. In J. M. Sharpley and A. M. Kaplan (ed.) Proc. Third Intl. Biodegradation Symp. Applied Science Publishers Ltd, London.

Zobell, C. E. 1973. Microbial Degradation of Oil: Present Status, Problems and Perspectives. In D. G. Ahearn and S. P. Myers (ed.), The microbial degradation of oil pollutants. Center for Wetland Resources, LSU, Baton Rouge, LA.

Pharmaceutical Manufacturing Wastes

W.D. Burge

The waste stream from the pharmaceutical industry is small compared to that of heavy industry and municipal wastewater treatment plants (Swan, 1979). Bewick (1977) states that the figures for the total quantity of wastes are difficult to obtain, because the industry is somewhat reluctant to divulge the data. Lederman et al. (1975) estimated United States production at about 1.6 x 10^6 metric tons per year based upon raw materials consumed and production losses. For one state, Connecticut, De Roo (1975) estimated the waste production to be about 52,000 tons per year containing 350 tons of nitrogen, an amount of waste equivalent to the sewage sludge generated by a city of 300,000 people. On this basis using the estimate by Lederman et al. (1975), the wastes generated by the pharmaceutical industry in the U.S. would be equivalent to the sludge generated by about 31 small cities or an amount equivalent to the sewage sludge produced by approximately 1.4 Greater Metropolitan Chicago Sewage Districts. Fortunately, unlike that of Chicago, the wastes are not concentrated in one place. According to Swan (1979) there are some 229 pharmaceutical plants in the U.S. with only about 54 of the facilities generating significant levels of waste. Therefore, the average plant generates considerably less than the amount equivalent to that of the sewage sludge from a small city.

Kinds of Waste Generated

The kinds of waste generated are related to the type of production. Struzeski (1977) has divided the industry into five major kinds of plants: (1) fermentation plants; (2) synthesized organic chemicals plants; (3) combined fermentation and synthesized organic chemicals plants; (4) biologicals production plants; and (5) drug mixing, formulation and preparation plants. The wastes as a result of the four functions carried out in these plants: fermentation; organic chemicals synthesis; biologicals production; and formulation, etc. have been described by Mayes (1974). The largest bulk of wastes are produced by the fermentation process. The process is utilized to produce antibiotics (penicillin, streptomycin, and aureomycin) and many of the steroids including cortisone. Organic acids (citric, itaconic, and ketoglutaric) are also produced (De Roo, 1975). The wastes include purification and cleanup wastes and the spent beer from which the product has been removed usually by an organic solvent. Spent beers contain residual nutrients such as sugars, starches, and vegetable oils; and the cells or mycelia of the organisms doing the fermenting. In addition microbial metabolic products and some residual antibiotics will be present (Bewick, 1977). Chemical analyses for nitrogen show it to be 2 to 4.3% (u = 2.6%) for organic acid fermentation residues (King and Vick, 1978; Nelson and Sommers, 1979; De Roo, 1975). Only a few analyses for heavy metals have been reported. De Roo (1975) found Pb, Cd, and Cr were not detectable for organic acid or antibiotic residues, but that Zn, Cu, and Ni were. Zn ranged from 75 to 25,000 ppm, Cu from 4.8 to 9.5 ppm, and Ni from 1 to 2 ppm.

According to Mayes (1974), the waste effluents from the organic synthesis process are the most difficult to treat. Examples of chemical reactions used in the process are nitration, halogenation, sulfonation, and alkylation. The wastes generated contain high COD, acids, bases, solvents, cyanides, suspended solids, bacteriostats, disinfectants and various organics, some of a refractory nature. Nitroanalines used in the synthesis of sulfanilamides and phenol mercury are examples of refractory organics (Struzeski, 1977). Solids and precipitates from the process are usually disposed of in designated landfills. Solvents are generally removed by stripping, not only because of their reuse value but to prevent fires in the waste stream. Residual tars and sludges remaining from the stripping process are usually incinerated (Mayes, 1974).

Biological processes utilize both animals and plants to obtain medicinals (Mayes, 1974). The production of gas-gangrene antitoxin, toxoid, typhus, and influenza vaccines require production of large volumes of blood serum, and the facilities to house the necessary animals may require several hundred acres of land. Sources of wastes from the production of animal preparations are mostly the animal wastes, including bedding and waste feeds. The crude animal isolates used for production of the desired products generate negligible effluent contaminants and at times may be removed from the waste stream and used in the production of animal feeds.

The processes for obtaining medicinals from plants have been discussed by Swan (1979). The pharmaceuticals include alkaloids such as quinine, and plant steroids and similar products. The bulk of the waste consists of the plant material remaining after extraction, usually with an organic solvent, to remove the desired components. The organic solvents are generally stripped from the wastes. In the production of stigmasterol, a steroid obtained from soybeans, the soybeans are fused after extraction and disposed of in a landfill.

The formulating and packaging process is the sole function conducted by a majority of the firms in the pharmaceutical business (Mayes, 1974). They take the drugs manufactured elsewhere and compound them to produce pills, powders, lotions etc. for the consuming public. The work is labor intensive and low in waste production. Ingredients are expensive and the degree of quality and raw materials control is reflected in company profits or losses. Plant effluents consist mainly of employee sanitation wastes and are usually introduced into an adjacent community's domestic-waste stream.

Land Application of Fermentation Wastes

The history of fermentation wastes in the U.S. and U.K. has been reviewed by Bewick (1977). Fermentation effluents can have BODs as high as 32,000 ppm. The official maximum BOD for disposal of effluents in water is 20 ppm in the U.S. The result has been to force the fermentation industry to look for alternatives other than disposal in rivers, lakes, and oceans. Use of the wastes as feed supplements for livestock was popular because of their growth promoting potential until the Swan Report in 1969 in the U.K. and the Antibiotics Task Force Report in 1972 in the U.S. cast grave doubts concerning the wisdom of this practice. These reports gave support to the fear that the use of these materials for feed supplements would increase the pool of antibiotic resistant disease organisms. With the options of dumping in adjacent or nearby bodies of water and utilization as feed supplements closed, manufacturers in addition to chemical or biological oxidation of their fermentation wastes are beginning to consider the agricultural utilization of wastes as a source of nitrogen. The high cost of petroleum which is used in the manufacture of nitrogen fertilizer has made the utilization of low nitrogen fertilizer, despite the costs involved in distribution and application, more reasonable from an economic standpoint.

The application of fermentation wastes to agricultural soils has only recently been studied. Bewick (1977), in reviewing the literature in preparation of a study with tylosin fermentation wastes, could find only a few publications. Those publications were newspaper reports of studies conducted by Pfizer Ltd., Groton, Connecticut on their use as a fertilizer on potatoes and an article by Uhliar and Bucko (1974) in Czechoslovakia on spraying lucern with penicillin and streptomycin wastes in which no information was given as to the nutritional status of the wastes or the rates used. Most studies carried out since then have been laboratory studies.

Laboratory studies-- Bewick (1978) applied tylosin fermentation wastes and the purified antibiotic to a mixture of peat, sand, and soil to determine their effect on microbial activity. The tylosin waste was added at rates of 0, 0.4, 1.2, and 2 metric tons per ha on a dry basis, while the antibiotic was added at rates that would have been equivalent to 4, 20, and 40 tons per ha of waste. From the waste, one-half of the carbon was mineralized in 10 weeks. The amount of nitrogen mineralized ranged from 31-38%. Ten to 32% of the tylosin from the waste was detected in leachates from the columns. Addition of the pure antibiotic caused a decrease in microbial respiration relative to the control.

In a study (Bewick and Tribe, 1980) on the influence of temperature on degradation, tylosin wastes were applied at 0, 2.56, 7.56, and 12.5 tons per ha dry weight and incubated in soil columns at temperatures of 4, 10, 15, and 23°C. Again, in 10 weeks despite the higher rates of application, approximately one-half the carbon was converted to CO_2. The rate of carbon release was related to soil temperature, but not to amount of material applied. With decreasing temperature there was an increase in the lag period. For 23 and 15°C the amount mineralized was the same at the end of 10 weeks, but at 10°C the amount mineralized was considerably less. Nitrification was inhibited at 23 and 15°C for 2-6 weeks. At 10 and 4°C, nitrification did not occur during the incubation period. Bewick and Tribe (1980) compared their results with those found by Mommick (1962) for straw at 5, 12, and 24°C and found that the tylosin decomposition rates were equal or greater.

Presscake is one form of waste from the production of citric acid by fermentation and consists of the microbial cells, $CaSO_4$, and a filter aid such as perlite. When applied to soil for laboratory incubation studies as small chunks or powder, nitrogen release and nitrate accumulation was relatively rapid (King and Vick, 1978). Most nitrate appeared in the first 8 weeks with little change thereafter. At the highest rate of application (446 ppm N), incorporation of the chunks resulted in 58% of the presscake nitrogen accumulating as nitrate as opposed to 28% for surface application and 32% for incorporated powder. The differences appeared to be due to gaseous losses that were respectively 31, 55, and 61%. The authors suggest that the form of material and method of application might be useful in determining whether the nitrogen becomes available for nutrient use or is converted to gaseous forms for disposal to the atmosphere. Carbon mineralization was not followed in the study.

DeRoo (1975) used a mixture of fermentation residues from the production of citric, itaconic and ketoglutaric acid and of the antibiotics terramycin and tetracycline as produced by Chas. Pfizer Co., Connecticut, in potting studies for the growth of chrysanthemums and junipers, in greenhouse studies for the growth of tomatoes and corn followed by oats, and in field plot studies for the growth of tobacco and corn. Analysis of the materials indicated a nitrogen content ranging from 0.37 to 4.3% (average 0.472%). The actual observed value for the mixture was 1.90%. Both junipers and chrysanthemums were successfully grown in containers in mixtures of 25 to 33% by volume of mycelium and wood fiber wastes. Some stunting and yellowing occurred during the early growth of chrysanthemums, but the final product was of good to excellent market quality. The junipers showed no harmful effects throughout their growth. During winter storage they remained greener than the controls, apparently as a result of the slow release of nitrogen.

Both tomatoes and corn grown in the greenhouse in a loamy sand soil amended with 12, 36, and 108 tons per acre (wet wt) of mycelium produced yields equivalent or better than the fertilizer controls, although the tomatoes showed early retarded growth and mottling of leaves. The corn showed some retardation at the highest application rate but no mottling. Barley planted in the tomato stubble did well indicating a slow release of nitrogen. Shade grown wrapper tobacco did not benefit from application of the mycelial waste in the container studies. In a field study the application of mycelial wastes at 100 tons per acre (wet wt) produced as good or better forage than the control to which commercial fertilizers had been applied.

The limiting factor in utilization of the waste as a nitrogen source was its total salt content. Toxic effects from the relatively high concentration of Zn were not evident.

Field studies--Only a very few field-study results have

been published. Volz and Heichel (1970) applied a fermentation waste, whose source they did not specify, at a rate of 75 dry tons per ha each of two years to a Yaleville sandy loam soil. In terms of N, the rates were 2, and 130 kg per ha the first year and 1,780 kg per ha the second. Yields of soybeans were significantly enhanced over that of the chemically fertilized controls. Populations of nitrifying and denitrifying organisms in the soils were increased by application of the fermentation residues. Calculation of a mass balance of N from the studies suggested that 17 to 19% of the N ended up in the grain stover, 35% was lost by leaching, and 48 to 78% was lost to the atmosphere as N_2O or N_2. Wright (1978) applying higher rates (112 and 224 dry metric tons per ha) of fermentation residues from the manufacture of "various antibiotic organic acids" to a silt loam field soil in Rhode Island also found extensive leaching of nitrates.

Nelson and Sommers (1979) have evaluated the use of fermentation wastes and sludge from the production of cephalosporin and tylosin production. Wastes were subsurface injected or surface broadcast. Attempted application rates were 224, 448, and 673 kg N per ha, but the actual rates achieved ranged fairly widely around those desired. Corn yields were in general increased both by application of antibiotic wastes and commercial fertilizers over untreated soil, but the soil fertility level was too high for the production of significant differences at the 95% probability level among any of the three treatments. Phytotoxicity was observed in the corn planted one month after application of the fermentation wastes activated sludge. Levels of phytotoxicity were lower in surface applied than in subsurface injected waste plots possibly because of higher NH_3 losses from the surface applied wastes. The authors felt nitrate was removed by leaching from the profile, but the problem was no worse for the fermentation wastes than for commercial inorganic fertilizers.

The results of the study are currently being used by Eli Lilly and Co. to aid in the land treatment of fermentation production wastes on 220 ha of company owned land and on 3,300 ha of farmer-cooperator land.

Fate and Effects on Soil Organisms

Antibiotic production wastes contain residual antibiotics. The fate of antibiotics in soil and their effects on the microflora and plants have been discussed in a review by Bewick (1977). Low concentrations of antibiotics added to soils may be metabolized, chemically decomposed, adsorbed, or leached. The inactivation in soil of penicillin, viridin, gliotoxin, frequentin, and albidin has been attributed to their chemical instability. The chemical nature of antibiotics affects their adsorptivity. Basic antibiotics such as streptomycin, dihydrostreptomycin, neomycin, kanamycin and erythromycin may be adsorbed by clay minerals. Amphoteric antibiotics such as bacitracin, aureomycin, and terramycin will act either as bases or acids depending on their isoelectric points. Since their isoelectric points are usually above the pH of most soils they may be adsorbed by the cation exchange charge. Neutral or acidic antibiotics, that is, penicillin and chloramphenicol are adsorbed only in small amounts. Antibiotics adsorption in some cases has been reversed by use of phosphate and citrate buffers. Although streptomycin was inactivated in sterile soil its disappearance was more rapid in nonsterile soil. Penicillin was rapidly degraded in soil. The addition of chloramphenicol to soil did not cause an increase in numbers of the organisms resistant to antibiotics, but an increase in numbers of resistant organisms was found for clavicin. In general it appears unlikely that the addition of low levels of antibiotics as present in fermentation residues would produce persistent levels of antibiotics in soils.

Although antibiotics in low concentrations in laboratory media have been shown to inhibit the growth of microorganisms, their action in soil is much less pronounced. In solution, streptomycin at 1 ppm inhibited nitrification in a laboratory broth, but 1,000 ppm was required to produce the same inhibition in sand and 10,000 ppm parts was required in soil.

Summary and Research Needs

The waste stream from the pharmaceutical industry in the U.S. is small. The total sludge generated is equivalent to only about 1.4 times that produced by the Greater Metropolitan Chicago Sewage District. There may be disposal problems, however, in local areas where some of the larger plants are located. The most hazardous wastes result from the synthesis of organic chemicals. They may contain acids, bases, solvents, cyanides, bacteriostats, disinfectants and various organic chemicals, some of which may be resistant to degradation. Currently the organic solvents in these wastes are reclaimed not only for their reuse value but also to prevent fires in the waste stream. Precipitates are land filled and still bottoms produced from the solvent reclaiming process are incinerated. Landfarming of these wastes might be an economic solution for some plants, but the overall size of the problem appears not to be large.

The remaining wastes from the industry appear not to be particularly hazardous. They consist mainly of sanitary wastes from the labor intensive formulating and packaging plants, animal and plant residues from the production of biologicals and the wastes from the fermentation process for the production of organic acids, antibiotics, and steroids. The fermentation wastes represent the greatest volume and are currently being applied to farm land in some areas in the country.

The fermentation wastes consist of residual nutrients supplied originally for the fermenting microbes, microbial metabolic products, including some residual antibiotics, and the cells or mycelia of the microorganisms. At one time these wastes were used for animal feed supplements, but concerns regarding the possible buildup of antibiotic-resistant pathogens has essentially halted this practice. The high cost of commercial fertilizers has begun to make their use in agriculture economical. Laboratory and field studies on their use as yet have been small in number, but those that have been conducted show that the major problems are the possible overloading with nitrogen and its release to the environment and the effect of their relatively high salt content on the inhibition of plant growth. Early studies in the 1950's and 1960's indicated that the residual antibiotics in these materials would have little if any influence on microbial processes in the soil and since they do not persist would be unlikely to constitute a source for developing antibiotic resistance in pathogenic microorganisms. Information is needed on application rates of pharmaceutical wastes to prevent nitrate leaching and the build up of excess salts which could inhibit plant growth. Pharmaceutical wastes do not appear to comprise a hazardous waste problem of any magnitude.

REFERENCES

Bewick, M. W. M. 1977. Considerations on the use of antibiotic fermentation wastes as fertilizers, a review of their past use and potential effects on the soil ecosystem. Commonwealth Bureau of Soils, Special Publ. No. 4.

Bewick, M. W. M. 1978. Effects of tylosin and tylosin fermentation waste on microbial activity of soil. Soil Biol. Biochem. 10:403-407.

Bewick, M. W. M., and H. T. Tribe. 1980. The decomposition of tylosin fermentation waste in soil at four temperatures. Plant Soil 54:249-258.

DeRoo, H. C. 1975. Agricultural and horticultural utilization of fermentation residues. Bull. No. 750. Connecticut Agricultural Field Station.

King, L. D., and A. L. Vick. 1978. Mineralization of nitrogen in fermentation residue from citric acid production. J. Environ. Qual. 7:315-318.

Lederman, P. B., H. S. Skovronek, N. J. Edison, and P. E. Des Rosiers. 1975. Pollution abatement in the pharmaceutical industry. Chem. Eng. Prog. 71:93-99.

Mayes, J. H. 1974. Characterization of wastes from the ethical pharmaceutical industry. EPA-670/2-74-057. U.S. Environmental Protection Agency, Cincinnati, Ohio. 65p.

Mommick, H. 1962. Mineral nitrogen immobilization and carbon dioxide production during decomposition of wheat straw in soil as influenced by temperature. Acta. Agric. Scand. 12:81-94.

Nelson, D. W., and L. E. Sommers. 1979. Recycling antibiotic wastes on cropland. Proc. 2nd Annual Conf. of Appl. Research and Practice on Municipal and Indus. Wastes. September 1979. Madison, WI.

Struzeski, E. J. 1977. Status of waste handling and waste treatment across the pharmaceutical industry and 1977 affluent limitations. pp. 1095-1110. In Proc. 30th Industrial Waste Conf. Purdue University, Lafayette, IN. Ann Arbor Science, Ann Arbor, MI.

Swan, R. 1979. Pharmaceutical industry sludge, drug makers face waste management headache. Sludge 2:21-25.

Uhliar, J., and M. Bucko. 1974. Moznost vyuzitia priemyslonjch odpadov 2 vyroby antibiotik v rostlinnej vyrobe. Rostlinno Vyroba 20:923-930.

Volz, M. G. and G. H. Heichel. 1979. Nitrogen transformations and microbial population dynamics in soil amended with fermentation residues. J. Environ. Qual. 8:434-439.

Wright, W. R. 1978. Laboratory and field mineralization of nitrogen from fermentation residues. J. Environ. Qual. 7:343-346.

Pulp and Paper Industry

L.J. Sikora

Size of the Industry

Pulp and paper production is a major industry in the United States as evidenced by $40.2 billion in sales in 1977, which accounted for over 3% of all manufacturing sales. The industry employs almost 1% of the total U.S. work force (676,000 people in 1976). It is the third largest industrial water user and the fourth largest industrial purchaser of fuel and electricity, even though it generates approximately 45% of its own heat and electricity (Energy Resources Co., Inc., 1980). Many of the companies are large, highly diversified establishments which produce pulp, paper, paperboard and building board, maintain their own timberland, and market their finished products.

The industry is also one of the leading generators of industrial solid waste. A Gorham International (1974) report on the solid waste practices of the pulp and paper industry indicated that approximately 600 pounds of solid waste were generated for each ton of finished product produced. Solid wastes include fly ash, bark, silt, general mill sweepings and trash, and inorganic precipitates formed during chemical makeup and recovery (e.g., grits, dregs, lime mud) which are normally collected at low moisture and incinerated and/or landfilled with little or no treatment (NCASI, 1979).

Waste Treatment

Waste treatment practices are vital aspects of the pulp and paper industry because of the large quantities of liquid, solid, and gaseous by-products from the manufacturing process, many of which are regulated by EPA-imposed discharge standards. Although many treatment technologies are utilized from plant to plant, a general description of the various liquid, solid, and gaseous waste controls is given here.

Sludges from the complex wastewater treatment facility must be disposed of; land application and incineration are the major disposal methods. Coal ash originating from power generation is usually landfilled. The bituminous ash has ten major constituents with SiO_2, Al_2O_3, and Fe_2O_3 making up the major portion at 46, 26, and 18%, respectively. The wastes accumulated within the stack scrubbers usually go to some type of land disposal.

Table 16 contains a summary of pulp and paper solid waste streams. Most of the solid waste streams can be readily identified as to source and method of generation. Others such as the dregs, rejects, and muds have not been discussed previously and are generated as part of the chemical recovery process.

Recovery of the spent cooking liquor (black liquor) in the Kraft pulping process involves a series of interdependent operations. Initially, the black liquor is concentrated in evaporators and then incinerated. The smelt resulting from the incineration is mixed with water and clarified. The liquid portion following clarification is termed green liquor, while the settled materials are labeled green liquor dregs.

Lime is then slaked in the green liquor to convert Na_2CO_3 to NaOH. Following the lime slaking, the liquor is again clarified to remove lime mud and the resulting solution (white liquor) is ready for reuse in the digester. The lime mud, $Ca(OH)_2$, is usually burned in a kiln to recover lime for reuse in the slaking process. However, if the lime kiln is not functioning, or an excess of lime mud accumulates, the mud is landfilled. Thus, four solid wastes are generated with the kraft liquor recovery: green liquor dregs, slaker rejects, lime kiln rejects and occasionally lime mud. Wastewater treatment sludges are the second largest solid waste generated by the paper and pulp industry. Table 17 outlines the basic treatment processes used to remove specific types of wastewater contaminants and, subsequently, lead to the generation of sludges for disposal.

Pretreatment of wastewaters includes pH adjustment (neutralization) or equalization to prevent treatment upsets from slugs of strong wastes and removal of grit and other materials which will damage pumps and clog conveyance

TABLE 16. SUMMARY OF PULP AND PAPER SOLID WASTE STREAMS (FROM ENVIRO CONTROL, 1979)

PROCESS	WASTE STREAM	CONSTITUENTS	FATE	HAZARD EVALUATION	ESTIMATED WET TON GENERATION (Nationwide)
WOOD PROCESSING	Wood waste	Bark, dirt, sand, wood fragments	Incinerated/energy source, landfilled or composted	Non-hazardous based on existing data - further testing due to reported leachate toxicity	2.0×10^7
PULPING	Pulp rejects	Undigested woodchips, cellulose	Repulped or landfilled	Non-hazardous	6.0×10^5 includes mechanical and chemical
PAPER MAKING	White water dregs	Cellulose fines, dyes	Landfilled	No data- testing to determine chemical constituents advised	
	Wastepaper rejects	Staples, glue, fiber-glass tape, string	Landfilled	Non-hazardous	
AUXILIARY PROCESSES Chemical Recovery Kraft	Green liquor dregs	Unburned carbon, refractory materials, sodium hydroxide, sodium sulfide	Sewered or occasionally lagooned	Non-hazardous based on available data - further characterization advised.	6.0×10^5

TABLE 16. (cont'd)

PROCESS	WASTE STREAM	CONSTITUENTS	FATE	HAZARD EVALUATION	ESTIMATED WET TON GENERATION (Nationwide)
	Slaker rejects	Calcium carbonate	Landfilled	Non-hazardous	
	Lime kiln rejects	Calcium, silica, aluminum	Landfilled	Non-hazardous	
	Lime mud	Calcium carbonate,	Recovery kiln, occasionally landfilled	Non-hazardous based on available data - further characterization advised	
Sulfite					
Calcium bisulfite	Spent calcium bisulfite liquor		Combustion/energy source, lagooned or recovered for non-captive use	Further testing due to lack of data	
Magnesium bisulfite	Spent magnesium bisulfite liquor	Wet liquor ash	Landfilled or lagooned		
Ammonium bisulfite	Ammonium bisulfite spent liquor	Ammonium oxide and heavy metals	Combustion/energy source, feed supplement or lagooned		
Sodium bisulfite	Sodium bisulfite spent liquor				

TABLE 16. (cont'd)

PROCESS	WASTE STREAM	CONSTITUENTS	FATE	HAZARD EVALUATION	ESTIMATED WET TON GENERATION (Nationwide)
Semi Chemical NSSC	Green liquor dregs	Unburned carbon, refractory materials	Sewered or lagooned		
Sulfurless	Green liquor dregs	Silica, carbon char, sodium hydroxide	Sewered or lagooned		
Power Generation	Coal fly ash	Unburned inorganics	Lagooned	Further testing advised due to heavy metals content	1.8×10^6 - all coal ash
	Coal bottom ash	Unburned inorganics, SiO_2, Al_2O_3 & Fe_2O_3			
	Bark fly ash	Unburned inorganics			
	Bark bottom ash	Unburned inorganics CaO and silica			
	Boiler water treatment sludge	Lime & minerals	Landfilled	Non-hazardous	
	Boiler blowdown	Particle buildup	Sewered	Shows up in water treatment sludges	

TABLE 16. (cont'd)

PROCESS	WASTE STREAM	CONSTITUENTS	FATE	HAZARD EVALUATION	ESTIMATED WET TON GENERATION (Nationwide)
Wastewater Treatment	Primary sludge	Short fibers, fillers coating clays, starch, metal	Landfilled or incinerated	Non-hazardous based on available data - further testing advised due to variation from mill to mill	2.0×10^6
	Secondary sludge	Organic material, heavy metals	Landfilled, incineration land injection, or recovery for fertilizer		2.0×10^5
MISCELLANEOUS WASTE STREAMS	Water treatment sludge	Algae, organic debris, precipitated metals	Landfilled	Non hazardous based on available data	
	Miscellaneous trash	Paper, steel strapping, corrugated boxes, scrap lumber, cement	Landfilled		
	Waste lubricating oils	Hydrocarbons	Recovered, burned as fuel, drained and landfilled	Hazardous	
	Spent cleaning solvents	Organic solvents, miscellaneous solid wastes	Drummed and land filled	Hazardous	

TABLE 17. BASIC TREATMENT PROCESSES USED TO REMOVE SPECIFIC TYPES OF WASTEWATER CONTAMINANTS. (MODANI AND HOLLEY, 1973)

Contaminants	Pretreatment and primary treatment	Secondary treatment	Tertiary treatment
Suspended solids (SS)	Screening; flotation; sedimentation	Secondary clarification; activated sludge; aerated lagoon; stabilization basin	Filtration; holding pond; chemical precipitation
Excess acidity or alkalinity	Equalization; neutralization		
Biodegradable organics (BOD)	Sedimentation; flotation	Aerated lagoon; activated sludge; trickling filter; anaerobic lagoon; stabilization basin; oxidation pond; secondary clarification	Filtration holding pond
Refractory organics (COD, TOC) and color turbidity	Sedimentation; flotation	Aerated lagoon; activated sludge	Carbon adsorption; massive lime treatment; reverse osmosis; filtration; holding pond
Dissolved inorganic solids			Adsorption and coprecipitation; ion exchange; electrodialysis; reverse osmosis

systems. Primary treatment removes solids that can be filtered (sand filters), those that float to the surface (grease and oil), suspended solids with specific gravities greater than water which can be buoyed up by contact with air bubbles (flotation), and solids which settle out (sedimentation). Primary clarification can also reduce the fraction of BOD_5 associated with suspended solids (Gehm, 1973). The size of that fraction depends on the source of fiber, pulping process used, and type of paper produced. Effluents from fine paper and tissue mills are relatively low in dissolved organic solids and thus show a high BOD_5 reduction after primary treatment. Low BOD_5 reductions are obtained from pulpmill and wastepaper effluents on settling since they are high in dissolved organics. Typical BOD_5 reductions are almost zero for sulfite wastes, 10% for sulfate, 15 to 20% for newsprint, 20 to 25% for book mill and 35 to 65% for tissue (Billings and DeHass, 1971).

The most common treatment for suspended solids is mechanical clarification. The clarifier is a large tank (usually round) equipped with a skimmer for surface debris and a scraper on the bottom that collects and removes settled sludge (Billings and DeHass, 1971). Clarification results in a 2 to 6% solids sludge which can be further thickened by filtration, centrifugation, mechanical pressing and/or drying.

Secondary treatment is the biological oxidation of soluble and suspended organics in the effluent stream with subsequent clarification and disposal of microbial residue (secondary sludge). Secondary treatment methods are those common to treatment of municipal sewage. All methods are based on the growth and maintenance of a healthy microbial population that is acclimated to the wastewater as an energy substrate. Secondary treatment results in oxidation of oxygen-demanding material (BOD_5) and destruction of organic compounds which are toxic to stream organisms. Resin acids, chlorinated resin acids, unsaturated fatty acids, chlorinated phenolics, diterpene alcohols, juvabiones and certain other lignin degradation products of pulping and bleaching are detoxified by biological treatment (Leach and Thakore, 1976; Leach, Mueller, and Walden, 1976; Walden and Howard, 1977). Much of the turbidity remaining after primary treatment is also removed in biological treatment (Carpenter and Janis, 1969; Gove and McKeown, 1975). Turbidity, which is the scattering of incident light, is caused by the presence of finely divided particles in the effluent. The most common sources of these particles are kaolinite clays, titanium dioxide, starches, other fillers and coating used in papermaking, and cellulose fines. High rate BOD_5 treatment methods were found more efficient for turbidity removal due to the higher concentration of biomass. Adsorption, flocculation and biological destruction of peptizing starches were given as the active reactions involved.

Pulp and papermill wastes are normally low in N and P (NCASI, 1977). An insufficient nutrient content limits bacterial growth and metabolism, which in turn limit the breakdown of carbonaceous material in the waste. Deficiencies of these elements are also found to affect sludge settling and dewatering properties. Requirements for N and P were found to be greater in high rate activated sludge processes than in longer-term treatments. Because of costs of secondary treatment, mills have been able to obtain permission to send their primary effluents to municipal treatment plants (Schoemaker and Dickinson, 1979). Together with the relatively nutrient rich municipal effluent, the paper mill effluent is treated biologically at the sewage treatment plant with no addition of N and P necessary (Clingenpeel and Jones, 1973). The average BOD:N:P ratio for treatment systems varied from 100:4.6:1.0 for activated sludge to 100:2.8:0.6 for aerated stabilization. In some cases the addition of N and P in waste treatment is supplemental to N and P compounds wasted during the manufacturing process. Two ammonia-based sulfite mills examined by the EPA were discharging 6 to 9 kg of N per ton of production.

Miscellaneous wastes--Various wastes from routine plant maintenance add to the solid waste load and these include water treatment sludge, miscellaneous trash, and waste lubricating oils. The water treatment sludge is generated when the water supply of the mill is a river or lake and the water is treated with alum (aluminum sulfate) to remove suspended matter and organics. The flocs slowly settle out forming a sludge. Miscellaneous trash is that generated from personnel (office waste paper, lunch wrappings, and newspapers), manufacturing services (packaging scraps), and construction wastes (scrap lumber, cement, and metal).

Waste lubricating oil arises from routine machine maintenance involving periodic oil changes and subsequent disposal. Some oil spillage ends up in the washwater and necessitates the incorporation of an oil skimmer into the wastewater treatment system.

Spent cleaning solvents also arise from hardware maintenance in which solvents are used to remove residual buildup on operating parts. In most cases, a solution of more than one organic solvent is used to remove the chemicals and wood substances adhering to the machines. An analysis of spent solvent indicated that methylethylketone, toluene, and cyclohexane were constituents (Enviro. Control, Inc., 1979).

Current Practice of Disposing of Solid Wastes in the Paper and Pulp Industry--

The paper industry disposes of its waste largely by the methods of landfilling, lagooning, and incineration. Table 18 contains the results of a survey of 163 mills and their current disposal practices for sludges.

Landplaced as defined by McKeown (1978) is placement on land without a soil cover. Landfilling and landplacing account for 72% of the wastes being disposed of in this survey. Lagoon disposal, generally of dilute sludges, accounts for 13% of the wastes disposed of.

TABLE 18. DISTRIBUTION OF INDUSTRY DISPOSAL METHODS (FROM McKEOWN, 1978)

METHOD	NUMBER OF MILLS IN SURVEY	PERCENT OF TOTAL WASTES COVERED
Incinerator	19	10
Landfilled	40	33
Landplaced	53	39
Incinerator and landfilled	4	3
Incinerator and landplaced	4	2
Recycled	5	1
Sold	3	-
Lagoon	13	7
Municipal/contractor	3	-
Municipal and landplaced	7	1
Municipal and landfilled	3	-
Other	9	4
Totals	163	100

Landfilling--This has been the most acceptable disposal alternative for pulp and paper wastes (Blosser, 1960). Landfilling as opposed to incineration was suggested as the only feasible choice for disposal of sludges with ash contents greater than 50% (NCASI, 1979). The reason cited was the inability of these sludges to sustain self-combustion during incineration. One of the main objectives in landfill design has been to contain as much sludge as possible in the smallest volume. Paper and pulp sludges have low solids content and when placed in a landfill, the fill material will settle or decrease in volume by 30 to 50% (Zimmerman and Perpich, 1978). A technique used to add stability to a landfill and allow drainage is to layer soil or bark waste between layers of sludge or mix with refuse, a process called co-disposal (Reinhardt and Kohlberg, 1978).

All landfills unless lined generate leachate which is characteristic of material buried in the landfill. Reinhardt and Kohlberg (1978) compared the concentrations of several parameters in leachate and groundwater samples beneath pulp and paper mill landfills and municipal refuse landfills. The ranges for all parameters were comparable and they concluded that pulp and paper mill residuals have leaching potentials as great as municipal refuse.

Depending upon the geological and soil characteristics of a landfill site, different design concepts have been utilized. Some of these designs include clay or sludge lined cells with leachate collection systems and diked cells over clay soils. The highly fibrous nature of primary sludges make these sludges ideal for liners in sandy locations. Besides the leachate generated from below some pulp and paper mill landfill sites, leachate occasionally seeps through the dike walls which then endangers the quality of surface water (Energy Resources Co., 1980).

In the Energy Resources Co. report (1980), data were quoted from a study of paper industry landfills for the Canadian government. Wood waste disposal sites as opposed to sludge disposal sites were investigated and the data indicated that sites used for hog fuel disposal generated undesirable leachates which were acidic, high in BOD, and color. Because industries have gone to burning most of their wood wastes, hog fuels are being burned as opposed to landfilled.

Reinhardt and Kohlberg (1978) reported that the State of Wisconsin has permitted engineering modification of landfill sites which overcome hydrogeologic problems. These include bentonite cutoff walls, leachate collection systems and surface water control devices. The State feels that these improvements in design plus adequate monitoring of the site during and after operation will protect ground and surface waters near these sites.

Soil treatment--In the 1960's the pulp and paper industry pursued intensively the reduction of BOD, odor, aquatic toxicity, turbidity, and color in plant effluents by soil treatment. Experience has shown conclusively that spray irrigation of most effluents can be an effective and economical method of disposal where sufficient land is available in the proximity of the mill and where the overall volume of treated effluents is low (Blosser and Caron, 1965; Gehm, 1973). The principle of irrigation for the pulp and paper industry is to totally exclude effluents from receiving waters or remove oxygen-consuming materals by soil filtration and microbial decomposition prior to entry into receiving waters. The industry considers the following methods for irrigation use:

1. Seepage ponding. Wastewater flows into diked basins where the natural permeability of the soil allows seepage into underlying strata and eventually to ground or surface waters.

2. Application to fallow soil. Effluents are drained or sprayed onto bare soil where the soil layers act as a medium for biological breakdown and filtration.

3. Application to vegetated areas. Effluents are applied by surface or overhead irrigation to areas with natural or seeded plant cover. This procedure capitalizes on the added permeability afforded by root penetration and resistance to surface sealing and on increased volume reduction due to evapotranspiration. The primary purpose of these areas is for maximum disposal of the waste with little or no use of biomass produced.

4. Application to agricultural land. Effluents are applied by surface or overhead irrigation onto closely managed grazing land or cropland for the purpose of increasing biomass production in the process of disposing of the effluent. The irrigation water and recycled nutrients from the waste are expected to increase production of a marketable product. The greatest potential exists where lack of water normally limits crop production and quality irrigation water is limited in supply.

Gellman and Blosser (1959) reported on a survey conducted of several mills which had either pilot scale or full scale irrigation systems for effluent (Table 19). A number of different methods were employed and, in some cases, crops were used. In summarizing the results, the authors stated that: a) land requirements were approximately 0.4 to 0.6

TABLE 19. SURVEY OF COMPANIES WHICH LAND APPLY EFFLUENT FROM PULP AND PAPER MILLS (ADAPTED FROM GELLMAN AND BLOSSER, (1959)

Company	Effluent	Application Rate	Frequency of Application	Crops	Field Size (ha)	Comments
Ohio Boxboard	Paperboard white water-Sprinkler	5.33 cm/appl	Every 5 days	Oats, reed canary grass	.5	Pilot study - no fiber matting observed
Weston Paper	Paperboard white water-Sprinkler	4.8 cm/appl	Every day	Alfalfa, legumes, oats	-	Pilot study - no fiber matting observed
Forest Fibre Prod.	Effluent from defibration and board forming	8.6 cm/appl	Every 10 days	Fescue, clover, orchard grass,	41	Full scale permanent facility
U.S. Gypsum	Same as above	5.1 cm/appl	Every 10 days	Same as above	122	Full scale permanent facility, gradual soil crusting occurred
American Boxboard	Spent cooking liquor-semichem-sprinkler	26.7 cm/appl	Every 3 years	No cover crop	2	Full scale installation-ceased due to declining seepage rate
Sonoco Prod.	White water-semichem - sprinkler	10.0 cm/cycle	Every 1.5 days	No cover crop	3.3	Full scale installation
Diamond-Gardner	Groundwood white water - seepage pond	20.0 cm/cycle	Continuous	No cover crop	2.0	Full scale installation
Kimberly-Clark Flambeau Nekoosa Badger Paper	Spent sulfite liquor - sprinkler - contour ditches - seepage pond	.1 to 3.8 cm/cycle	Continuous	No cover crop	2.8 to 6.5	Full scale operation

TABLE 19. (cont'd)

Company	Effluent	Application Rate	Frequency of Application	Crops	Field Size (ha)	Comments
Crown-Zellerbach	Spent sulfite liquor	0.8 cm/appl	1 day/month	corn, beans cabbage, potatoes, carrot, fescue clover, alfalfa	.4	Pilot
Southland Paper East Texas Pulp and Paper Western Kraft Springfield Mill	Craft effluent - -sprinkler -furrow	.75 to 9.1 cm/appl	1 day/week to 1 day/2 weeks	no cover crop	.4 to 2.5	Pilot
Union-Bag-Camp Franklin	Combined digester and evaporator condensates, cauticizing dregs and washings	9.1 cm/appl	1 day/week	corn, peanuts	28	Full scale

ha/ton of pulp capacity; b) all irrigation sites should have groundwater monitoring wells; c) use of storage ponds in connection with irrigation systems is desirable, permitting removal of nozzle-plugging particles and providing needed flexibility for relocation of irrigation equipment; d) undiluted spent sulfite or NSSC liquor was toxic to vegetation; e) a plant cover was desirable even if not harvested because it improves soil infiltration properties and provided for atmospheric release of some liquid by evapotranspiration.

These findings were reasserted in more recent studies. Guerri (1973) found that full-strength NSSC spent liquor (10% solids) containing 16% Na would destroy soil structure and lower treatment efficiency when applied at an average rate of 0.6 cm/day without a crop cover. Blosser and Owens (1964) provided guidelines and considerations on field application of pulp mill effluents. They pointed out that efficient BOD removal (95%) occurred when BOD loads were less than 224 kg BOD/ha/day. For normal crop growth an effluent pH between 6.0 and 9.5 was necessary. Liming the soil would aid, but below pH 4.5, plant growth is impaired. Effluents high in Na but low in divalent cations could displace Ca and Mg to reduce permeability. At levels greater than 134 kg Na/ha/day, effluents greater than a SAR(Sodium Adsorption Ratio) of 8 would impair permeability. Reduction of color in soil leachates varied with soil characteristics (texture, pH, and CEC), the effluent applied, and the season of the year. Color is retained by clays more than sands and color bodies, if persistant, may leach out after irrigation with subsequent infiltration.

Narum, Michelson, and Roehne (1979) recently reported on the Simpson Paper Co. irrigation project in Anderson, CA. Using an automatic flood irrigation system (much like ridge-furrow irrigation), nearly 5.3 million cubic meters of treated, secondary effluent were applied on a 162 ha field in 20 months. Yields of several crops grown including wheat, oats, corn, alfalfa, and beans were equal to or greater than the averages for those crops in California. The effluent percolate which eventually enters the Sacramento River was essentially devoid of suspended solids, BOD_5, COD, color and toxicity components. Soil column experiments conducted along with the field studies indicated that limestone would be needed occasionally to balance the levels of cations in the effluent.

Sludge Disposal/Utilization--There are two types of sludges or residuals that accumulate in the industry during treatment of wastewaters. Primary sludges arise from sedimentation and floatation basins that are used to remove suspended solids, neutralize excessive acidity or alkalinity, and to partially treat organics. Secondary sludges or biological sludges arise from high rate biological waste treatment systems such as aerated lagoons, activated sludge systems, trickling filters, anaerobic lagoons, and oxidation ponds.

The primary sludges are generally very fibrous and are easily dewatered (Frederick, Grace, and Joyce, 1980). The solids in primary sludge can contain fibers, filler, and coating clays (Enviro Control Inc., 1979). Deinking process sludges, for example, may contain a great variety of inks, dyes, and coating constituents. Some primary sludges arising from an acid neutralization system contain large amounts of calcium carbonate and have been used as a lime substitute. Panda and Das (1971) reported increases in pH and Bray's extractable P with addition of a primary lime sludge from a paper mill to an acid soil. Corey (1977) reported that the application rate of lime sludges depended on the $CaCO_3$ content of the sludge, the particle size of the sludge solids, the type of crop being grown, and the rate of N application. In general, from 3 to 8 tons of $CaCO_3$ equivalents per acre can be applied.

The non-lime primary sludges contain a modest amount of N which, if the sludge is applied to soils, becomes plant available only after a sizeable portion of the carbonaceous portion is decomposed. This topic is discussed in the process section dealing with C:N ratios. Joyce, Webb, and Dugal (1979) determined the composition of primary sludges of 10 pulp and paper mills and found that the ash contained largely titanium dioxide, coating clays, and other fillers.

Aspitarte et al. (1973) reported on a study investigating ways of disposing of primary treatment plant sludge. They included incineration, burning in hog fuel boilers, incorporation into soil as an amendment and use as a mulch for soil stabilization. They reported on the economics of incineration and burning for fuel. In the soil incorporation study, they reported some difficulty in spreading and incorporating very high sludge amendments. The annual 400 ton/acre application ended when, after 2 years, the plots were spongy and marsh-like and would not support equipment. They concluded that after 600 tons/acre application, a year of fallow was necessary in order to obtain adequate crop yields. Crop yields in fresh mixtures of low sludge application rates (100 to 200 tons/acre) were satisfactory, provided that adequate N was added. Some of the crops tested in the soil amendment study included tomato, mushroom, raspberry, strawberry, corn, and beans. Sludge alone or in combination with bark was competitive as a hydromulch material in establishing grass stands on steep embankments.

Secondary sludge is the settled biomass that results from biological treatment of oxygen demanding solids. It contains fibers, micro- and macroorganisms, and micro- and macroelements. Ganczarczyk (1972) analyzed five sulfate secondary sludges and found that each had a considerable amount of lignin, from 22 to 39%. Because secondary sludges are composed mainly of biomass, the balance of macroelements in the sludge is better than in primary sludges. This characteristic enables the sludge to have a wider range of applications. For instance, the secondary sludge produced at the Mehoopenny, PA. plant of Proctor and Gamble Paper Products Company contains adequate N and P to be marketed as a slow-release turfgrass fertilizer (Eberhardt et al., 1978). Preliminary studies of the same sludge have also indicated that the sludge may be a suitable fish food (C. A. Barton, personal communication Procter and Gamble Co., Cincinnati, OH Sept. 1980).

Jacobs (1978) reported on a two year field study using secondary sludge as a soil amendment. The paper mill was one producing corrugated medium paperboard. The sludge was applied utilizing a liquid manure tank wagon which the farmer filled from a storage container located on his property. This concept proved fruitful because it allowed the farmer to be flexible and use the sludge when he desired. Jacobs concluded from his studies using kidney bean and potato crops that application of the paperboard wastewater sludge at agronomic rates was not likely to change the plant composition or soil properties for heavy metals or nutrients other than N or P. The same paper mill is now utilizing all of its secondary sludge for conditioning and fertilizing land on which trees will be grown for future harvesting. The sludge contains 5% solids and is applied at a rate of 4 dry t/a. The nutrient level of the sludge is 6% N, 2% P, and 1% K. The sludge is injected into the fields the year before the spring planting of the trees. Although the mill has an incinerator, they have land applied the sludge as much as possible for economic reasons. The cost of incineration is $112/ton while the cost for land applying is $70/ton (Personal communication, Roger Smith, Packaging Corp. of America, Filer City, MI, Sept. 1980). The sludge is also used as a soil amendment at the company's intensive tree farm.

Huettl (1979) reviewed the various commercial applications of paper mill sludges. They include use as a feed supplement for ruminant animals, as cattle bedding, as an absorbent for oil spills, and as a component in clay bricks, cement tiles, fiber building board, construction paper, and potting mix.

Buchanan (1978) reported on the subsurface injection of secondary sludge from a plant producing corrugated medium. Analysis of the sludge is shown in Table 20. The company's major concern in using the sludge for agricultural purposes is the B content. The application rate is nearly 30 dry tons/acre which is injected on three passes through the field. Crops grown on amended fields are corn, soybeans, and winter wheat. The state permitting process requires that each farmer requesting sludge be visited by a state agricultural agent and, if approved, a permit is written for a one time/one season application. The dictating factors include soil type, proximity to neighbors and domestic wells, and crops to be grown. Previous season (fall) injection is recommended. Local response by farmers in the area of the mill has been extremely positive.

A commercial operation has been successful in obtaining permits from states to spray or inject lagoon sludges from paper mills located in the state. They canvass local farms to find farmers interested in having their fields amended with paper mill sludges. They then sample the soils, analyze soil and sludge samples, submit the data and samples to the state and file for permits. If approved, they then return to the farmer and schedule application dates. In most instances, the application is performed in the fall after harvest in preparation for the following spring planting. The only problem experienced has been compaction of fields by the heavy application equipment when the soils are wet.

TABLE 20. ANALYSIS OF SECONDARY SLUDGE UTILIZED IN STUDY REPORTED BY BUCHANAN (1978)

COMPONENT	CONCENTRATION (mg/kg, dry weight)
Boron	148
Copper	42.1
Nickel	105
Zinc	174
Chromium	68.4
Mercury	0.31
Cadmium	26.3
Lead	111
Calcium	11,000
Magnesium	2790
Potassium	4740
Phosphorus	4210
Nitrate	10.7
Total Kjeldhal Nitrogen	18,600

Composting of Solid Wastes--Composting of organic wastes is a method of stabilizing, drying, and reducing the volume of materials, making it ideal in the treatment of many pulp and paper industry wastes. Attempts have been made to compost these wastes with the intention of marketing the compost or utilizing it in-house for plant propagation. P.H. Gladfelter has undertaken a 60 ton/day composting process in which dewatered primary-secondary sludge mixture is mixed with bark wastes in a ratio of 1:1 (volume basis) and composted by the aerated-pile method (Epstein et al., 1976). Greenhouse studies are being conducted utilizing the unscreened final product by the University of Maryland (F. R. Gouin, Dept. of Horticulture, University of Maryland, Sept., 1980 personal communication). Weir (1977) reported on a proposed composting project utilizing wastewater treatment plant sludge and paper mill sludge. The paper mill sludge, dewatered below 50%, will be used as the bulking agent for mixing with the wastewater treatment plant sludge. The final product will be tested as a humus soil conditioner. In a separate composting pilot study, sludge and hog fuel were mixed and composted by the Beltsville Aerated Pile Method (D. W. Marshall, NCASI, New York, Sept., 1980 personal communication). The variables tested were addition of N and P to the sludge prior to composting, and the addition of N and P to the compost. The conclusions were that N and P addition did not aid in the composting process but did aid in plant production when the nutrients were added to the compost.

The general consensus of the industry on composting is that if other methods of disposal and utilization of the wastes such as land spreading or landfilling are prohibited for some reason, composting may be beneficial if the final product can be utilized or marketed. Because pathogenic organisms are not a concern in pulp and paper wastes, the necessity for high temperature treatment is absent. Therefore other means of drying or volume reduction, if more cost-effective, will be chosen over composting.

Hazardous Wastes in the Pulp and Paper Industry

Enviro Control Inc. (1979) performed a study to produce a list of hazardous or possibly hazardous wastes that are generated in the pulp and paper industry. Table 16 includes the conclusions on the waste solids listed. Several wastes were listed as inconclusive because testing had not been completed. The only wastes listed as bonafide "hazardous" materials were waste lubricating oils and spent cleaning solvents, both of which are generated in small quantities in most plants.

Aside from the hydrocarbons and organic solvents, the

underlying concern of regulatory agencies and industry personnel is the level of heavy metals or chlorinated hydrocarbons in various waste streams. In contrast to the heavy metal concentration of many industrial and municipal sludges, pulp and paper mill sludges consistently contain very low levels of heavy metals (Gehm, 1973). Zinc is the only heavy metal used to any extent. Zinc hydrosulfite is used at a few mills for bleaching groundwood pulp, but at levels low enough to prevent toxic levels in receiving waters. The three possible sources of heavy metals in the mill, pulping chemicals, papermaking additives and equipment corrosion, have not contributed significant levels to plant effluents. Problems experienced with Hg, once used in slimicides and also used in Hg cell-Cl-caustic plants have been corrected. A study by NCASI (NCASI, 1979) showed that of 31 solid wastes representing a cross section of the industry's manufacturing-derived residuals, none were found to be hazardous in terms of the currently proposed criteria for metals analyzed for and including Cd, Cr, Pb, Ni, Cu, and Zn. Table 21 presents the range of metals found in primary and secondary sludges. The only sludges that contain relatively high levels of metals are the deinking sludges.

Polychlorinated biphenyls (PCB) are no longer in use by the pulp and paper industry and regulations concerning PCB-containing transformers in mills producing packaging for food items deem that these transformers be removed. Dugal (1977) found trace amounts of PCB present in several wastewater deinking mills which is not entirely surprising since these mills deal with wastepaper. Work performed at

TABLE 21. CONCENTRATION OF HEAVY METALS IN PRIMARY AND SECONDARY SLUDGES (FROM NCASI, 1979)

Metal	No. Samples	Range	Median	Mean
		ug/g		
Cd	7	0.09 to 2.4	0.5	0.9
Cr	12	22 to 210	54	72
Cu	16	3.9 to 200	52	70
Ni	14	1.3 to 99	13	26
Pb	15	1 to 880	52	192
Zn	16	13 to 1400	218	439

the Institute of Paper Chemistry showed that PCB's which were once used in the manufacture of printers' ink dyes, paper coatings (plasticizers) and carbonless carbon paper, are sorbed on cellulose fibers which are mostly removed in primary clarification. The concern then is the fate of PCB in the disposal of the primary sludge. Dugal (1977) reported that PCB leaching from pulp and paper mill landfills would be negligible.

Organic compounds exhibiting a toxic effect on fish have long been known to exist in pulping and bleaching wastes. Types of compounds found in pulp mill effluents and contributing to effluent toxicity include resin acids, chlorinated resin acids, unsaturated fatty acids, other acidics, and chlorinated phenols (Easty, Borchardt, and Wabers, 1978). Some of the specific compounds identified in each group are:

Resin Acids
abietic, dehydroabietic, isopimaric, pimaric

Chlorinated resin acids
monochlorodehydroabietic, dichlorodehydroabietic

Unsaturated fatty acids
oleic, linoleic, linolenic

Unsaturated fatty acid derivatives
9,10-epoxystearic acid
9,10-dichlorostearic acid

Chlorinated phenolics
3,4,5-trichloroguaiacol
3,4,5,6-tetrachloroguaiacol

Halomethane
chloroform

Although these toxic compounds are significantly removed from effluents by biological treatment, they are still found in varying quantities in outfalls from plants (Keith, 1977). In conjunction with the land disposal of sludges from biological treatment the question arises as to the degradation and mobility of these compounds in soils. Leach and Thakore (1977) have suggested sources of these compounds and possible degradation mechanisms. They stated that resinous wood extractives and the products of treatment are the sources of most toxicants. Only the chlorinated compounds were probably somewhat resistant to microbial attack.

Recently, bleach effluents have been tested for mutagenicity and both softwood and hardwood krafts were mutagenic (Eriksson, Kolar, and Kringstad, 1979). Using the Ames test it was determined that biological treatment which included exposing the effluent to an adapted microflora for 48 hrs resulted in loss of mutagenicity, indicating that the toxic component was biodegradeable. Therefore, sludges from secondary treatment processes should not contain mutagenic materials. Bjorseth, Carlberg, and Moller (1979) analyzed effluents from different bleaching states in a sulphite and a sulphate plant. Most effluents were mutagenic as determined by the Ames test and contained several organohalogenated compounds including dimethyl-propyl-naphthalenes and alkylated catechols.

Data which demonstrate the presence of chlorinated organics along with other data which asserts their known persistence in the environment raise questions whether a hazard is present when the sludges from bleach kraft plants are land applied. There are insufficient data available concerning the degradation of these known compounds and research data should be generated to this effect in light of the mutagenic character of the compounds.

Summary and Research Needs

The pulp and paper industry generates a major amount of solid wastes which have been scrutinized carefully by industry, university, and government personnel as to their toxic or hazardous nature. The wastes known to be hazardous, the oils and solvents, are generated in relatively small quantities and in most cases, are disposed of in secured landfills. The waste which is "technically" considered hazardous because of the high pH is lime mud. Neutralization of this material may result in H_2S evolution which is another hazardous substance. Data should be collected on the application of the mud to soils before neutralization to determine if, with proper application, this material can be used for liming of acid soils.

Because little information is available on the fate of the chlorolignins and other chlorinated materials in the biological treatment of bleaching effluents, a research effort should be made to quantify whether degradation or flocculation or both are occurring in treatment process. If chlorinated materials still remain in the sludges, information on quantities and fate after sludges are placed on land or in landfills should be obtained. Finally, testing of these materials for mutagenic agents should be performed in response to recent findings that mutagenic factors have been detected in secondary sludges.

REFERENCES

Aspitarte, T. R., A. S. Rosenfield, B. C. Smale, and H. R. Amberg. 1973. Methods for pulp and paper mill sludge utilization and disposal. U. S. Environmental Protection Agency, EPA-R2-73-232. 139 p.

Billings, R. M. and G. G. DeHass. 1971. Pollution control in the pulp and paper industry. pp. 18-1 to 18-28. In Industrial pollution control handbook. McGraw-Hill Book Co.

Bjorseth, A., G. E. Carlberg, and M. Moller. 1979. Determination of halogenated organic compounds and mutagenicity testing of spent bleach liquors. The Science of the Total Environ. 11:197-211.

Blosser, R. O. 1960. Sludge disposal. Pulp and Paper Magazine of Canada (March) pp. T 195-T 199.

Blosser, R. O., and A. L. Caron. 1965. Recent progress in land disposal of mill effluents. TAPPI. 48:43-46A.

Blosser, R. O., and E. L. Owens. 1964. Irrigation and land disposal. Pulp Paper Mag. Can. p. T-263-T-267.

Buchanan, B. 1978. Land application of secondary sludge. pp. 37-45. In Proc. NCASI, Central Lake States Regional Meeting.

Carpenter, W. L., and J. R. Janis. 1969. Removal of white water effluent turbidity by biological treatment processes. pp. 209-211. In Proc. 24th Indust. Waste Conf. Purdue Univ. Lafayette, IN.

Clingenpeel, W. H., and M. K. Jones. 1973. Design for joint treatment of municipal and paper mill waste at Lynchberg, Virginia. pp. 109-116. In Proc. 28th Indust. Waste Conf., Purdue University, Lafayette, IN.

Corey, R. B. 1977. Agricultural application of pulp and paper industry waste treatment sludges. pp. 89-91. In Proc. NCASI Central Lake States Regional Meeting. Special Report 77-02.

Dugal, H. S. 1977. Analysis, distribution and control of PCB in the pulp and paper industry. pp. 112-118. In Proc. 41st Executives Conf., Instit. of Paper Chem., Appleton, WI.

Easty, D. F., L. G. Borchardt, and B. A. Wabers. 1978. Removal of wood-derived toxics from pulping and bleaching wastes. U. S. Environ. Prot. Agen., Wash., D.C. EPA-600/2-78-031.

Eberhardt, W. A., J. L. Lewis, R. A. Scharp, and C. A. Barton. 1978. Conversion of sulfite pulping waste to an agricultural product. J. Water Pollut. Control Fed. 50:1893-1904.

Energy Resources Co. Inc. 1980. Disposal Practices for Selected Industrial Solid Wastes. Final Report submitted to U.S. Environmental Protection Agency. 655 p.

Epstein, E., G. B. Willson, W. D. Burge, N. C. Mullen, and N. K. Enkiri. 1976. A forced aeration system for composting wastewater sludge. J. Water Pollut. Control Fed. 48:688-694.

Erikkson, K-E., M-C. Kolar, and K. Kringstad. 1979. Studies on the mutagenic properties of bleaching effluents. Svensk. Papper. 4:95-101.

Enviro Control Inc., 1979. Hazardous Waste Listing, Pulp and Paper Industry. Final Report submitted to U. S. Environmental Protection Agency, 55 pp.

Frederick, W. J., T. M. Grace, and T. W. Joyce. 1980. Disposal of secondary sludge in the kraft recovery system. Environ.Conf. of TAPPI. pp. 43-47.

Ganczarczyk, J. 1972. Fate of lignin in activated sludge treatment of kraft effluents. pp. 256-269. In Proc. 27th Indust. Waste Conf. Purdue Univ. Lafayette, IN.

Gehm, H. 1973. State-of-the-Art Review of Pulp and Paper Waste Treatment. Environ. Prot. Agency, Wash., D.C. EPA-R2-73-184.

Gellman, I., and R. O. Blosser. 1959. Disposal of pulp and papermill waste by land application and irrigational use. pp. 479-494. In Proc. 14th Indust. Wast. Conf. Purdue University, Lafayette, IN.

Gove, G. W. and J. J. McKeown. 1975. Current status of paper reprocessing effluent characteristics and disposal practices. TAPPI 58:121-126.

Gorham International, Inc. 1974. Study of Solid Waste Management Practices in the Pulp and Paper Industry. U.S. Environmental Protection Agency. NTIS No. PB-234-944. 143 p.

Guerri, E. A. 1973. Industrial land application spray field disposal and wildlife management can coexist. TAPPI 56:95-97.

Huettl, P. J. 1979. The production and disposal of wastewater and wastewater sludges in the pulp and paper industry. A literature review submitted in partial fulfillment of the requirements for a PhD degree in the Dept. of Soil Science, Univ. of Wisconsin, Madison. WI. 123 p.

Jacobs, L. W. 1978. Utilizing paperboard wastewater sludge on agricultural soils. pp. 509-516. First Annual Conf. of Applied Research and Practice on Municipal and Industrial Waste.

Joyce, T. W., A. A. Webb, and H. S. Dugal. 1979. Composition of pulp and paper mill primary sludges. TAPPI 62:83-84.

Keith, L. H. 1977. GC/MS analyses of organic compounds in treated kraft paper mill wastewaters. pp. 671-707. In L. H. Keith (ed.) Identification and Analysis of Organic Pollutants in Water. Ann Arbor Science Publishers, Ann Arbor, MI.

Leach, J. M., J. M. Mueller, and C. C. Walden. 1976. Identification and removal of toxic materials from kraft and ground pulpmill effluent. Proc. Biochem 11:7-10.

Leach, J. M., and A. N. Thakore. 1976. Toxic constituents in mechanical pulping effluents. TAPPI 59:129-132.

Leach, J. M., and A. N. Thakore. 1977. Compounds toxic to fish in pulp mill waste streams. Prog. Water Tech. 9:782-798.

McKeown, J. J. 1978. Sludge dewatering and disposal practices- a review of the U.S. paper industry. TAPPI 62:97-104.

Modani, N. K., and W. H. Holly. 1973. The impact of water pollution abatement on competition and pricing in the Alabama paper industry. Water Resour. Res. Inst. Bull. 13. Auburn Univ. 147 p.

Narum, Q. A., D. P. Mickelson, and M. Roehne. 1979. Disposal of an Integrated Pulp-Paper Mill Effluent by Irrigation. U. S. Environmental Protection Agency, EPA-600/2-79-033. 118 pp.

NCASI. 1977. Pulp and paper mill effluent nitrogen and phosphorus requirements for biological treatment and residuals after treatment and residuals after treatment. Nat'l. Council of the Paper Industry for Air and Stream Improvement, Inc., New York. Tech. Bull. No. 296. 36 p.

NCASI. 1979. Nature and environmental behavior of manufacturing--derived solid wastes of pulp and paper origin. Natl. Council of the Paper Industry, for Air and Stream Improvement, Inc., New York. Tech. Bull. No. 319. 90 pp.

Panda, N., and J. C. Das. 1971. Evaluating of lime sludge from paper mills as an amendment for acid soil. In Proc. Internatl. Symp. on Soil Fertility Evaluation. Vol. 1. pp. 781-787. New Dehli (India).

Reinhardt, J., and N. F. Kohlberg. 1978. Pulp and paper mill sludge disposal practices in Wisconsin. pp. 253-264. Third Annual Conf. on Treatment and Disposal of Industrial Wastewaters and Residues. Information Trans. Inc., Silver Spring, MD.

Shoemaker, G. H., and R. H. Dickinson. 1979. A method for disposal of sludge from pulp and paper waste treatment. pp. 17-19. In Proc. of Environ. Conf. of TAPPI.

Walden, C. C. and T. E. Howard. 1977. Toxicity of pulp and paper mill effluents - A review of regulations and research. TAPPI 60:122-125.

Weir, D. R. 1977. Composting industrial and municipal wastes, paper mill and city treatment plant. Compost Sci. 18:27.

Zimmerman, E. R., and W. M. Perpich. 1978. Design and construction of pulp and paper mill residue disposal sites. pp. 495-504. In Proc. First Annual Conf. Applied Research and Practice on Municipal and Industrial Waste, Madison, WI.

Soap and Detergent Industry Wastes

R.H. Fisher and P.B. Marsh

Introduction

The soap and detergent industry produced 5.8 billion kg (12.8 billion lb) of products as of 1973, with virtually all of it eventually ending up in municipal wastewater treatment plants. Less than 5% of the manufacturing process water is treated by the industry itself, with the large remainder being discharged to municipal wastewater treatment plants. The process effluents contain readily degradable organic compounds and are considered non-toxic (EPA, 1974). There is little information on the land application of wastewater and sludges from the soap and detergent industry, probably because so little of it is treated by the manufacturer.

Background

Synthetic surfactants were first produced and added to detergents in the 1930's but it was not until the 1950's that they came into widespread usage. These surfactants were mainly of the alkylbenzenesulfonate (ABS) type, which has a branched alkyl chain not completely biodegradable in normal wastewater treatment processes. This led to the production of foam in wastewater treatment processes. In addition, in some areas that depended on septic tanks for waste treatment, ABS was found in groundwater. Even though ABS is considered non-toxic to humans at the levels which caused foaming, the foaming produced by low concentrations of ABS was considered undesirable.

In the early 1960's, technological improvements allowed the production of straight alkyl chain surfactants called linear alkylbenzenesulfonates (LAS), which replaced ABS by 1965. These are much more readily biodegradable and caused little problem in wastewater treatment.

It was also during the 1960's that the phosphate builders in detergents were blamed for causing the eutrophication of lakes and streams by supplying phosphates to algae. This led to the banning of phosphate-containing detergents in some states and municipalities and caused a search for replacements to the phosphate builders. The prime candidate was nitrilotriacetic acid (NTA), which was added to detergents during the late 1960's, but reports of its carcinogenicity and teratogenicity and concern about the ultimate fate of NTA in the environment led to a temporary removal. At the present time, it is allowed in detergents with restrictions. The search for replacements for NTA also started during the late 1960's and is continuing at this time.

The future for the soap and detergent industry is one of _status quo_ with no radically new developments in surfactants expected and a gradual reassessment of existing products due to health and environmental concerns. In a majority of the country, phosphate builders will still be present in detergents because of their superior performance over substitutes (Schneider, 1977).

Chemical Characteristics of Soaps and Detergents

In general, a surfactant is composed of a strongly hydrophilic group and a strongly hydrophobic group in the same molecule, which gives surfactants their unique properties. The hydrophobic group generally has a 10-20 carbon atom chain for the desired physical properties. Fatty acids derived from plants or animals are used in soaps while paraffins, olefins, alkylbenzenes, long chain alcohols, and alkylphenols from the petrochemical industry are used for the manufacturing of synthetic surfactants. The hydrophilic group can be anionic (sulfonate, sulfate, or carboxylate), cationic (quaternary ammonium compounds), or nonionic (polyoxyethylene, sucrose, or polypeptide) (Swisher, 1970).

Besides the surfactant, which makes up 15-30% of detergents, the other major component is the phosphate builder, which is usually composed of sodium tripolyphosphate and can comprise up to 60% of the detergent. The main purpose of the phosphate builder is to soften the water by complexing with Ca and Mg ions, with prevention of fabric discoloration by metal ions, dispersion of soil particles, and maintenance of proper pH occurring also. Other components such as non-phosphate builders like NTA, coloring, fragrances, enzymes, and fluorescent whitening agents may also be present (Devey and Harkness, 1973).

General Manufacturing Process

The soap and detergent industry tends to purchase the alcohols, alkylates, and other hydrophobic components of the surfactants from the petrochemical industry instead of manufacturing them itself. They do, however, for the most part sulfate or sulfonate these components themselves and formulate the detergents and soaps by adding builders, coloring compounds, etc. In the manufacture of soaps, the fats and oils are purified and then treated with caustic and heat for the reactions to make soap. All byproducts from this process are either recycled back into soap making or sold as in the case of glycerin. Depending on the manufacturing process used, there may be no wastewater produced except that needed for equipment washdown and spill removal. Even those processes that create wastewater do so only to a small extent (30 gal/1000 lb soap produced) for highly caustic impure soap wastes.

The sulfation and sulfonation of alkylbenzenes, fatty alcohols, etc., requires no process water, so the only waste flow from this step is cleanup of spills and washout of unreacted chemicals from equipment. In preparation of powdered detergents, again there is no process wastewater except that arising from spill cleanups and equipment washout since during the drying of the components of the detergent mix, the water comes off as water vapor. Compared to other industries, the volume and concentration of wastewater from the soap and detergent industry is small (EPA, 1974).

Waste Treatment Practices by the Soap and Detergent Industry

The Jeffersonville, IN plant of the Colgate-Palmolive Company is one of the few plants in the soap and detergent industry that treats its own waste. The wastewaters are treated by an aeration basin followed by a clarification stage. A BOD reduction of 92% was obtained with 90% loss in surfactant levels. Temporary high concentrations and flows of wastewater and low wintertime temperatures reduced operating efficiency (Herin, Marlow, and Stigger, 1970). The construction of an equalization basin and a chemical pretreatment system with lime has helped improve treatment. The acclimation of the microorganisms in the aeration basin was found to be aided by the addition of soap scraps to obtain better surfactant degradation (Brownell, Busch, and Herrin, 1975).

In 1976, the chemical sludge started to be disposed of on a land site. After a few earlier preliminary attempts had been unsuccessful, deep injection on a 20 ha (50 ac) field was tried. In one year 5700 m^3 (1.5 million gal) of sludge had been cultivated into the field, mainly during the summer months. The high pH (12.4) of this lime sludge was desired by the farmer owning the field as an alternative to the addition of lime to the soil as a means of raising soil pH. Although initially shallow subsurface injection of the sludge at a depth of 20 cm (8 in.) was used, state regulatory agencies ordered a change to injection at 40 cm (16 in.). At this greater depth, anaerobic degradation of sludge can occur with possible odor and leaching problems, although disking a few weeks after application was practiced. The sludge was applied at a rate of 285 m^3/ha (30,000 gal/ac) with a solids loading of 28,000 kg/ha (25,000 lb/ac). The chemical sludge contains high levels of P, B, and Ni, although concentrations of the latter two are being reduced through product reformulation. Higher levels of pH, Ca, P, and Na were found in the soil after sludge application. The field was replanted with corn the next year with no visible reduction in the crop. The future for this site seems to be an expansion to a 60 ha (150 ac) field with application of chemical sludge and farming to occur in alternate years. Aside from reducing soil acidity, there appears to be no real benefit from the addition of the chemical sludge, which is low in nutrients, to the farmland, aside from the disposal of the sludge (Phung et al., 1978b).

The Texize Chemicals Company in Greenville, SC compared an aeration lagoon system to a rotating biological contactor (RBC) for the treatment of liquid detergent manufacturing wastes and found that although the RBC was superior in treating the wastewater, it was subject to many operating difficulties that made the aeration lagoon system better overall (Lense, Mileski, and Ellis, 1978).

Basu (1967) reported on the treatment of various vegetable oil wastes used in the production of soap in India and found that the high concentration of oils reduced aerobic treatment efficiency and that anaerobic digestion resulted in a better treatment of the effluent. McCarty et al. (1972) also investigated the treatability of vegetable oils and animal fats used in soap manufacture and found that they were readily treatable separately or when combined with municipal wastes under both aerobic and anaerobic conditions.

Biodegradation of Soaps and Detergents

Before 1965, there was considerable research done on the incomplete biodegradation of branched chain ABS surfactants in natural waters, sewage treatment processes, and soils. For instance, Swisher (1963) found only 70% of an ABS surfactant was degraded in two months in river water and Cook (1968) found removal of ABS was from 30% to 70% in activated sludge systems. However, in simulated septic tank drain-fields, degradation of ABS reached 98% and biodegradation occurred best with proper dosing levels (Robeck et al., 1963). This higher level of biodegradation in soils is thought to be due to the longer contact time of surfactants with the soil microorganism populations which allowed a more complete biodegradation (Robeck et al., 1964). The addition of organic material from sludge and adsorption of ABS onto soil particles also aided in the biodegradation, although it must be noted that the complete mineralization of the ABS to biomass and CO_2 was not determined.

Klein, Jenkins, and McGauhey (1963) investigated the uptake of ABS by plants and determined that there was inhibition of the growth of sunflowers and barley at levels of 10 mg/l ABS or above, even though the concentration of ABS in the plants was low (0.01% on a dry weight basis) and was mainly concentrated in the roots. Seed germination of lupine was increased by low concentrations of ABS. Cationic surfactants, which are generally used for their bactericidal properties can be degraded in activated sludge units as shown by Fenger et al. (1973) and Gerike, Fischer, and Jasiak (1978) and nonionic surfactants could be degraded at concentrations of up to 500 mg/l in activated sludge (Cook, 1979). McClelland and Mancy (1969) concluded that the presence of surfactants can interfere with microbial metabolism of other C sources through enzyme inhibition and interference in mass transport in the aqueous phase.

The biodegradation of surfactants has been reviewed by Huddleston and Allred (1967), Willets (1973), Cain (1976), and Overcash and Pal (1979) and has been the subject of a number of papers on the relationship of molecular structure to biodegradation in pure culture studies and in natural systems such as activated sludge and river water. The following generalizations concerning structure and biodegradability can be made:

1. The biodegradability of C_6-C_{14} linear alkyl chains increases with chain length. However, at chain lengths of C_{18} and above, degradability decreases markedly due to low solubility in water. As the distance between the hydrophilic group and the terminal C of the alkyl group increases, there is a reduction of steric interference in the enzyme-substrate complex. This principle is valid for alkyl groups in anionic and nonionic surfactants and also for the presence of phenyl groups in phenylalkanes.

2. Branching of the alkyl chain causes a reduction in degradation with simple branching being easiest to degrade, multiple branching on more than one C of the alkyl chain or branching of the side chain itself being much harder to degrade, and the presence of quaternary branching on the terminal C of the alkyl chain being the most resistant to degradation.

3. In nonionic polyethoxylate surfactants, an increase in the ethylene oxide chain length from zero to ten ethylene oxide units had no effect on the rate or completeness of degradation, but when more than ten units were present, the degradability was severely reduced. Nonionic alcohol ethoxylate surfactants were more degradable if the alkyl chain was derived from primary or secondary alcohols as opposed to more complex structures. Polyethoxylated alkylphenols were more readily degradable if the phenol ring had a primary attachment rather than secondary attachment. The most resistant structure is when the phenol group is located four C's from the terminal C (Steinle, Myerly, and Vath, 1964; Osburn and Benedict, 1965; Patterson, Scott, and Tucker, 1970).

The most common metabolic pathway for surfactant biodegradation is the beta-oxidation sequence which is also used for fatty acid metabolism in all organisms. The low specificity of the enzymes of this pathway allows the degradation of a wide variety of compounds that either occur naturally such as fats and hydrocarbons or are man-made as in the case of surfactants. In anionic surfactants, removal of the hydrophilic sulfonate or sulfate group by an alkylsulphatase usually precedes beta-oxidation resulting in the production of sulphite and free sulfate groups. The desulphonation usually occurs at the cell membrane to prevent denaturation of intracellular proteins by the intact surfactant, although some strains of Pseudomonas can degrade the surfactant totally without prior desulphonation. The beta-oxidation of the alkyl chain is initiated by oxidation of the terminal C to a carboxylate group followed by a progressive removal of two C units as acetyl-CoA. Linear alkyl chains and those with simple methyl branching can be broken down with this mechanism. Alkyl chains with more complex branching can be degraded by alpha-oxidation, which removes one carbon unit at a time and is found only in a few strains of Micrococcus. Branched alkyl chains may also be degraded by valine or pantothenate-type oxidation, but these are also relatively rare metabolic pathways.

If an aromatic ring is present in the surfactant, it is usually degraded after the alkyl chain has been removed. The ring cleavage is generally achieved through an 'ortho' cleavage mechanism common to many microorganisms or more rarely through a 'meta' cleavage. Bird and Cain (1974) reported that ring cleavage by Alcaligenes occurred before alkyl chain removal in compounds with C_4 or smaller alkyl chains although this is not a common mechanism. It has also been shown that when long alkyl chains are degraded, there is a temporary buildup of C_4-C_6 alkyl chain compounds when there is a phenyl group attached to the chain. This is thought to be due to a lower level of acetyl-CoA synthases specific for C_4-C_6 alkyl chains than for longer alkyl chains. Also when C_4 alkyl chains are present, beta-oxidation is not used. Instead a dehydrogenation and alpha-oxidation of the side chain occurs at a slower rate than beta-oxidation of the longer alkyl chains.

The alkyl chains of nonionic polyethoxylated surfactants are also degraded by beta-oxidation and are usually removed before the remaining polyethylene glycol is attacked. An alternative mechanism is an initial cleavage of an ether bond leaving a long chain alkanol and polyethylene glycol which are then degraded separately. However, in alkylphenol ethoxylates, there is a beta-oxidation of the alkyl chain followed by an oxidative or hydrolytic breakdown of the polyethoxylate with the formation of ethylene glycol. There are a wide variety of bacteria that can break down polyethylene glycol and ethylene glycol and metabolize it in several different pathways.

Even the bacteriostatic cationic quaternary ammonium surfactants can be broken down by an initial demethylation which eliminates the charged nature of these compounds. This is followed by beta-oxidation of the alkyl side-chains eventually ending up with the formation of the amino acid glycine which is readily utilized. Some bacteria have been found that degrade the alkyl chains with beta-oxidation first, with demethylation of the quaternary ammonium occurring last.

The enzymes of the beta-oxidation mechanism are constitutive since they are a normal part of bacterial fatty acid metabolism, whereas the enzymes involved in desulphonation, ring cleavage, and other reactions are generally inducible. Swisher (1970) states that these inducible enzymes are probably involved with the early stages of degradation such as transport of the surfactant across the cell membrane, desulphonation, and initial oxidation of terminal C of alkyl group. The amount of acclimation time needed to degrade different surfactants is probably dependent mainly on how well these enzymes are induced by the various isomers of the surfactant present. Willets (1973) found a wide range of desulphonating enzyme induction depending on the position of the phenyl group on the alkyl chain with the terminal position causing the most induction. It is thought that there is a regulator gene that recognizes the various isomers to different degrees that controls a single structural gene for the production of the desulphonating enzyme. The induction of other regulator genes for other enzymes probably operates in a similar manner. Furthermore, the ability to degrade surfactants may be plasmid-associated genes similar to those for the degradation of camphor, octane, and other compounds in *Pseudomonas*.

With the continual application of surfactants to a soil system, there would be a gradual change in microorganism population favoring those organisms capable of degrading the surfactants more rapidly and to a greater extent than the original soil microorganisms. The great diversity of microorganisms present would also tend to enable the soil system to degrade an extensive range of organic compounds that would only be partially degraded by any single species.

The biodegradation of other components that go into making soaps and detergents has also been studied especially in the area of replacements for phosphate builders. NTA has been the subject of the most study, with reports on its almost complete biodegradation under aerobic conditions (Thompson and Duthie, 1968; Thom, 1971; Ferguson et al., 1971; Stoveland, Lester, and Perry, 1979) and anaerobic conditions (Moore and Barth, 1976). However, an acclimation period of more than a week was needed and very poor biodegradation occurred at less than 10° C.

Initially there was concern that NTA levels would keep rising in receiving waters, especially in colder climates, but Maylaiyandi, Williams, and O'Grady (1979) showed that there was no buildup in Canadian surface or ground waters except in cases where septic tank drainfields were not working properly. However, NTA can cause problems in sewage treatment since it chelates heavy metals during primary sedimentation and does not permit them to settle out with other sludge solids. Instead, they pass on to the secondary treatment where the NTA is biodegraded and the metals are released and subsequently are discharged at levels much higher than in the absence of the NTA (Stoveland, Lester, and Perry, 1979). Also, NTA is capable of solublizing heavy metals from sediments in lakes and rivers (Gregor, 1972; Banat, Forstner, and Muller, 1974) and could present problems in land disposal of this compound. Since Etzel et al. (1975 a,b,c) and Doemel and Brooks (1975) have shown that the banning of phosphate-containing detergents did not reduce the amount of phosphates entering receiving waters to levels low enough to prevent algae blooms, perhaps replacement of phosphate builders by substitutes will be slowed. Devey and Harkness (1973) estimated the proportion of phosphate in sewage due to detergents to be from 15 to 65%, with the remainder contributed by human excreta and some industrial wastewaters. Non-point sources of phosphate are runoff from urban areas and agricultural areas and other sources of erosion. While some lakes and streams definitely do show signs of eutrophication due to phosphorus additions, the banning of phosphates in detergents will probably not be the solution to this problem. Even advanced wastewater treatment to remove phosphates and other compounds may not be sufficient if the major source of phosphates is from land runoff (Lee, 1973).

Substitutes for phosphate builders beside NTA have been studied for their biodegradability and environmental impact. Among these are sodium citrate, sodium carbonate, and metasilicates (Jenkins et al., 1974; Ashforth and Calvin, 1973; Shannon, 1975). These compounds are thought to be of little concern at expected concentrations. Other new proposed builders that have been investigated and found biodegradable are carboxymethyloxysuccinate (Klein and Jenkins, 1972; Singer et al., 1978) and carboxymethyl-tartronate (Gledhill, 1978). Other components of soaps and detergents have also been shown to be readily treatable. Bacterial proteases which are present in enzyme-containing detergents were shown to be completely degraded in activated sludge (Crawford and Bouchard, 1971) and an antibacterial compound used in soaps, 3,4,4'-trichlorocarbanilide (TCC), was also found to be readily degraded by many bacteria when in low concentrations (Gledhill, 1975). Although the fluorescent whitening agents (FWA) used in many detergents are not very biodegradable, they are removed during primary sedimentation. In land application of sludges containing FWA's, only slight uptake of them by cover crops was observed with no detrimental effects observed (Ganz et al., 1975).

Possibilities for Land Application of Soap and Detergent Wastes

The high biodegradability and low toxicity of all components of the soap and detergent industry make it an excellent candidate for the land disposal of wastewaters and sludges. Only the fact that so few of the manufacturers treat their own wastes seems to stand in the way. Robeck et al. (1963, 1964) showed that the more recalcitrant ABS surfactants were degraded adequately in properly maintained soil systems.

Overcash and Pal (1979) reviewed the application of surfactants to soils and its possible effects mainly in greenhouse and small-scale field studies. The presence of surfactants increased the water retaining capability of the soil to various degrees depending upon the soil type and concentration and type of surfactant applied. Leaching of surfactants under conditions of unsaturated flow was thought to be minimal at low surfactant concentrations. This was due to the adsorption of surfactants onto the soil particles and subsequent biodegradation. The level of surfactant in soil that reduces microbial activity also varied depending on soil type and concentration and type of surfactant. Generally, at surfactant levels of greater than 100 ppm (approx. 240 kg/ha) in the soil, there is a reduction in microbial activity.

Overcash and Pal (1979) also reviewed information on the effect of surfactant addition on plant growth. The level of surfactant added to the soil that was thought to decrease crop yield was 10 kg/ha for cationic surfactants, 15-3,000 kg/ha for nonionic surfactants and 300 kg/ha for anionic surfactants. Surfactant applications in the range of 10-20 kg/ha would probably be best for maximum microbial degradation and minimum reduction in crop yield.

Since the soap and detergent industry is continuing to improve the biodegradability of surfactants, there should be no problems concerning land cultivation of these wastes if proper precautions are taken. The amount of wastewaters and sludges available from the soap and detergent industry was estimated to be 12-14 x 10^6 m^3 in 1975, increasing to an estimated 20-22 x 10^6 m^3 in 1985 (Phung et al, 1978a).

Summary and Research Needs

Although there is little literature available on land application of wastes from the soap and detergent industry, it appears that there would be relatively few problems associated with this method of disposal. The high biodegradability and low toxicity of many of the organic constituents has been well established in aquatic systems, but there is considerably less information concerning the rate and extent of degradation in soil. Research into determining the biodegradation of these compounds in soil should provide data as to possible application rates and methods for disposal of soap and detergent wastes on land.

REFERENCES

Ashforth, G. K., and G. Calvin. 1973. Safety evaluation of substitutes for phosphates in detergents. Water Res. 7:309-320.

Banat, K., U. Forstner, and G. Muller. 1974. Experimental mobilization of metals from aquatic sediments by nitrilotriacetic acid. Chem. Geol. 14:199-207.

Basu, A. K. 1967. Treatment of effluents from the manufacture of soap and hydrogenated vegetable oil. J. Water Poll. Control Fed. 39:1653-1658.

Bird, J. A. and R. B. Cain. 1974. Microbial degradation of alkylbenzene sulphonates. Metabolism of homologues of short alkyl-chain length by an *Alcaligenes* sp. Biochem. J. 140:121-134.

Brownell, R. P., P. L. Busch, and J. L. Herrin. 1975. Chemical-biological treatment of surfactant wastewater. p. 1085-1094. *In* Proceedings of the 30th Industrial Waste Conference. Purdue University, Lafayette, IN, Ann Arbor Science Publishers, Inc., Ann Arbor, MI.

Cain, R. B. 1976. Surfactant biodegradation in wastewaters. p. 283-327. *In* A. G. Callely, C. F. Forster, and D. A. Stafford (ed.). Treatment of industrial effluents. John Wiley and Sons, New York.

Cook, K. A. 1979. Degradation of the non-ionic surfactant Dobanol 45-7 by activated sludge. Water Res. 13:259-266.

Cook, R. 1968. The bacterial degradation of synthetic anionic detergents. Water Res. 2:849-876.

Crawford, J. G. and E. F. Bouchard. 1971. Environmental considerations in the manufacture of detergent enzymes. Dev. Ind. Microbiol. 12:25-35.

Devey, D. G. and N. Harkness. 1973. The significance of man-made sources of phosphorus: detergents and sewage. Water Res. 7:35-54.

Doemel, W. N. and A. E. Brooks. 1975. Detergent phosphorus and algal growth. Water Res. 9:713-719.

E.P.A. 1974. Development document for effluent limitations guidelines and new source performance standards. Soap and detergent manufacturing point source category. EPA-440/1-74-018-a, U.S. Environ. Prot. Agency, Washington, D.C. 202 p.

Etzel, J. E., J. M. Bell, E. G. Lindermann, and C. J. Lancelot. 1975a. Detergent phosphate ban yields little phosphate reduction, Part I. Water Sewage Works 122(9):91-93.

Etzel, J. E., J. M. Bell, E. G. Lindermann, and C. J. Lancelot. 1975b. Detergent phosphate ban yields little phosphate reduction, Part II. Water Sewage Works 122(10):91-93.

Etzel, J. E., J. M. Bell, E. G. Lindermann, and C. J. Lancelot. 1975c. Detergent phosphate ban yields little phosphate reduction, Part III. Water Sewage Works 122(11):68-70.

Fenger, B. H., M. Mandrup, G. Rohde, and J. C. K. Sorensen. 1973. Degradation of a cationic surfactant in activated sludge pilot plants. Water Res. 7:1195-1208.

Ferguson, B. C., R. L. Todd, H. W. Holm, J. D. Pope, Jr., and H. J. Kania. 1971. Environmental fate of NTA. p. 271-283. In Proceedings of the 26th Industrial Waste Conference. Purdue University, Lafayette, IN, Ann Arbor Science Publishers, Inc., Ann Arbor, MI.

Ganz, C. R., C. Liebert, J. Schulze, and P. S. Stensby. 1975. Removal of detergent fluorescent whitening agents from wastewater. J. Water Poll. Control Fed. 47:2834-2849.

Gerike, P., W. K. Fischer, and W. Jasiak. 1978. Surfactant quaternary ammonium salts in aerobic sewage digestion. Water Res. 12:1117-1122.

Gledhill, W. E. 1975. Biodegradation of 3,4,4'-trichlorocarbanilide, TCC, in sewage and activated sludge. Water Res. 9:649-654.

Gledhill, W. E. 1978. Microbial degradation of a new detergent builder, carboxymethyltartronate (CMT), in laboratory activated sludge systems. Water Res. 12:591-597.

Gregor, C. D. 1972. Solubilization of lead in lake and reservoir sediments by NTA. Environ. Sci. Technol. 6:278-279.

Herin, J. L., L. H. Marlow, and C. T. Stigger. 1970. Development and operation of an aeration waste treatment plant. p. 420-426. In Proceedings of the 25th Industrial Waste Conference. Purdue University, Lafayette, IN. Ann Arbor Science Publishers, Inc., Ann Arbor, MI.

Huddleston, R. L., and R. C. Allred. 1967. Surface-active agents: biodegradability of detergents. p. 343-370. In A. D. McLaren and H. Peterson (ed.). Soil biochemistry. Marcel Dekker, Inc., New York.

Jenkins, D., W. J. Kaufman, P. H. McGauhey, A. J. Horne, and J. Gasser. 1974. Environmental impact of detergent builders in California waters. Water Res. 7:265-281.

Klein, S. A. and D. Jenkins. 1972. The fate of carboxymethyloxysuccinate in septic-tank and oxidation pond systems. SERL Report No. 72-10. Univ. of California at Berkeley, Berkeley, CA.

Klein, S. A., D. Jenkins, and P. H. McGauhey. 1963. The fate of ABS in soils and plants. J. Water Poll. Control Fed. 35:636-654.

Lee, G. F. 1973. Role of phosphorus in eutrophication and diffuse source control. Water Res. 7:111-128.

Lense, F. T., S. E. Mileski, and C. W. Ellis. 1978. Effects of liquid detergent plant effluent on the rotating biological contactor. EPA-600/-2-78-129, U.S. Environmental Protection Agency, Cincinnati, OH. 58 p.

Malaiyandi, M., D. T. Williams, and R. O'Grady. 1979. A national survey of nitrilotriacetic acid in Canadian drinking water. Environ. Sci. Technol. 13:59-62.

McCarty, P. L., D. J. Hahn, G. N. McDermott, and P. J. Weaver. 1972. Treatability of oily wastewaters from food processing and soap manufacture. pp. 867-878. In Proceedings of the 27th Industrial Waste Conference. Purdue University, Lafayette, IN. Ann Arbor Science Publishers, Inc., Ann Arbor, MI.

McClelland, N. I., and K. H. Mancy. 1969. The effect of surface active agents on substrate utilization in an experimental activated sludge system. pp. 1361-1384. In Proceedings of the 24th Industrial Waste Conference, Purdue University, Lafayette, IN. Ann Arbor Science Publishers, Inc., Ann Arbor, MI.

Moore, L. and E. F. Barth. 1976. Degradation of NTA acid during anaerobic digestion. J. Water Poll. Control Fed. 48:2406-2409.

Overcash, M. R. and D. Pal. 1979. Design of land treatment systems for industrial wastes - theory and practice. Ann Arbor Science Publishers, Inc., Ann Arbor, MI. 684 pp.

Osburn, Q. W. and J. H. Benedict. 1965. Polyethoxylated alkyl phenols: relationship of structure to biodegradation mechanism. J. Amer. Oil Chem Soc. 43:141-146.

Patterson, S. J., C. C. Scott, and K. B. E. Tucker. 1970. Non-ionic detergent degradation. III. Initial Mechanism of the Degradation. J. Amer. Oil Chem. Soc. 47:37-41.

Phung, T., L. Barker, D. Ross, and D. Bauer. 1978a. Land cultivation of industrial wastes and municipal solid wastes: State-of-the-Art Study. Vol. I. Technical Summary and Literature Review. EPA-600/2-78-140a, U.S. Environmental Protection Agency, Cincinnati, OH. 206 pp.

Phung, T., L. Barker, D. Ross, and D. Bauer. 1978b. Land cultivation of industrial wastes and municipal solid wastes: State-of-the-Art Study. Vol. II. Field investigations and case studies. EPA-600/2-78-140b. U.S. Environmental Protection Agency, Cincinnati, OH. 157 pp.

Robeck, G. G., T. W. Bendixen, W. A. Schwartz, and R. L. Woodward. 1964. Factors influencing the design and operations of soil systems for waste treatment. J. Water Poll. Control Fed. 36:971-983.

Robeck, G. G., J. M. Cohen, W. T. Sayers, and R. L. Woodward. 1963. Degradation of ABS and other organics in saturated soils. J. Water Poll. Control Fed. 35:1225-1236.

Schneider, H. J. 1977. Detergents-1985. Soap Cosmet. Chem. Spec. 53(3):40,42,121.

Shannon, E. E. 1975. Effects of detergent formulation on wastewater characteristics and treatment. J. Water Poll. Control Fed. 47:2371-2383.

Singer, E. J., E. Mones, A. S. Rothenstein, and J. M. Weaver. 1978. CMOS: Environmental and human safety. Soap Cosmet. Chem. Spec. 54(8):38-44.

Steinle, E. C., R. C. Myerly, and C. A. Vath. 1964. Surfactants containing ethylene oxide. Relationship of structure to biodegradability. J. Amer. Oil Chem. Soc. 41:804-807.

Stoveland, S., J. N. Lester, and R. Perry. 1979. The influence of nitrilotriacetic acid on heavy metal transfer in the activated sludge process. I. At constant loading. Water Res. 13:949-965.

Swisher, R. D. 1963. The chemistry of surfactant biodegradation. J. Amer. Oil Chem. Soc. 40:648-656.

Swisher, R. D. 1970. Surfactant biodegradation. Marcel Dekker, Inc., New York. 496 pp.

Thom, N. S. 1971. Nitrilotriacetic acid. A literature survey. Water Res. 5:391-399.

Thompson, J. E. and J. R. Duthie. 1968. The biodegradability and treatability of NTA. J. Water Poll. Control Fed. 40:306-319.

Willetts, A. J. 1973. Microbial aspects of the biodegradation of synthetic detergents: A Review. Int. Biodetn. Bull. 9(1-2):3-10.

Textile Wastes

S.B. Hornick and D.D. Kaufman

Textile mill wastes contain either natural impurities such as dirt, grease, etc. from the fiber or detergents, dyes, oils, etc. from chemical processes. Therefore, the wastes produced by the textile industry are usually characterized by the fiber utilized and the process employed to manufacture the products. The fibers used can be wool, cotton, or synthetic fibers such as rayon, nylon, dacron, cellulose acetate, and polyester. The processes utilized by various textile manufacturers are (1) wool scouring, (2) wool finishing, (3) woven fabric dyeing and finishing, (4) knit fabric dyeing and finishing, (5) carpet dyeing and finishing, (6) stock and yarn finishing, (7) nonwoven manufacturing, and (8) felted fabric processing (EPA, 1979).

Unlike wool, the making of cotton and synthetic fibers into yarn or unfinished cloth is a dry process. Cotton is much cleaner than wool and can be opened, picked, and combed mechanically to remove impurities.

Synthetic fibers are basically pure chemicals and have no impurities. Synthetics are either cellulosic, produced from cellulose, or non-cellulosic, synthesized from organic materials. Rayon and cellulose acetate are the major cellulosic fibers with polyester, nylon, and acrylics comprising the majority of non-cellulosic fibers.

Dry processes use little or no water and usually precede wet operations. Some of the major dry processes, are spinning, tufting, knitting, weaving, slashing, adhesive processes and functional finishing. Spinning converts the fiber into yarn or thread; tufting is used in carpet manufacturing to mechanically pull yarn through jute or polypropylene backing when forming carpet pile; knitting is a major fabric manufacturing method which applies oils to the yarn for lubrication during the high-speed processing; and slashing coats warp yarns with sizing compounds to prevent breakage during weaving. Common sizing agents are starch, polyvinyl alcohol (PVA), carboxymethyl cellulose (CMC) and polyacrylic acid (PAA) (EPA, 1979).

Some adhesive processes are bonding, laminating, and coating. Bonding joins two materials in order to extend or improve performance or appearance. The adhesive used in bonding is either a water-based acrylic compound or urethane foam. Laminating bonds fabric with non-textile materials such as foam with latex adhesive. A process related to laminating is carpet backing. Coating applies a continuous film to the fabric by using polyvinyl chloride (PVC).

Functional finishing utilizes various chemicals to impart certain properties upon a fabric such as wrinkle resistance, water-repellancy, stain resistance, and moth resistance. Chemicals such as synthetic resins, silicones, silicofluorides, chromium fluorides, chlorinated phenols and various metallic salts are used in the finishing processes.

The major wet processes or high water use processes are desizing, scouring, mercerizing, bleaching, dyeing, and printing. Several additional processes specific for wool handling are raw wool scouring, carbonizing, and fulling.

Raw wool can contain 30 to 70% impurities such as dirt, grease, and suint depending on the breed and habitat of the sheep which have been sheared. After shaking, sorting, and blending, the wool must be scoured and dried before further processing. The wool is usually treated with a hot soap-alkali (pH 9.5-10.5) and then followed by the hotter neutral (pH 7) detergent baths. Scouring emulsifies the dirt and grease. In many mills, recovery of grease and oil for lanolin is practiced since the substance is not readily biodegradable (Nemerow, 1978). Carbonizing removes burrs and vegetative or plant material to ensure equal adsorption of dyes. In the carbonizing process the wool is soaked in dilute sulfuric acid, baked to oxidize the cellulosic contaminants, crushed so the charred material can be removed by mechanical agitation, neutralized with sodium carbonate solution, and rinsed. The thick feel and appearance of woolen cloth is the result of fulling. Alkali fulling uses soap or detergent, sodium carbonate, and sequestering agents. Acid fulling utilizes dilute sulfuric acid, hydrogen peroxide and metal catalysts such as Cr, Cu, and Co (EPA, 1979).

Desizing removes the compounds which were added in the slashing operation by solubilizing the size with dilute sulfuric acid or enzymes (vegetable or animal). Scouring of cotton fabric removes the natural impurities, such as pectins, by a hot alkali soap or detergent followed by a water rinse. Since synthetic fibers contain very few impurities, mild baths containing weak alkali, anti-static agents, and lubricants are used before the dyeing process (EPA, 1979).

In mercerization, the tensile strength of cotton is increased by treatment with 15 to 30% cold NaOH. The alkali is rinsed off and in many plants recovered for further use.

Bleaching removes color and can remove impurities such as sizing compounds from the fibers. The wool is bleached with sodium or hydrogen peroxide and requires careful pH control. Cotton is usually washed and treated in several steps with sodium hydroxide, hydrogen peroxide, and hot water rinses. Cellulosic synthetic fibers are bleached similarly to cotton; non-cellulosic synthetics are bleached only when blended with natural fibers. While stock, yarn, fabric, or any state of goods can be dyed, most of the dyeing is performed on fabric. A detailed listing of the chemical constitution of the dyes is found in the Colour Index (Society of Dyers and Colourists, 1971), but only a general discussion of the properties of dye classes will be presented here.

The various types or classes of dyes are vat, direct, azoic or naphthol, reactive, sulfur, basic, mordant or chrome, acid, and disperse (EPA, 1979; DOI, 1967). The use of a particular dye depends on the fabric, the finish of the fabric and the desired color. Besides the dye, additional chemicals or carriers may be needed to give the desired results.

An example of one dye class is the azoic or naphthol dyes used in the dyeing of cellulosic fibers which involves a two stage dyeing process. The first stage involves the adsorption of insoluble pigments by using a soluble azoic coupling component and then treating with a diazonium salt. Naphthol is the usual coupling component but there are over fifty coupling components and 50 bases which can be diazotized to react with the coupling compound (EPA, 1979). Another example is disperse dyes which are very slightly soluble organic compounds used to dye hydrophobic cellulose acetate. With the various dyes, additional chemicals such as butyl benzoate carriers, chlorobenzene, and diethyl phthalate may be used to give satisfactory results.

Printing differs from dyeing in that only small or selected areas of fabric are colored or dyed. This is accomplished by applying the print paste on or through rollers or screens which have the desired designs for the planned pattern.

Each year numerous new dyes come on the market because of the demand for better performance. Desired qualities in dyes are resistance to ozone, light, hydrolysis, and other environmental factors which will cause degradation. Of course these qualities are not desirable when the dyes end up in the waste stream.

In wool production, the scouring, dyeing and washing processes contribute the most to the waste stream. A large amount of dirt, grease (high BOD), high temperature, and alkalinity come from scouring; colour, high BOD, acid, and possibly toxic waste from dyeing; and high oil and BOD, and high temperatures from the wash after fulling. With cotton, a high BOD, neutral pH, and high total solids comes from desizing; scouring gives high BOD, high pH, high total solids, and high temperature; bleaching wastes have high BOD, alkaline pH, and high solids; with mercerizing, a low BOD, alkaline pH, and low solids result; the dyeing and printing process gives a waste with high pH, high solids content and neutral to alkaline pH (Clemson University, 1971)

Each fiber used in synthetic fiber production contributes different waste components. Depending on the desired end product and dyes used, these wastes can vary considerably from mill to mill.

Although some of the priority pollutants have been suspected or detected in some textile wastewater effluent (EPA, 1979), these compounds may not be adsorbed into wastewater sludges due to limited contact of the wastewater with solids. Some mills have no sludge due to dilution of separated solids with wastewater.

If textile wastes were land-applied, metals such as Cr if present in high quantities should be rapidly converted to less toxic and soluble form if the soil is aerobic and limed. Similarly, metals such as Zn, Cu, As and Cd will precipitate in the soil profile if the pH is maintained at 6.5 or above. Specific loading rates for these metals in organic wastes have been cited (Chaney, 1974). Not enough is known about the essentiality of metals such as Sr to recommend a maximum loading rate. High levels of metals or salts in textile wastes could be avoided by chemical substitution.

Since dye manufacturers want their products to be resistant to degradation, it is not surprising that research has for the most part confirmed their success under aerobic conditions (Porter and Snider, 1976). The exact combinations of environmental factors, if any, that would lead to dye degradation in cropped soils needs further investigation.

The persistence of finishing compounds has also been investigated. Some studies have shown that polyvinyl alcohol (PVA) can be degraded. According to Porter and Snider (1976) discrepancies in PVA's degradability have arisen since the hydrolysis of polyvinyl acetate is not complete and acetate groups remain on PVA. The degradability varies with the number and location of the residual groups.

Carriers such as chlorinated benzenes, o-phenylphenol, phenylmethylcarbinol, and benzoic or salicylic acids, which are used in polyester dyeing, are usually recovered in waste treatment. These carriers contribute to the varying and usually high BOD's found in dye wastes, e.g., salicylic acid, 25,000 ppm; o-phenylphenol, 6,060 ppm. Toxic carriers such as chlorinated benzenes are very toxic to humans and are specially vented.

Very little is actually known about the photodegredation of many of the chemicals used in textile manufacture. Certain of the textile carrier compounds are chlorinated benzenes. The volatile nature of these compounds makes them likely candidates for photodegradation in air or on moist surfaces. Replacement of the chlorine with either hydrogen or hydroxyl groups to yield benzene or phenols, respectively, would be the most likely reactions. Obvious candidates for photodegradation studies are the dyes. Their stability within the fabric is a most desirable feature which can be in basic conflict with their susceptibility to degradation in the environment as a waste. Porter (1973) examined the photodegradation of 36 dyes selected from those most used by the textile industry. Most of the dyes were quite resistant to photodegradation and had an average of only 40% color loss after 200 hours of exposure to artificial light in water.

A comparison of artificial light and natural sunlight effects on Basic Green 4 and Direct Blue 76 showed that these dyes degraded at least 10 times slower in natural sunlight (Porter, 1973). This means that a minimum of 80 24 hr-days would be required to produce appreciable degradation of these two dyes in the environment. These tests were conducted in the laboratory, however, with the dyes dissolved in water. It is well known that water in streams or lakes, etc. may contain sensitizers which facilitate the degradation of otherwise resistant compounds. It should also be recognized that the light intensity used in the laboratory may have had a higher concentration of the specific wave lengths necessary to degrade these dyes.

Dyes such as Basic Green 4 photodegrade to a point where the products could be more susceptible to biodegradation. The coupling of photodegradative mechanisms with biodegradative mechanisms needs further investigation. Other dyes such as Acid Black 52 appeared resistant to photodegradation (Porter, 1973).

Porter and Snider (1976) conducted 30-day BOD studies on several textile chemicals and dyes. The results were compared with the chemical oxygen demand (COD) of the sample, which in most cases measures its maximum oxygen combining power. The results were also compared with the BOD test performed on a standard glucose-glutamic acid reference solution used for BOD calibration. The BOD/COD ratio with the glucose-glutamic acid solution was 99%, which indicates that the glucose and glutamic acid were almost completely oxidized. The BOD/COD ratio of the textile chemical butyl benzoate was also comparatively high. The BOD was low (less than 10%) for all dyes except Disperse Yellow 144. This was attributed to the larger concentrations of degradable diluents such as dextrin or lauryl sulfate used for the formulation of this dye. These and other degradable chemicals were used for the dye formulation and account for the BOD loading.

These data illustrate several points of primary concern. While the BOD-COD methods may be useful as a general indication of the fact the same oxidation of the formulation has occurred and a general cause of pollution may be averted, they do not assure that the key, and possibly hazardous constituents, have been degraded. Although there was an immediate and rapid uptake of oxygen during the degradation of this formulation, there was essentially no change in the optical density measurements used to record degradation or dissipation of the dye itself. Thus, there presumably was not alteration or degradation of the dye, which in this formulation may well be the chemical of greatest concern. The results obtained with the textile dyes, Reactive Red 21 and Direct Blue 80, illustrate an additional phenomenon. In these cases there is no immediate elevation of oxygen consumption, but definite lag periods occur before oxygen consumption begins. This may be the result of: (1) the period of time required for the development of microbial population effective in degrading the chemicals in question due to the unique nature of the chemicals or microbes involved; or (2) the presence of an inhibitory factor in the formulation which must either first be eliminated, or a microbial population resistant to its toxic effects must develop. Again, regardless of the BOD/COD expression there was essentially no change in the concentration of the dye as determined by optical density.

Garrison and Hill (1972) examined the dissipation of nine

dye carriers exposed individually to aerated mixed cultures of aerobic microorganisms to test their resistance to bacterial action. Mono, o-di-, and p-dichlorobenzene volatilized completely in less than one day from the aerated culture, but 1,2,4-trichlorobenzene persisted up to nine days. Concentrations of biphenyl and o-phenylphenol decreased asymptotically in 4-6 days to 5-50% of their original concentrations. Acetophenone was formed during the degradation of α-methylbenzyl alcohol, but was also subsequently degraded.

Many commercial dyes are characterized by the presence of an azo group (-N=N-) in their molecular structure. The azo group is not widespread in naturally occurring products although it has been identified in degradation of aromatic amine and nitro based pesticides. There is strong evidence that azo reduction occurs within the liver of higher animals with the subsequent formation of amines and phenolic metabolites (Robinson and Wright, 1964). Some microorganisms contain the same enzyme system found in higher animals and may thus be able to successfully degrade azo dyes.

Literature on the degradation of azo dyes by microorganisms is sparse. The discovery of sulfa drugs was actually the result of an investigation of the degradation of azo dyes (Flege, 1968). Prontosil is enzymatically reduced to sulfanilic acid which inhibits microbial growth. Sulfanilic acid is a primary dye intermediate or building block for the manufacture of many dyes. Consequently, it may also be a primary metabolite resulting from the degradation of many dyes.

Dyes for textiles are deliberately manufactured to resist strong oxidizing and reducing chemicals used in commercial laundries. Thus, a product susceptible to degradation in simple chemical or biological systems would fail to quality as a dye for commercial usage.

Flege (1968) examined the degradation of several disperse dyes in a digester. Both Disperse Orange 5 and Disperse Red 5 were degraded by the same mechanism to similar products. The degradation scheme illustrated here indicates that the azo bond of the parent dye molecule is reduced to a hydrazo bond. This bond is subsequently cleaved to yield two different aromatic amines which themselves are further degraded. This degadation pathway is of particular interest because it leads to the formation of the soil fungicide DCNA or dicloran. While DCNA is used commercially as a fungicide, its possible toxicity to further degradation of other textile chemicals is unknown. Several conflicting reports on DCNA metabolism exist in the literature. Some of these indicate a rapid microbial degradation of DCNA with liberation of the chloride ion and

$^{14}CO_2$ from ring-labeled DCNA. The results of other investigations indicate that DCNA is only slowly degradable by microbiological processes.

In actuality it is conceivable that numerous other degradation pathways could exist for Disperse Orange 5. Reduction of the nitro group to an amine could occur at any point during its metabolism. Removal of either the methyl or ethanolic groups (dealkylation), such as happened in the formation of the p-phenylenediamine, could also occur at any point during the metabolism of the parent molecule.

Summary and Research Needs

With the advent of wastewater treatment plants in the last few years, textile manufacturers have a larger amount of sludge and wastewater to dispose of or utilize. Generally, textile wastes are low in most metals which can affect crop production or animals. Overall, textile sludges are lower in metals than municipal sludges. Occasionally one element such as Zn, Cu, B, etc., can be elevated and become a limiting factor in the consideration of land applying a textile waste. Usually the P and K contents of textile wastes are minimal (0.02 to 0.1% on a dry weight basis). Nitrogen, on the other hand, can be sufficient to provide some fertilizer benefits (1.5%). If land application of textile wastes were based strictly on inorganic constituents, many wastes could safely be applied to cropland.

The unknown factor in utilizing textile waste for land application is the persistence and quantity of organic constituents. The actual reactivity of these compounds with the soil and with the microorganisms necessary to maintain plant life is uncertain at this time. Possible pretreatment processes such as composting, which might degrade some of the harmful organics, or product recovery would render the wastes more suitable for land application.

Further research conducted on waste quality, adsorption of metals and organics in the soil profile, the persistence or mobility of organics and their degradation products in the environment, and the effect of waste constituents on soil microbes, will allow the land application of textile wastes to become an acceptable and safe practice.

REFERENCES

Chaney, R. L. 1974. Recommendations for management of potentially toxic elements in agricultural and municipal wastes. USDA-ARS-NPS, Beltsville, MD.

Clemson University. 1971. State of the art of textile waste treatment. USEPA Water Pollution Control Research Series, 12090 ECS. USGPO Wash., D. C.

D.O.I. 1967. The cost of clean water. Volume III. Industrial Waste Profiles No. 4 - Textile Mill Products. USGPO, Wash., D.C.

E.P.A. 1974a. Development document for effluent limitations guidelines and standards for the textile mills. EPA-440/1-79-022b. U. S. Environ. Prot. Agen., Washington, D.C.

Flege, R. F. 1968. Determination, evaluation, and abatement of color in textile plant effluents. Completion Report: OWRR Project No. B-012-GA. Water Resources Center, Georgia Institute of Technology, Atlanta, GA.

Garrison, A. W., and D. W. Hill. 1972. Organic pollutants from mill persist in downstream waters. American Dyestuff Reporter 61:21-25.

Nemerow, N. L. 1978. Industrial water pollution - origins, characteristics, and treatment. Addison-Wesley Pub. Co., Reading, MA.

Porter, J. J. 1973. A study of the photodegradation of commercial dyes. Environmental Protection Technology Series, EPA-R2-73-058. Office of Research and Monitoring, U.S. Environmental Protection Agency, Wash. D.C. 94 pp.

Porter, J. J., and E. H. Snider. 1976. Long-term biodegradability of textile chemicals. J. Water Pollut. Control Fed. 48:2198-2210.

Robinson, A. J., and S. E. Wright. 1964. J. Pharmaceuticals and Pharmacology 16: Suppl. 80, T-82T.

Society of Dyers and Colourists and the Amer. Assoc. of Textile Chemists and Colourists. 1971. Colour Index. 3rd edition. Volume 3. The Society of Dyers and Colourists, Bradford, England.

Wood Preservative Wastes

L.J. Sikora

The Wood-Preserving Process

Wood is a major material of construction and appears likely to remain in this category for the foreseeable future. In the absence of special precautions, however, its usefulness for certain purposes is limited by its susceptibility to attack by insects and fungi. The chief precaution used is to treat it with preservatives.

Damage to untreated wood is sometimes deep-seated and sometimes superficial. Thus, an untreated pine fence post may have a useful service life of only two years before it has rotted below ground level. Lumber in outdoor storage may be disfigured and reduced in economic value by staining fungi. In both cases proper preservative treatment will provide a significant degree of control of the problem.

The three most common types of preservatives for wood are creosote, pentachlorophenol, and arsenicals as none are satisfactory for all wood products. Creosote has been used mainly for railroad ties, utility poles, and piling and pentachlorophenol (PCP) for utility poles, cross arm posts, and lumber, while arsenicals have been used on lumber, plywood, and poles (Table 22).

Discussion of the details of the wood-preservative treatment process is beyond the intended purpose here but a few essentials may be mentioned. The process consists of two basic steps: (a) conditioning the wood to reduce its moisture level and increase its permeability to preservatives and (b) the actual impregnation with the preservative.

TABLE 22. ESTIMATED PRODUCTION OF TREATED WOOD, 1978[a].

Products	Treated With			
	All Preservatives[b]	Creosote Solutions	Penta	CCA ACA FCAP*
	1,000 cu. ft.			
Crossties and switchties[c]	106,085	103,138	449	2,498
Poles	46,179	18,237	41,905	4,038
Crossarms	1,685	41	1,615	29
Piling	12,090	9,993	1,154	943
Lumber and timbers	105,305	10,779	21,209	73,317
Fence posts	20,028	4,584	10,983	4,461
Other products[d]	18,113	7,815	2,681	7,616
All products	327,485	158,587	79,996	92,903

*CCA: chromated-copper-arsenate, ACA: ammonical copper arsentate, FCAP: fluor-chrome-arsenate phenol

a Volume reported for 1977 (AWPA), plus volume reported by respondents to Assessment Team Survey, plus volume estimated for nonrespondents.

b Creosote, Penta, and CCA/ACA/FACP only.

c Includes landscape ties.

d Includes plywood.

Note: Components may not add to totals due to rounding.

In conventional or steam conditioning, unseasoned or partially seasoned stock is subjected to direct steam impingement at elevated pressure in a retort. Steam condensate that forms in the retort is conducted to oil-water separators after which removal of emulsified oils requires further treatment. In closed steaming, the steam is generated in the retort with water from a reservoir and the oils are removed periodically from the water, which is recycled in the system. Alternatively, a vapor drying process may be used in which the wood is exposed to hot vapors of xylol, naphtha, or Stoddard solvent, causing the water in the wood to vaporize.

Following conditioning, the wood is immersed in preservative chemicals, either at ambient or elevated temperature, either with or without the use of pressure. Table 23 gives an indication of the various process schemes. Ninety-five percent of all wood treated with preservatives is done via pressure processes.

TABLE 23. PARTIAL RESULTS OF SURVEY TO DETERMINE CHARACTERISTICS OF WOOD PRESERVING PLANTS IN U.S. (ADAPTED FROM DEVELOPMENT DOCUMENT, 1979).

Conditioning Process	Preservative	Number of Plants	Raw Flow	Prod
			-------m^3 or l/day-------	
Steaming	C+	4	13,000 - 94,000	76 - 248
Steaming	C,P	11	7,500 - 236,000	85 - 461
Steaming	C, inorg.	2	6,400 - 161,000	96 - 515
Steaming	P, inorg.	1	950	65
Steaming	C,P, inorg.	3	7,500 - 52,000	110 - 212
Vapor Drying	C	1	94,600	198
Boulton	P	1	28,400	62
Boulton	P, inorg.	1	8,330	78
Boulton	C,P, inorg.	3	18,900 - 57,900	142 - 308

+ C = Creosote, P = pentachlorophenol, inorg = inorganic salts.

There are approximately 600 wood preserving plants in the United States, about 200 of these being small. Half use inorganic salts and half use combinations of PCP and creosote. There are several large commercial plants which handle poles and process 3000 to 5000 cu ft of lumber per day. (Personal communication, Mr. Thomas Marr, Koppers Co., Pittsburg, PA. and Dr. Warren Thompson, Miss. State Univ., State College, Miss. Sept. 1980).

Origin of Wastes

Steaming (preconditioning) is the main source of wastewater from wood treating plants, while some wastewaters also arise when the treated wood product is removed and allowed to drain. The steam condensate contains oil, phenols, suspended solids, and dissolved organic matter. Since the chemicals used for treatment are biocidal in nature, it is not surprising that the concentrated wastewaters are not amenable to biological waste treatment processes.

Wastewater Treatment Processes

The three major wastewater treatment processes used by the wood preserving industry include (a) primary treatment and pH adjustment to render the wastewater acceptable to the municipal treatment plant, (b) primary treatment followed by biological treatment (secondary treatment) and spray irrigation of the non-recycled water, and (c) primary treatment followed by biological treatment in ponds and then spray evaporation. The procedure for treating the water-borne preservatives (CCA) wastewater was adapted from the metal plating industry, and requires the reduction of chrome from the hexavalent to the trivalent form, followed by precipitation of the chrome, copper, and arsenic (Teer and Russell, 1977).

The methods used for reclaiming creosote and PCP from the treatment solutions are based upon their physical characteristics. The solutions are conducted into tanks or ponds, after which the creosote, because of its greater density, settles to the bottom, PCP oil rises to the surface, and water comprises the middle layer. The PCP can then be recovered by skimming and the creosote with submerged pumps (Middlebrooks, 1968).

Dust and Thompson (1972) discussed the chemical and physical methods for treating wastewater from the wood preservative industry. Primary treatment involves flocculation and sedimentation. Because oil-water emulsions are not broken by mechanical oil-removal procedures, chemical flocculation is required to reduce the oil content of wastewaters containing emulsions. Lime, alum, ferric chloride, and polyelectrolytes have been used with lime being the most successful.

Some polyelectrolytes have been successful when used in combination with lime. Reduction in oil content by flocculation averaged 95%, PCP removal was 90%, and the phenol concentration did not change. Sludge from the flocculation treatment is removed by a gravity separator and dewatered in sludge drying beds. Effluent from the drying beds is conducted to the city sewer or into holding lagoons for secondary treatment.

Biological treatment is the principal secondary treatment method but other treatment methods employed at some wood preservative plants are carbon adsorption, membrane filtration systems, and oxidation by chlorine, hydrogen peroxide, and ozone. Many of these treatments are experimental in the wood preserving industry and further research is necessary to sufficiently determine the effectiveness of these treatments. Biological treatment has been shown to be quite effective in decreasing the concentrations of phenols, oil and grease, PCP, and organic toxic pollutants from wood preserving wastewaters. The effectiveness of biological treatment depends on influent concentration, detention time in the biological system, extent of aeration, and type of biological system employed. The reduction in concentration of PCP and other high molecular weight toxic pollutants is thought to occur by adsorption upon the biomass, rather than actual degradation (Development Document, 1979).

Trickling filters, aerated lagoons, oxidation ponds, and activated sludge systems are all used by plants throughout the industry. Several plants utilize land treatment systems for biological treatment in which treated wastewater is applied by sprinkler or by gravity irrigation. At some sites, runoff is collected in catchment basins and leachate in underdrains and the liquid reused or left to evaporate.

Approximately 10 wood preserving plants currently use spray irrigation as a final treatment step. Fisher and Tallon (1971) reported on 4 plants of which one plant was located in a large metropolitan area with access to city sewers. The wastewater from creosote treatment of crossties was deoiled, placed in a holding tank, and gradually released to the city sewer along with cooling waters and plant sewage. A second plant stored decanted wastewater from a creosote process in five lagoons with a residence time of one year. Water from the lagoon eventually seeped into a swampy area, then to a storm ditch where sampling and analysis indicated removal of phenol and a decrease in N, P, BOD, and COD. A third plant was located in a dry climate and used both creosote and salt treatments. Wastewaters flowed through a series of two ponds where significant evaporation and degradation occurred. Phenol removal was reported as excellent. A fourth plant produced 15,000 gallons of wastewater a day from creosote treatment. The wastewaters were placed in an equalization tank with N and P salts added, and aerated for 96 hours which decreased the BOD by 90% and removed 99% of the phenols. After clarification the wastewaters were conducted to a lagoon for further treatment followed by spray irrigation.

Myers et al. (1979) analyzed samples taken at various points in the processes and operations of two wood preserving plants. The treatment process for one plant (plant 10) consisted of a flocculation tank, an aerated lagoon to which was added N and P, and a lagoon from which effluent was applied to land by spray irrigation. Evaporation was the final wastewater disposal process. A second plant (plant 11) was similarly sampled and its wastewater treatment process involved a settling basin where oil was skimmed, a large storage basin, an aerated lagoon to which nutrients were added, a spray irrigation site planted to fescue, and a pond to collect runoff from the irrigation field.

Myers et al. (1979) analyzed the raw wastewater, the bottom sediment, and the effluent from the two wood preserving plants described previously. In most instances the effluent levels at plant 10 were below detectable levels for metals except arsenic (20 ug/l), polynuclear aromatics (10 ug/l) and priority pollutants except 2,4,6-trichlorophenol (116 ug/l) and PCP (663 ug/l). At plant 11, zinc was detected at 100 ug/l, phenol at 2 ug/l, 2,4-dimethylphenol at 26 ug/l, 2,4,6-trichlorophenol at 37 ug/l, and PCP at 105 ug/l. Both of these plants are classified as "zero discharge" plants.

The quantity of research data on the fate of the constituents in wood preservative wastewater in soils is relatively small. Thompson and Dust (1973) monitored total phenol content of water collected at various soil depths following irrigation of land with untreated creosote wastewater applied at a rate of 3500 gallons/acre/day (32,800 l/ha/day). Removal of phenols equaled or exceeded 99% at all soil depths with the range of 1 to 4 feet. In a laboratory study drums containing 60 cm of a heavy clay soil were loaded at rates of 3,500, 5,250, and 8,750 gallons/acre/day (32,800; 49,260; and 82,000 l/ha/day). Sufficient N and P were added to the waste to provide a COD:N:P ratio of 100:5:1. Influent COD and phenol concentrations were 11,500 and 150 mg/l, respectively. Reductions of more than 99% in COD content of the wastewater were observed from the first week in the case of the two highest loadings and from the fourth week for the lowest loading. A breakthrough occurred during the 22nd week for the lowest loading rate and during the fourth week for the highest loading rate. The COD removal steadily decreased thereafter for the duration of the test. Phenol removal showed no such reduction with time, but instead remained high throughout the test. Removal of COD exceeded 99% prior to breakthrough and averaged over 85% during the last week of the test.

Because of the relatively low volumes of wastewater generated by wood preserving plants, evaporation is a feasible and widely used principle for achieving zero discharge status. Based on the large number of plants which have adopted evaporation technology to achieve zero discharge status, this appears to be the method of choice for many wood preserving plants to comply with effluent guideline limitations and standards (Development Document, 1979).

Three types of evaporative systems are common in the industry. They are (a) spray evaporation, (b) cooling tower evaporation, and (c) thermal evaporation. All evaporation systems are preceded by flocculation to reduce the emission of hydrocarbons (Shack and Reynolds, 1977).

Incineration of wastewater sprayed over chips or wood shavings is a disposal method practiced by a small percentage of plants. Approximately 8% of the plants surveyed in Canada use incineration (Shields and Stranks, 1978). The nature of the organic compounds in wood preservatives require that incineration process be carefully controlled in order that emission of toxic constituents does not occur.

Sludge Volumes and Constituents

In the treatment of the wastewater before evaporation or irrigation, residuals or sludges are generated. The three most common wastewater treatment schemes in use are: 1) gravity oil-water separation by chemical flocculation and sand-bed drying, 2) gravity oil-water separation followed by biological treatment, and 3) zero discharge evaporation systems. The average sludge generation rate for the three processes is 0.018, 0.015, and 0.016 cu yd/1000 cu ft production respectively (Development Documentation, 1979). Although sludge generation rates at most plants are small, sludge is often allowed to accumulate for months or even years before removal and disposal. The amount of creosote-oil emulsion and penta-oil emulsion sludge produced annually by the entire industry is only 239 to 930 and 600 metric tons, respectively. (Listing Background Document, Wood Preserving, 1980).

Ultimate disposal of these sludges in the past has been in landfills. Tables 24 and 25 indicate that a number of toxic organics are present at high levels in the sludges including polynuclear aromatics and other constituents of creosote (Lorenz and Gjovik, 1972). Some of the aromatic hydrocarbons of creosote are known to be metabolized by various microorganisms including the bacterium *Pseudomonas creosotensis* (Drisko and O'Neill, 1966). Laboratory studies of oil contamination demonstrate that biodegradation is greatly enhanced if essential microbial nutrients are added. Some higher molecular weight compounds in creosote, particularly benzo(a)pyrene and benzo(a)anthracene are susceptible to photooxidation (USDA, State's EPA Preserv. Chem. Assess. Team, 1980).

Information regarding the degradation of PCP originates largely from its use as a wood preservative or as an herbicide in rice paddies. There is some evidence that wood-rotting and wood-staining fungi are capable of degrading PCP (Lyr, 1963; Duncan and Deverall, 1964; Cserjesi, 1967; Ingols and Stevson, 1963). Although no metabolites have been recovered or identified, changes in UV spectra, the liberation of chloride ions, and the appearance of colored products all point to a degradation of the parent molecule. While

TABLE. 24. CONCENTRATION OF CONSTITUTENT IN SLUDGES OF PLANT 10. (FROM MYERS et al., 1979)

	Bottom Sediment Dry Weight Aerated Lagoon (ug/kg)	Final Pond (ug/kg)
POLYNUCLEAR AROMATICS:		
Benzo(a)anthracene	3,700	149
Chrysene	4,500	2,060
Fluorene	17,600	210
Pentanthrene/anthracene	19,500	3,390
Pyrene	5,300	4,140
PHENOLICS:		
Phenol	9,030	16,000
2,4-Dimethylphenol	4,398	3,418
2-Chlorophenol	396,000	1,200
2,4,6-Trichlorophenol	No data	25,000
Pentachlorophenol	302,000	58,000
TOTAL METALS		
Arsenic	9,300	7,600
Selenium	4,500	1,600
Cadmium	2,100	1,300
Beryllium	1,900	3,200
Copper	40,000	4,300
Antimony	3,700	1,100
Chromium	5,600	3,100
Nickel	19,000	18,000
Zinc	310,000	48,000
Silver	2,100	690
Thallium	6,300	690
Lead	2,100	27,000
Mercury	136	3.4

TABLE. 25. CONCENTRATION OF CONSTITUENTS IN SLUDGES OF PLANT 11 (FROM MYERS et al., 1979)

Weight	Bottom Sediment Dry Aerated Lagoon (ug/kg)
POLYNUCLEAR AROMATICS:	
Benzo(a)anthracene	1,250
Benzo(a)pyrene	5,980
Chrysene	9,280
Acenaphthylene	1,400
Fluorene	547
Phenanthrene/anthracene	43,700
Pyrene	4,250
Acenaphthene	1,840
PHENOLICS:	
Phenol	4,500
2-chlorophenol	300
Pentachlorophenol	4,800
TOTAL METALS	
Arsenic	51,000
Selenium	940
Cadmium	200
Beryllium	2,500
Copper	99,000
Antimony	900
Chromium	56,000
Nickel	18,000
Zinc	280,000
Silver	910
Thallium	910
Lead	20,000
Mercury	20

photodecomposition is believed to be the major degradation process in paddy fields, most of the PCP applied to the soil surface in flooded paddy fields is known to infiltrate into the soil with percolating water (Kuwatsuka, 1972). Based on experiments involving soil sterilization, soil temperature, and PCP degradation products, it is believed that PCP degradation proceeds by both biological and chemical means (Kaufman, 1978; Kuwatsuka, 1972). It appears, however that chemical and microbial degradation are closely interrelated, since degradation was not observed in soil with a very low content of organic matter. Murthy et al. (1979) followed the degradation of PCP in aerobic and anaerobic soils and found that pentachloroanisole was a major by-product. Sikora et al. (1982) found that PCP degraded during composting at temperatures above 70 C which indicated that the degradation was being performed by a thermophilic organism. Published data indicated that PCP persistence in soil is about 9 months (USDA, State's EPA Preserv. Chem Assem. Team, 1980). Photodecomposition of PCP has been shown with the end products being colored quinones and aromatic ring cleavage products. Dioxins, found as minute impurities in PCP, are probably also photodegraded.

Hilton and Yuen (1963) compared oil absorption of PCP to the soil adsorption of a number of substituted herbicides. They found that adsorption of PCP was the highest of all compounds studied indicating that its mobility in soil may be limited. Further discussion of the degradation of PCP and other chlorophenols is located in a previous section.

Hazardous Nature of Sludges and Wastewater from Wood Preserving Industries

In 1980, the U.S. Environmental Protection Agency concluded that wastewater from wood preserving processes using creosote or pentachlorophenol and the resulting bottom sediment sludges from wastewater treatment (EPA, 1978) may pose an immediate or potential hazard to human health or the environment when improperly treated, stored, disposed of or otherwise managed. This conclusion is based on the fact that wastewaters and sludges contain several toxic constituents such as polynuclear aromatics which are known to be toxic, mutagenic, and carcinogenic and phenolics which are toxic and in some cases, bioaccumulative and carcinogenic (Listing Background Document, 1980). Landfills are used most often in the disposal of the sludges which raises the possibility of migration of toxic sludge constituents to domestic groundwater sources if the landfill is improperly designed or operated.

Incineration, which is an occasional disposal method for some wastewaters (with woodchips), is suspected of releasing hazardous vapors to the atmosphere. Because the degradation rates and migration rates of the components are unknown or are in dispute because of conflicting data (Listing Background Document, 1980), EPA questions the treatment processes that remove the toxic constituents. The EPA has utilized damage incidents to demonstrate either the mobility of these toxic constituents or their relative resistance to degradation. Also, because the observed human toxicity levels for many of these constituents are exceeded several times in treated effluents, the disposal of this material falls under Subtitle C of RCRA (Costle, 1980).

Summary and Research Needs

Wood preserving is a significant industry in the commerce and construction field. Although the total wastewater volume of the industry is large, the volume of wastewater for a single individual site is relatively small. Treatment of the wastewaters includes primary and secondary treatment with ponds used either for storage area or for a treatment system (aerated lagoons). Reduction in the levels of preservative constituents is substantial for most treatments but because of the toxicity of several of the constituents, the residual levels in the effluents are still above presently acceptable standards. For this reason, EPA has proposed that such wastewaters may be hazardous if improperly disposed of because of the presence of constituents which are mutagenic, carcinogenic, or bioaccumulative.

Sludges from the various treatment processes are classified as hazardous and are generally landfilled. Their volumes are relatively small, from 840 to 1530 dry metric tons per year for the entire industry. The sludges have a much higher concentration of the hazardous constituents than the wastewaters.

Data describing the fate of wood preservative constituents in biological treatment systems are generally incomplete and at times inconclusive. Although several studies have shown removal of constituents from effluent, the question remains whether the constituent was degraded or flocculated and became a portion of the sludge. Laboratory studies have shown that the chemical and biological degradation of PCP does occur and some of the end products have been identified. The mobility of the wood preservatives and degradative byproducts in soils is incompletely understood but damage incidents indicate mobility of these materials may be substantial.

Land treatment of wood preservative wastes has been successfully achieved at a number of plants. Thus extension of the land treatment approach to other waste streams and possibly sludges appears feasible and warrants further research. High priorities for research include (a) assessing different methods of waste application, such as subsurface injection to minimize volatilization losses, (b) composting of sludges prior to land application which may allow initial degradation of the more toxic constituents to avoid extensive biocidal effects on the soil microflora, and (c) investigating the mobility or leaching potential of the various chemical preservatives and their degradation products. Studies should also be undertaken to assess the long-term effects on human and animal health, and on the immediate soil/water ecosystem, from the land treatment of wood preservative wastes.

REFERENCES

Costle, D. M. 1980. Hazardous Waste Management System: General. Fed. Register Vol. 45. No. 98:33066-33073.

Cserjesi, A. J. 1967. The adaptation of fungi to pentachlorophenol and its biodegradation. Can. J. Microbiol. 13:1243-1249.

Development Document for Proposed Effluent Limitations Guidelines, New Source Performance Standards and Pretreatment Standard for the Timber Products Processing - Point Source Category. 1979. Office of Water and Waste Management, U.S. Environmental Protection Agency, Washington, D. C. EPA 440/1-79/023b. 427 pp.

Drisko, R. W., and T. B. O'Neill. 1966. Microbiological metabolism of creosote. Forest. Prod. J. 16:31-34.

Duncan, C. G. and F. J. Deverall. 1964. Degradation of wood preservatives by fungi. Appl. Microbiol. 12:57-62.

Dust, J. V., and W. S. Thompson. 1972. Pollution control in the wood-preserving industry. III. Chemical and physical methods of treating wastewater. Forest Prod. J. 22:25-30.

E.P.A. 1978. Federal Register, May 19, 1980, Part III; 33123.

Fisher, C. W. and G. R. Tallon. 1971. Wood preserving plants wastewater problems - some solutions. Proc. 67th Amer. Wood-Preservers Assoc. Meeting. 67:92-96.

Hilton, H. W., and Q. H. Yuen. 1963. Adsorption of several pre-emergence herbicides by Hawaiian sugarcane soils. J. Agr. Food Chem. 11:230-234.

Ingols, R. S. and P. C. Stevenson. 1963. Biodegradation of the carbon-chlorine bond. Res. Engr. 18:4-8.

Kaufman, D. D. 1978. Degradation of pentachlorophenol in soil and by soil microorganisms. pp. 27-39. In K. Rango Rao (ed.) Pentachlorophenol: chemistry, pharmacology and environmental toxicology. Plenum Press, N.Y.

Kuwatsuka, S. 1972. In Environmental Toxicology of Pesticides. Academic Press, New York. p. 385.

Listing Background Document - Wood Preserving. 1980. Title 40 CFR, Part 261. Subpart D-Lists of Hazardous Wastes, Appendix VII- Basis for Listing. 42 p.

Lorenz, L. F. and L. R. Gjovik. 1972. Analyzing creosote by gas chromatography: Relationship to creosote specification. Proc. Amer. Wood Preserv. Assoc. 68:32-39.

Lyr, H. 1963. Enzymatische detoxification chlorierter penole. Phytopathol. Z. 47:73-83.

Middlebrooks, E. J. 1968. Wastes from the preservation of wood. Jour. of the Sanit. Eng. Div. (ASCE) 94:41-54.

Murthy, N. B. K., D. D. Kaufman, and G. F. Fries. 1979. Degradation of pentachlorophenol in aerobic and anaerobic soil. J. Environ. Sci. Health. 14:1-14.

Myers, L. H., T. E. Short, Jr., B. L. DePrater, F. M. Pfeffer, D. H. Kampbell, J. E. Matthews. 1979. Indicator Fate Study. U. S. Environmental Protection Agency, Ada, Oklahoma. EPA-600/2/79-175. 82 pp.

Shack, P. A., and T. A. Reynolds. 1977. An evaluation of atmospheric evaporation for treating wood preserving wastes. Proc. Nat'l. Conf. on Treat. and Disposal of Indust. Wastewaters and Residues. pp. 29-33.

Shields, J. K., and D. W. Stranks. 1978. Wood preservatives and the environment. Proc. Tech. Trans. Sem. on the Timber Industry. Rep. No. EPS 3-WP-78-1. Environmental Protection Service, Canada.

Sikora, L. J., D. D. Kaufman, G. B. Willson, and J. F. Parr. 1982. Degradation of pentachlorophenol and pentachloronitrobenzene in a laboratory composting system. Proc. of the Eight Ann. Res. Symp. on Land Disposal of Hazardous Waste. EPA-600/9-82-002. pp. 372-382.

Teer, E. H., and L. V. Russell. 1977. Wastewater treatment for the small wood preserver. pp. 75-80. Proc. 32nd Indust. Waste Conf., Purdue Univ.

Thompson, W. S. and J. V. Dust. 1973. Pollution control in the wood preserving industry. IV. Biological methods of treating wastewater. Forest Prod. J. 23:59-66.

U.S. Dept. of Agriculture, State's EPA Preservative Chemical Assessment Team. 1980. The biological and economic assessment of pentachlorophenol, inorganic arsenic, and creosote. Draft 487 pp.

Summary

The objective of this report is to review critically and evaluate existing data relevant to the land treatment of hazardous wastes. The report discusses this information, identifies management controls to improve the effectiveness and environmental safety of land treatment, and suggests research activities that would improve the technology of land treatment and minimize environmental risks from this practice.

Soils provide a unique medium for treatment of wastes. They can serve effectively as filters, thereby protecting groundwater against contamination. The clay and organic matter content of soils adsorb both the organic and inorganic waste chemicals, thereby reducing their effective concentration in the system. Soil microorganisms possess a sufficiently wide range of physiological capabilities to initiate the degradation of potentially toxic organic waste chemicals. Careful selection of land treatment sites, plus management of waste composition, placement in the soil, rate and timing of applications, soil pH and fertility, soil erosion, and the crop grown can optimize chemical photo- or biodegradation while protecting air and water resources and the food-chain as well.

When wastes contain volatile constituents, they may have to be applied to land treatment systems by subsurface injection, or perhaps composted before land application. Although most toxic organic chemicals are biodegradable, some are relatively recalcitrant in most mineral soils. Management of these compounds may require pretreatment by UV-ozonation, composting, alternate aerobic/anaerobic sequencing, or other techniques to initiate degradation. Microorganisms may yet be isolated or genetically engineered which can biodegrade these relatively recalcitrant toxic organic chemicals and thus provide for inoculation compost and land treatment systems designed for rapid and sustained degradation.

Many hazardous wastes are actually bulky biological sludges from the treatment of industrial wastewaters; they contain only low levels of toxic compounds. Segregation of wastes provides an added dimension in that acceptable treatment/disposal alternatives other than land treatment can be selected for especially toxic or recalcitrant compounds (e.g., thermal oxidation, UV-ozonation, or anaerobic/aerobic sequencing).

Land treatment sites may be managed with or without cover crops, or with crops produced for sale. It may be possible to produce feed grains safely, and with minimal food chain impact, on soils treated with low levels of toxic organic chemicals and heavy metals. Crop removal at a land treatment site also removes nutrients and even plant-absorbed organic chemicals that could undergo leaching and subsequently contribute to groundwater pollution.

During the operation of a land treatment site, residues of toxic organic compounds and heavy metals could be sufficiently high to cause adverse health effects in livestock or humans if the land were immediately converted to pastures or home gardens. Cattle and sheep can ingest substantial amounts of soil containing chemical industrial sludges in their normal grazing activities. Children can and do ingest soil during outside activity. Improperly managed land treatment sites can allow undue human and animal exposure to toxic organic compounds and heavy metals, which must be prevented by proper closure procedures. During the closure period, residual levels of toxic organic chemicals can be degraded and their associated risks alleviated.

Selective removal of certain heavy metals from soils at land treatment sites may be possible using the principle of phyto-extraction and involving rare hyperaccumulator plants. The application of some metals and toxic organic chemicals may have to be limited by regulation, as is now the case for Cd.

Many research needs are identified for land treatment systems in general, and for land treatment of selected industrial wastes. Research can provide the scientific basis for the safe and effective design and management of land treatment sites, and for permitting this practice. A conclusion of this study is that land treatment systems offer a potentially cost effective, relatively low-risk, and environmentally acceptable method for the management of a wide range of hazardous wastes.

Recommendations for Research

During the course of this study, a number of deficiencies were identified in our knowledge and understanding of the behavior of certain hazardous wastes in the soil ecosystem. To ensure the safe, economic, and effective management of hazardous wastes in land treatment systems, the editors have compiled the following list of research needs. Included are the development and utilization of soil and crop management practices, composting technology, special bioassay and analytical techniques, and aesthetically acceptable waste management practices.

Specific recommendations:

1. Examine the influence of the composting process in degradation of hazardous wastes.

Composting of chemical industrial wastes or co-composting certain chemical wastes with sewage sludges or other organic materials may provide an important option for initial detoxification and degradation of toxic constituents prior to land application of some hazardous wastes. This pre-treatment approach could greatly minimize undesirable biocidal effects on the soil microflora from direct land application. Another approach involving composting is that of amending or preconditioning the soil at a land treatment site with sewage sludge compost prior to direct application of chemical wastes. Both of these approaches should be thoroughly investigated in the development of land treatment systems.

2. Analyze and characterize the organic constituents in industrial sludges that may be landspread.

If the organic fractions of the wastes could be analyzed and characterized, degradation of many wastes might be found to benefit from special management. Research on accurate and rapid characterization industrial sludges is needed.

3. Examine the factors governing the competition of clays and organic matter for the sorption of metals and organic components from certain types of hazardous wastes.

Chelation and sorption reactions determine the mobility and/or persistence of waste constituents in soil and their availability to plants. After wastes have been applied to land, soil matrix reactions which occur with time may change the availability of waste-applied constituents. Studies conducted on lands which have received industrial wastes for extended periods would enable scientists to predict the potential long-term benefits or problems with landspreading certain industrial wastes. The development of methods for regulating the movement of chemicals in soil, so that they might be contained within specific soil zones for degradation, may facilitate land treatment of chemicals previously considered as unsuitable for land farming.

4. Examine the degradation and persistence of selected hazardous wastes in soil.

The pattern of degradation, metabolism, and persistence of many hazardous wastes in soil is not known. Research is needed to characterize the degradative mechanism, the pathway, products formed, and persistence of selected waste chemicals. This should be done first under laboratory conditions, and subsequently, in the greenhouse and the field.

Although some industrial wastes contain priority pollutants, many can be degraded or fixed permanently in the soil matrix under aerobic conditions. Whether such degradation and fixation will occur, and whether the degradation products and fixed compounds will be held harmlessly in the soil should be thoroughly investigated.

5. Investigate the degradation of hazardous wastes in selected combinations.

The degradation of any single compound in soil may be influenced by other wastes and by soil chemical-physical characteristics. The degradation of some wastes may be enhanced, unaffected, or inhibited when applied to soils in combination with other wastes.

6. Examine the residual capacity and persistence of waste degrading soil microorganisms.

Soil microbial populations can be manipulated to become efficient at degrading many wastes. The ability to develop and manipulate these populations in soil is critical. Such questions must be answered as how often waste treatments should be applied to a soil population in order to: (a) safely dispose of maximum amounts of waste; (b) allow development and maintenance of an active waste degrading soil; (c) avoid destroying or losing an active waste degrading population by improper loading of the soil, or by cross contamination with other less desirable wastes which might interfere with degradation of the waste being treated.

7. Investigate possible pretreatment methods of "softening" persistent or hard to degrade wastes, to facilitate their later degradation in soil.

Subjecting hazardous wastes to specific treatments such as anaerobic or aerobic composting, light, heat, chlorination, UV-ozonation, chemical hydrolysis, etc., may render them more suitable for land application and subsequent degradation in soil. Pretreatment of persistent hazardous wastes has received limited attention and should be explored in greater depth in relation to degradation during land treatment.

8. Examine methods of regulating volatilization of chemicals from soil.

Many wastes contain chemicals which are considered both hazardous and volatile. Mechanisms for containing the residue at the site of application are needed. Subsurface application by injection, trenching, temporary flooding, or incorporation have been used. But other methods such as temporarily tarping or sealing the soil surface may serve to hold chemicals in soil until suitably degraded.

9. Examine the use of adapted inoculum, special enzymes, or other additives for enhancing degradation of chemicals in soil.

While the use of specially developed inocula, enzyme preparations or genetically engineered microorganisms for degradation of some wastes has been explored, their full potential for degrading environmental pollutants has not been recognized or achieved. Exploration of these mechanisms, or the use of other additives (adsorbants, stimulants, alternate substrates, etc.) may enhance the degradation of some chemicals in soils.

10. <u>Study the nature of metal interactions in microbial degradation</u>.

Greenhouse and laboratory studies can confirm which metals are beneficial or toxic to microorganisms necessary to carry out normal soil processes, and reveal the metal-microbe combinations that can accelerate or inhibit degradation processes.

11. <u>Determine the bioavailability of Cd, Zn, Mn, and Co in forages and leafy vegetables</u>.

Studies are needed to determine whether, in a worst case situation, leafy vegetables or forages grown on strongly acidic land soils at land treatment sites could absorb sufficiently high concentrations of toxic elements (e.g., Co, Zn, Cu, Ni) to cause adverse effects on animals after ingestion. We need to know the extent to which metals interact with other plant constituents and are rendered less bioavailable. The bioavailability of potentially toxic soil contaminants (Pb, F, Cu, Fe, As, Zn) to animals in ingested soil is virtually unknown and should be studied.

12. <u>Determine the phytotoxicity and bioaccumulation potential of Zn, Cu, Ni, Co, Cr^{3+}, Be, V, etc. for a number of agricultural and horticultural crops</u>.

Phytotoxicity to, and metal accumulation by, crops at both recommended and worst case management conditions should be established. Known metal-tolerant strains potentially useful for land treatment sites should also be tested. The relationship of metal concentrations in soil to specific and visible plant injury symptoms is not well defined. Phytotoxic responses should be correlated with acceptable/unacceptable risks to the food chain.

Some plants have an unusually high capacity to accumulate certain metals. A study of the potential of these so-called hyperaccumulator plants to extract metals from contaminated soils is needed. Specific soil and crop management practices necessary for culturing these plants should also be investigated. The principle of "phytoextraction" for removing metals from contaminated soils and at land treatment sites, and their subsequent recovery from the plant biomass may provide an economically feasible means for commercial recycling.

13. <u>Examine the nutrient and water availability in waste-amended soil under stress conditions</u>.

The mobility or persistence of waste constituents can be

altered under stress conditions such as drought or excessive rainfall or temperatures. Soil physical measurements in the field on existing landspreading sites should be performed to characterize changes or trends.

14. Select crops which are tolerant of high C:N ratios and/or excessive hydraulic loadings.

Since many industrial wastes which are applied to the soil have high BOD levels, plant varieties or species which tolerate high C:N ratios and/or rapid biodegradation of organic matter could be selected to grow on these sites to better utilize waste nutrients and immobilize waste constituents.

15. Evaluate the applicability and adaptability of mathematical models for predicting the detoxification and degradation of chemical wastes in land treatment systems.

Although models have been developed for predicting the rate and extent of degradation of crop residues, animal manures, and sewage sludges in soil, and the nutrient transformations therefrom, no such models have been reported for the detoxification and degradation of hazardous and toxic industrial chemical wastes in land treatment systems. The applicability and adaptability of existing models to hazardous waste chemicals in land treatment systems should be thoroughly explored since use of these models may optimize the safe, effective, and efficient management of these wastes.

16. Identify mutagenic constituents in hazardous wastes, develop rapid and reliable bioassay procedures for their characterization, and investigate methods and procedures for their inactivation in soils at land treatment sites.

Land treatment of chemical industrial wastes increases the probability that mutagenic constituents will be involved. An important research priority to ensure safe and effective management of such wastes will be the identification of mutagenic compounds and development of reliable analytical procedures. A study of the factors affecting the degradation and inactivation of mutagenic compounds in soil is needed.

17. Determine the interaction of waste characteristics and soil factors that influence the movement of pathogenic bacteria and viruses through soils.

Some wastes are defined as hazardous because of the presence of pathogenic bacteria and viruses. The movement of such organisms through soils at land treatment sites after waste application may constitute a potential disease hazard. Little is known about the movement of pathogenic organisms through soils. Research is needed to determine those waste and soil factors and their interactions that influence the movement of disease organisms through soils.

Index

Other Noyes Publications

HAZARDOUS WASTE INCINERATION ENGINEERING

by T. Bonner, B. Desai, J. Fullenkamp, T. Hughes, E. Kennedy, R. McCormick, J. Peters, and D. Zanders

Monsanto Research Corporation

Pollution Technology Review No. 88

The engineering guidelines contained in this book are a compendium of the available literature on current state-of-the-art technology for hazardous waste incineration. They are intended to be used as a source of information for operational decisions and as a reference in the preparation of permit applications for hazardous waste incineration facilities.

A sizable fraction of the millions of tons of industrial waste material generated in the United States each year is considered hazardous (approximately 57 million metric tons in 1980). Incineration has recently emerged as an attractive alternative to other hazardous waste disposal methods such as landfilling, ocean dumping, and deep-well injection.

The advantages of incinerating hazardous wastes are several: toxic components can be converted to harmless or less harmful compounds; volume can be greatly reduced; heat recovery is possible as a means of saving energy; and incineration provides ultimate disposal, thereby eliminating the possibility of problems resurfacing at a later date. Because of these advantages, incineration may become a principal technology for hazardous waste disposal in the near future.

The various chapters of the book detail waste characterization, current commercial technology as well as emerging technology, incinerator design, and overall facility design. Listed below is a condensed table of contents giving **chapter titles and selected subtitles.**

ISBN 0-8155-0877-8 (1981)

385 pages

Other Noyes Publications

HOW TO DISPOSE OF OIL AND HAZARDOUS CHEMICAL SPILL DEBRIS

Edited by A. Breuel

Pollution Technology Review No. 87

This book describes various techniques which can be used to dispose of the collected debris from oil and hazardous chemical spills. It is based on research prepared by *SCS Engineers* and *CONCAWE* (the European oil companies' international study group for conservation of clean air and water).

Engineering constraints and equipment requirements for handling and disposal are evaluated. Debris management aspects considered are storage, transport, treatment, reprocessing, and disposal. Hardware and processing systems are identified and conceptual transport/disposal plans are developed. A literature review and several case studies are included. U.S. and European technology are covered.

Disposal is a complex problem. Spill incidents which have occurred in recent years have highlighted the difficulties in dealing with the large quantities of emulsions and debris which are collected during control and cleanup operations. Close cooperation between authorities and industry is a necessity at every stage, to ensure disposal in an environmentally acceptable, cost-effective, and energy-conserving manner.

Each case is different and no specific rules can be formulated, but guidance is given on the choice of options and their state of development. The condensed table of contents listed below gives **chapter titles and selected subtitles.**

ISBN 0-8155-0876-X (1981) **420 pages**

Other Noyes Publications

OIL SPILL CLEANUP AND PROTECTION TECHNIQUES FOR SHORELINES AND MARSHLANDS

Edited by A. Breuel

Pollution Technology Review No. 78

When a major oil spill occurs, it usually involves contamination of coastal or inland shorelines and marshlands, which can result in serious environmental and economic damage. Such damage can be significantly reduced if proper protection and cleanup actions are taken promptly.

This book provides a systematic easy-to-apply methodology for assessing the threat or extent of contamination and choosing the most appropriate protection/cleanup procedures for each shoreline or marshland contamination event.

Oil spill on-scene coordinators and local officials, as well as petroleum, shipping and chemical industry personnel should find this book directly applicable during prior planning and oil spill operations.

A condensed table of contents of the three parts of the book, including **important subtitles,** is given below.

ISBN 0-8155-0848-4 (1981) **404 pages**

Other Noyes Publications

HANDBOOK OF TOXIC AND HAZARDOUS CHEMICALS

by Marshall Sittig

"Protecting against potential public health hazards requires widespread knowledge about commercial chemicals. We need to know more about the effects they will have, and, most importantly, how we can minimize the risks posed by them."—*from the Foreword by Bill Bradley, U.S. Senator from New Jersey.*

This handbook presents concise chemical, health, and safety information on about 600 toxic and hazardous chemicals, so that responsible decisions can be made by chemical manufacturers, safety equipment producers, toxicologists, industrial safety engineers, waste disposal operators, health care professionals, and the many others who may have contact with or interest in these chemicals due to their own or third party exposure. This book will thus be a valuable addition to industrial and medical libraries.

Essentially the book attempts to answer six questions about each compound (to the extent information is available):

(1) What is it?
(2) Where is it encountered?
(3) How much can one tolerate?
(4) How does one measure it?
(5) What are its harmful effects?
(6) How does one protect against it?

Included in the book are **all** of the substances whose allowable concentrations in workplace air are adopted or proposed by the American Conference of Governmental Industrial Hygienists (ACGIH), **all** of the substances considered to date in the Standards Completion Program of the National Institute of Occupational Safety and Health (NIOSH), **all** of the priority toxic water pollutants defined by the U.S. Environmental Protection Agency (EPA), and **most** of the chemicals in the following classifications: EPA "hazardous wastes," EPA "hazardous substances," chemicals reviewed by EPA in Chemical Hazard Information Profiles (CHIPS) documents, and chemicals reviewed in NIOSH Information Profile documents.

The necessity for informed handling and controlled disposal of hazardous and toxic materials has been spotlighted over and over in recent days as news of fires and explosions at factories and waste sites and groundwater contamination near dump sites has been widely publicized. In late 1980 the EPA imposed regulations governing the handling of hazardous wastes—from creation to disposal. Prerequisite to control of hazardous substances, however, is knowledge of the extent of possible danger and toxic effects posed by any particular chemical. This book provides these prerequisites.

The chemicals are presented alphabetically and each is classified as a "carcinogen," "hazardous substance," "hazardous waste," and/or a "priority toxic pollutant"—as defined by the various federal agencies, and explained in the comprehensive introduction to the book.

Data is furnished, to the extent currently available, on any or all of these important categories:

Chemical Description
Code Numbers
DOT Designation
Synonyms
Potential Exposure
Incompatibilities
Permissible Exposure Limits in Air
Determination in Air
Permissible Concentration in Water
Determination in Water
Routes of Entry
Harmful Effects and Symptoms
Points of Attack
Medical Surveillance
First Aid
Personal Protective Methods
Respirator Selection
Disposal Method Suggested
References

An outstanding and noteworthy feature of this book is the Index of Carcinogens.

ISBN 0-8155-0841-7 (1981) **729 pages**

Other Noyes Publications

HAZARDOUS CHEMICALS DATA BOOK 1980

Edited by G. Weiss

Environmental Health Review No. 4

Instant information for decision-making in emergency situations by personnel involved with chemical accidents is the prime purpose of this compilation of about 1,350 hazardous chemicals. The book, prepared in clear, concise, easy-to-locate format, should become an invaluable source on any library or laboratory shelf; is intended for use by scientists, engineers, managers, transportation personnel, or anyone who may have contact or require data on a particular chemical.

There is a large amount of pertinent data provided for each chemical, and examples of a few of the headings are:

Toxicity and Health Hazards
Hazards Relating to Water Pollution
Spill and Leak Procedures
Special Protective Equipment
Flammability and Explosion Hazards
Chemical Reactivity
Extinguishing Media
Special Precautions

The volume is set up in two sections, alphabetically arranged. The first section is based on the Department of Transportation's Chemical Hazard Response Information System. The second section is based on Material Safety Data Sheets obtained from the Oak Ridge National Laboratory. The use of a convenient all-inclusive index furnishes the responsible individual with the desired information in a minimum of time.

A partial list of 54 hazardous chemicals taken from the index of about 1,350 chemicals, is shown below.

Acetal
Acetaldehyde
Acetamide
Acetic Acid
Acetic Anhydride
Acetone
Acetone Cyanohydrin
Acetonitrile
Acetonylacetone
Acetophenone
Acetylacetone
Acetyl Bromide
Acetyl Chloride
Acetylene
Acetyl Methylamine
Acetyl Methylurea
Acetyl Peroxide Solution
Acetyl Thiourea
Acetyl Urea
Acridine
Acrolein
Acrylamide
Acrylic Acid
Acrylonitrile
Adipic Acid
Adiponitrile
Aldrin
Alkylbenzenesulfonic Acids
Allyl Alcohol
Allylamine
Allyl Bromide
Allyl Chloride
Allyl Chloroformate
Allyltrichlorosilane
Aluminum Chloride
Aluminum Fluoride
Aluminum Nitrate
Aluminum Sulfate
p-Aminoazobenzene
p-Aminoazobenzene Hydrochloride
2-Amino-1,4-Dimethylbenzene
4-Amino-1,3-Dimethylbenzene
5-Amino-1,3-Dimethylbenzene
o-Aminodiphenyl
Aminoethylethanolamine
2-Aminopyridine
Ammonia, Anhydrous
Ammonium Acetate
Ammonium Benzoate
Ammonium Bicarbonate
Ammonium Bifluoride
Ammonium Carbonate
Ammonium Chloride
Ammonium Citrate
plus about 1,300 other chemicals

ISBN 0-8155-0831-X 1188 pages

Other Noyes Publications

PROTECTIVE BARRIERS FOR CONTAINMENT OF TOXIC MATERIALS 1980

Edited by R. Fung

Pollution Technology Review No. 67

The Resource Conservation and Recovery Act of 1976 created federal and state authority over solid and hazardous wastes. The projected regulations which become fully effective in 1980 will have profound effects, particularly on disposal practices for toxic industrial wastes.

This book describes an aspect of waste treatment which is receiving increasing attention—the use of protective, impermeable barriers to prevent or minimize the escape of toxic pollutants from disposal sites. Manmade and natural lining and cover materials for lagoons and landfills, and solidification/stabilization techniques are detailed, as well as listings of actual installations.

The book provides information for the engineer or land impoundment operator on physical methods for toxic waste containment. A condensed table of contents is given below.

ISBN 0-8155-0804-2 **288 pages**

0